The Interaction Between Medium Energy Nucleons in Nuclei–1982

(Indiana University Cyclotron Facility)

What mystery is it? The morning as rare
    As the Indian Summer may bring!
A tang in the frost and a spice in the air
    That no city poet can sing!
The crimson and amber and gold of the leaves,
    As they loosen and flutter and fall
In the path of the park, as it rustlingly weaves
    Its way through the maples and under the eaves
    Of the sparrows that chatter and call.

This is how James Whitcomb Riley, in the fall of 1882, described the
countryside in Indiana. The site of the Workshop was the Canyon Inn,
McCormick's Creek State Park, Spencer, Indiana.

AIP Conference Proceedings
Series Editor: Hugh C. Wolfe
Number 97

The Interaction Between Medium Energy Nucleons in Nuclei–1982

(Indiana University Cyclotron Facility)

Edited by
H. O. Meyer
Indiana University

American Institute of Physics
New York   1983

L.C. Catalog Card No. 83-70649
ISBN 0-88318-196-7
DOE CONF- 821050

Proceedings of the

Workshop on the Interaction

Between Medium Energy Nucleons in Nuclei

Indiana University Cyclotron Facility

Bloomington, Indiana

October 28-30, 1982

PREFACE

For the second time McCormick's Creek State Park and Indiana's beautiful fall colors have provided the backdrop to an IUCF Workshop: clearly reason to hope that this marks the beginning of a long series of productive and stimulating annual meetings on current topics in physics.

This year's subject pertains to an area in nuclear physics where recently significant theoretical advances have been achieved and where, at the same time, considerable experimental evidence has been collected.

The problem of describing the nuclear many-body problem in terms of an underlying, "elementary" process or rather, the question in which way this elementary process is modified by the presence of the nuclear medium, can be approached in many ways. However, there seems to be a rough division into the two groups of non-relativistic, selfconsistent theories on one hand, and relativistic mean-field models on the other. This somewhat controversial situation was mediated by the workshop.

In the past few years, the level of the theoretical activity in both fields has received an impetus by the advent of experimental data that clearly bear on, at least certain aspects of, the effective nucleon-nucleon interaction in the nuclear medium. Specifically, these measurements include proton-nucleus elastic scattering and proton induced reactions such as (p,p') and (p,n) with bombarding energies in the neighborhood of 200 MeV. In this energy region the impulse approximation becomes more reliable, which greatly simplifies the theoretical task. Further increasing the bombarding energy, say, beyond 400-500 MeV enhances the absorptivity of the nucleus and as this happens, the nuclear interior is more and more obscured. This creates a "window" in bombarding energy (from about 100 MeV to 500 MeV) for the study of medium effects in the nucleon-nucleon interaction.

The purpose of this workshop was to assess the theoretical as well as the experimental status of the field and to elicit an exchange

of ideas between groups working on different aspects of the same
physics problem. This is reflected in particular in the choice of
invited participants. In the initial planning stages the
international advisory board (see list following this preface) was
most helpful. In particular I appreciate the concrete advice I
obtained from W. Bertozzi, G.E. Brown, C. Glashausser, W.G. Love, M.
H. Macfarlane, J.M. Moss, I. Sick, G.E. Walker and many others.

The scientific program was structured into a number of sessions,
each devoted to a particular sub-topic. Each session was organized
and chaired by a pair of "Reviewers" whose task it also was to
stimulate a free discussion. Thanks to the concern and involvment of
these reviewers this scheme succeeded.

Paramount to the success of any meeting are also the
non-scientific affairs demanding an enormous amount of work behind the
scene. Here, I acknowledge the contributions of R.D. Bent, K.E.
Berglund, C. Foster, and R.P. Thompson from the IU Physics Department
and the Cyclotron and of R. Claussen from the IU Conference Bureau. I
was also very pleased with the active role played by the physics
graduate students, M. Cantrell, W. Estrada, M. Fatyga, B. Flanders, W.
Fox, M. Green, K. Pitts, M. Price, and T. Throwe; their help is
greatly appreciated.

From the early planning stages to the process of editing the
proceedings, i.e. for a time of well over a year, the two IUCF
secretaries Diana McGovern and Becky Westerfield dedicated a sizable
fraction of their time to this workshop. I would like to commend them
for their enthusiasm and competence with which they performed their
central role.

In the present proceedings contributions are grouped and ordered
in the same way as they were presented at the workshop. In addition,
some important contributions are printed here although time
constraints during the meeting made an oral presentation impossible.
The discussion between presentations was recorded and used in various
ways. However, only the discussion during Session E - unanimously
judged the culmination of the meeting in terms of free exchange of
ideas - was transcribed in full. This transcript was prepared and
edited by A. Gattone and E.T. Cooper, both nuclear theorists at
Indiana University. Parts of the discussion in Session C have been
transcribed by G.T. Emery.

My final thanks go to reviewers, speakers, the summary speaker H.
Feshbach and all participants for their active role that made the
workshop worth the effort.

Bloomington, Indiana   Hans-Otto Meyer
January, 1983

SPONSORS

U.S. National Science Foundation

U.S. Department of Energy

Indiana University Cyclotron Facility

Department of Physics, Indiana University

The President's Office, Indiana University

SCIENTIFIC ADVISORY BOARD

W. Bertozzi, Massachusetts Institute of Technology, Cambridge, MA

G.E. Brown, State University of New York, Stony Brook, NY

H.V. von Geramb, University of Hamburg, Hamburg, WEST GERMANY

C. Glashausser, Rutgers State University, Rutgers NY

H. Kamitsubo, University of Tokyo, Tokyo, JAPAN

W.G. Love, University of Georgia, Athens, GA

C. Mahaux, Universite de Liege, Liege, BELGIUM

J.M. Moss, Los Alamos National Lab, Los Alamos, NM

I. Sick, University of Basel, SWITZERLAND

R. Vinh Mau, Nuclear Physics Institute, Orsay, FRANCE

CONFERENCE PARTICIPANTS

R. Amado, University of Pennsylvania, Philadelphia, PA 19104, USA
N. Anantaraman, Michigan State Univ., East Lansing, MI 48824, USA
L. Antonuk, Universite de Neuchatel, CH2000 Neuchatel, Switzerland
N. Austern, University of Pittsburgh, Pittsburgh, PA 15260, USA
S. Austin, Michigan State University, East Lansing, MI 48824, USA
A. Bacher, Indiana University, Bloomington, IN 47405, USA
P. Barnes, Carnegie-Mellon University, Pittsburgh, PA 15213, USA
R. Barrett, University of Surrey, Surrey GU2 5xH, United Kingdom
R. Bent, Indiana University, Bloomington, IN 47405, USA
W. Bertozzi, MIT, Cambridge, MA 02139, USA
H. Bhang, University of Maryland, College Park, MD 20747, USA
G. Brown, SUNY, Stony Brook, NY 11794, USA
D. Bugg, Queen Mary College, London E1 4NS, United Kingdom
M. Cantrell, Indiana University, Bloomington, IN 47405, USA
T. Carey, Los Alamos National Lab, Los Alamos, NM 87544, USA
N. Chant, University of Maryland, College Park, MD 20742, USA
G. Ciangaru, University of Maryland, College Park, MD 20742, USA
B. Clark, Ohio State University, Columbus, OH 43210, USA
T. Clegg, University of North Carolina, Chapel Hill, NC 27514, USA
J. Comfort, Arizona State University, Tempe, AZ 85287, USA
J. Conte, Indiana University, Bloomington, IN 47405, USA
T. Cooper, Indiana University, Bloomington, IN 47405, USA
M. Coz, University of Kentucky, Lexington, KY 40506, USA
C. Davis, TRIUMF, Vancouver, B.C., V6T 2A3, Canada
R. DeVito, Indiana University, Bloomington, IN 47405, USA
F. Dietrich, Lawrence Livermore Natl. Lab, Livermore, CA 94550, USA
N. DiGiacomo, Los Alamos National Lab, Los Alamos, NM 87545, USA
M. Dillig, University Erlangen, D-852 Erlangen, West Germany
T. Drake, University of Toronto, Toronto, Ontario, M52 1A7 Canada
R. Eisenstein, Carnegie-Mellon Univ., Pittsburgh, PA 15213, USA
G. Emery, Indiana University, Bloomington, IN 47405, USA
W. Estrada, Indiana University, Bloomington, IN, USA
M. Fatyga, Indiana University, Bloomington, IN, 47405, USA
H. Feshbach, MIT, Cambridge, MA 02139, USA
B. Flanders, Indiana University, Bloomington, IN 47405, USA
C. Foster, Indiana University, Bloomington, IN 47405, USA
W. Fox, Indiana University, Bloomington, IN 47405
J. Friar, MS D283, Los Alamos, NM 87544, USA
G. Garvey, Argonne National Lab, Argonne, IL 60439, USA
A. Gattone, Indiana University, Bloomington, IN 47405, USA
C. Glashausser, Rutgers University, Piscataway, NJ 08854, USA
C. Glover, Indiana University, Bloomington, IN 47405, USA
C. Goodman, Indiana University, Bloomington, IN 47405, USA
B. Goulard, University of Montreal, Montreal, Quebec, Canada
R. Grace, Carnegie-Mellon University, Pittsburgh, PA 15213, USA
M. Green, Indiana University, Bloomington, IN 47405, USA
W. Haider, University of Oxford, United Kingdom
D. Halderson, Western Michigan University, Kalamazoo, MI 49008, USA
J. Hall, Indiana University, Bloomington, IN 47405, USA
I. Halpern, University of Washington, Seattle, WA 98195, USA
K. Hecht, University of Michigan, Ann Arbor, MI 48109, USA
D. Hendrie, Department of Energy, Washington, DC 20545, USA

CONFERENCE PARTICIPANTS

B. Hichwa, Hope College, Holland, MI 49423, USA
C. Horowitz, MIT, Cambridge, MA 02139, USA
M. Hugi, Indiana University, Bloomington, IN 47405, USA
P. Hwang, Indiana University, Bloomington, IN 47405, USA
M. Ichimura, University of Tokyo, Komaba, Tokyo, Japan
K. Jackson, TRIUMF, Vancouver, B.C., V6T 2A3 Canada
W. Jacobs, Indiana University, Bloomington, IN 47405, USA
M. Jaminon, Universite de Liege, B-4000 Liege 1, Belgium
A. Johansson, Tandem Accelerator Laboratory, Uppsala, Sweden
C. Johnson, Oak Ridge National Lab, Oak Ridge, TN 37830, USA
W. Jones, Indiana University, Bloomington, IN 47405, USA
H. Karwowski, Indiana University, Bloomington, IN 47405, USA
J. Kehayias, Indiana University, Bloomington, IN 47405, USA
J. Kelly, MIT, Cambridge, MA 02139, USA
F. Khanna, Chalk River, Ontario, K0J 1J0 Canada
J. King, University of Toronto, M5S 1A7 Canada
N. King, Los Alamos National Lab, Los Alamos, NM 87545, USA
P. Kitching, University of Alberta, Edmonton, Alberta, Canada
A. Klein, University of Pennsylvania, Philadelphia, PA 19174, USA
K. Kwiatkowski, Indiana University, Bloomington, IN 47405, USA
C. Li, University of Pennsylvania, Philadelphia, PA 19174, USA
D. Lichtenberg, Indiana University, Bloomington, IN 47401, USA
R. Lindgren, University of Massachusetts, Amherst, MA 01003, USA
E. Lomon, MIT, Cambridge, MA 02139, USA
W. Love, University of Georgia, Athens, GA 30601, USA
D. Low, Indiana University, Bloomington, IN 47405, USA
W. MacDonald, University of Maryland, College Park, MD 20742, USA
M. Macfarlane, Indiana University, Bloomington, IN 47405, USA
C. Mahaux, Universite de Liege, B-4000 Liege 1, Belgium
Th. A. Maris, University F.R.G.S., Brazil
S. Martin, University of British Columbia, Vancouver, B.C., Canada
K. Maung, Kent State University, Kent, OH 44242, USA
H. McManus, Michigan State University, East Lansing, MI 48824, USA
J. McNeil, Villanova University, Villanova, PA 19085, USA
H. Meyer, Indiana University, Bloomington, IN 47405, USA
C. Miller, University of British Columbia, Vancouver, B.C, Canada
D. Miller, Indiana University, Bloomington, IN 47405, USA
G. Miller, University of Washington, Seattle, WA 98195, USA
J. Moss, Los Alamos National Lab ., Los Alamos, NM 87545, USA
D. Murdock, University of Colorado, Boulder, CO 80309, USA
H. Nann, Indiana University, Bloomington, IN 47405, USA
J. Noble, University of Virginia, VA, 22301, USA
C. Olmer, Indiana University, Bloomington, IN 47405, USA
V. Pandharipande, University of Illinois, Urbana, IL 61820, USA
F. Petrovich, Florida State University, Tallahassee, FL 32306, USA
M. Pickar, Indiana University, Bloomington, IN 47405, USA
W. Pitts, Indiana University, Bloomington, IN 47405, USA
M. Plum, University of Massachusetts, Amherst, MA 01003, USA
R. Pollock, Indiana University, Bloomington, IN 47405, USA
C. Poppe, Lawrence Livermore National Lab, Livermore, CA 94550, USA
M. Price, Indiana University, Bloomington, IN 47405, USA
J. Price, Indiana University, Bloomington, IN 47405, USA

CONFERENCE PARTICIPANTS

J. Rapaport, Ohio University, Athens, OH 45701, USA
G. Rawitscher, University of Connecticut, Storrs, CT 06268, USA
L. Ray, University of Texas, Austin, TX 78712, USA
E. Redish, University of Maryland, College Park, MD 20742, USA
M. Rho, Saclay/Stony Brook, Stony Brook, NY 11794, USA
P. Roos, University of Maryland, College Park, MD 20742, USA
M. Saber, Indiana University, Bloomington, IN 47405, USA
H. Sakaguchi, Kyoto University, Japan
P. Sauer, Hannover University, 3000 Hannover, West Germany
J. Schiffer, Argonne National Lab, Argonne, IL 60439, USA
P. Schwandt, Indiana University, Bloomington, IN 47401, USA
A. Scott, University of Georgia, Athens, GA 30602, USA
B. Serot, University of California, Santa Barbara, CA 93106, USA
C. Shakin, CUNY-Brooklyn College, Brooklyn, NY 11210, USA
J. Shepard, Univ. of Colorado, Boulder, CO 80309, USA
H. Sherif, University of Alberta, Edmonton, Alberta, T6G 2N5 Canada
P. Singh, Indiana University, Bloomington, IN 47405, USA
P. Sood, Banaras Hindu University, India
D. Sparrow, University of Pennsylvania, Philadelphia, PA 19174, USA
J. Speth, KFZ, Julich, West Germany
E. Stephenson, Indiana University, Bloomington, IN 47405, USA
T. Taddeucci, Indiana University, Bloomington, IN 47405, USA
P. Tandy, Kent State University, Kent, OH 44242, USA
T. Throwe, Indiana University, Bloomington, IN 47405, USA
S. van der Werf, Univ. of Groningen, Groningen, The Netherlands
S. Vigdor, Indiana University, Bloomington, IN 47405, USA
V. Viola, Indiana University, Bloomington, IN 47405, USA
H. von Geramb, University of Hamburg, Hamburg, West Germany
J. Walecka, Argonne National Lab, Argonne, IL 60439, USA
G. Walker, Indiana University, Bloomington, IN 47405, USA
M. Walker, Indiana University, Bloomington, IN 47405, USA
S. Wallace, University of Maryland, College Park, MD 20742, USA
R. Walter, Duke University, Durham, NC 27706, USA
T. Ward, Indiana University, Bloomington, IN 47405, USA
J. Wesick, University of Maryland, College Park, MD 20742, USA
W. Wharton, Carnegie-Mellon University, Pittsburgh, PA 15213, USA
H. Willard, National Science Foundation, Washington, DC 20050, USA
D. Wolfe, Kent State University, Kent, OH 44242, USA
S. Wong, University of Toronto, Toronto, Canada
L. Woo, Indiana University, Bloomington, IN 47405, USA

TABLE OF CONTENTS

SESSION C

THE NN INTERACTION IN REACTIONS

Reviewers: R.D. Amado and G.T. Garvey

Contributions:

SUMMARY

SESSION A
THE NUCLEON-NUCLEON
INTERACTION

THE FREE NN INTERACTION, 50–1000 MeV

D V Bugg

Queen Mary College, Mile End Road, London E1 4NS, UK

ABSTRACT

Phase shift solutions for pp elastic scattering are unique up to 800 MeV, and for np up to 500 MeV. There is excellent agreement between three analyses for all phase shifts δ and for elasticity parameters η of dominant inelastic partial waves ($^1D_2$, $^3F_3$ and $^3P_2$). For low partial waves, notably ^3P_o, $^3P_1$ and ^1S_o, there are some disagreements about η parameters, reflecting insensitivity of elastic data and some discrepancies amongst experiments. The need for further data is reviewed. Amplitudes for np→pn between 140 and 500 MeV are known accurately from phase shift analysis, and are almost purely real; they are useful for extracting nuclear matrix elements from (p,n) reactions.

The Paris potential is the best potential model fit to data up to 425 MeV, but there are some small systematic discrepancies with data. Whether or not dibaryon resonances exist at higher energies is still controversial. If they exist, they are associated with the onset of strong inelasticity, and are weakly coupled to both NN and πd channels. It seems likely that the structure observed in $\Delta\sigma_L$ and $\Delta\sigma_T$ at 550 MeV is due to weak coupling of NN to the large peak in the πd total cross section close to this energy.

I. INTRODUCTION

In this review, I shall attempt to achieve two conflicting objectives in each section, firstly an uncontroversial survey of the current situation suitable for the non-expert, secondly a glimpse of the fine detail still controversial amongst the experts. For further reference, there are several recent reviews: by myself [1] from a mainly phenomenological and experimental viewpoint, by Signell [2], relating theory and phenomenology of potentials, and by Kroll [3], who gives extensive numerical results and valuable detail of the formulae essential for exchange models and dispersion relation analysis. Vinh Mau [4] gives a detailed review of the Paris potential. On dibaryons, the enthusiast's viewpoint [5] may be compared with that of the conservative [6].

The notation used here for (coupled triplet) partial waves is

$$f_{\ell,J} = \left\{ \eta_{\ell,J} \cos 2\bar{\varepsilon}_J \exp\left(2i\delta_{\ell,J} - 1\right) \right\}/2i,$$
$$\eta = \cos^2\rho.$$

II. DATA AND ELASTIC PHASE SHIFT ANALYSIS

Data and amplitudes are in excellent shape for pp elastic scattering up to 800 MeV and for np from 140 to 500 MeV; further np data from LAMPF are likely to extend the analysis up to 800 MeV very shortly.

There are comprehensive pp data sets clustered around energies of 25, 50, (95), 140, 210, 325, 425, 515, 580, 650 and 800 MeV, and further data are beginning to appear from Leningrad at 970 MeV. At 5 energies from 445 to 580 MeV, the data set is so complete, due to measurements of 15 parameters at SIN by the Geneva group, that an amplitude analysis is possible [7] without recourse to partial wave analysis.

Four groups are active in phase shift analysis. Arndt and Verwest [8], the Saclay-Geneva group [9] and QMC-TRIUMF [10] all concentrate on the energy range 0 to 800 MeV. Hoshizaki [11] and collaborators tackle the energy range 500 to 3000 MeV. Although there are minor differences of detail (e.g. theoretical constraints on high partial waves, [10] and para-metrisations of energy dependence), all the first three agree closely on phase shifts δ, as is illustrated in Fig 1, and <u>even more closely</u> on amplitudes and the common observables. In detail, mixing parameters $\bar{\varepsilon}_2$ and $\bar{\varepsilon}_4$ at 800 MeV are rather suspect because of a change in energy

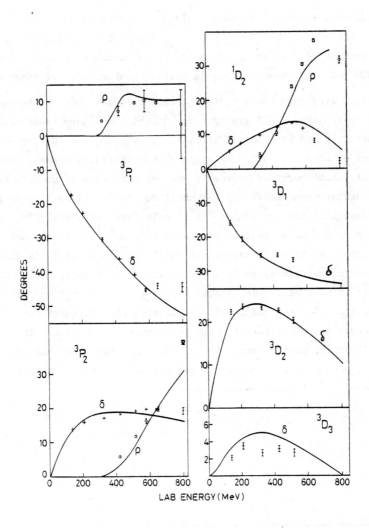

Fig. 1   A comparison of $^3P_1$, $^3P_2$, $^1D_2$, $^3D_1$, $^3D_2$ and $^3D_3$ phase shifts from Arndt and VerWest, ref. 8 (full curves) and Dubois et al., ref 10 (crosses δ, circles ρ).

dependence above 650 MeV; they could be improved by accurate (±0.02) measurements of either $D \equiv D_{NN}$ or $A' \equiv D_{LL}$ from 8 to $20°$ centre of mass. Measurements of A_{LL} and A_{SS} from $10°$ to $90°$ would be welcome at 650 and 800 MeV to reduce latitude in certain combinations of phase shifts.

For elasticity parameters η, the situation is not quite so happy in low partial waves. All groups agree roughly for those waves with large inelasticity and high multiplicity ($^3P_2$, $^1D_2$, $^3F_3$ and $^1G_4$), probably because of sensitivity in the shape of the diffraction peak. (Note, however, that there are few really good measurements of $d\sigma/d\Omega$ across the full angular range above 325 MeV until 800 MeV). There is also general agreement that inelasticities can be calculated accurately from OPE for $J \geqslant 4$ up to 800 MeV: results of the Paris group [12], Gruben and VerWest [13], and Silbar and Kloet [14] agree reasonably well. A detail is that the earliest calculations, by Green and Sainio [15], disagreed with the other authors in coupled triplet states, notably $^3F_4$ and $^3H_4$; a recent publication of Green and Sainio [16] locates the origin of the discrepancy and harmonises their results with the remainder. However, for low partial waves, particularly $^1S_0$, $^3P_0$ and $^3P_1$, there are disagreements about η parameters, to which elastic data show little sensitivity.

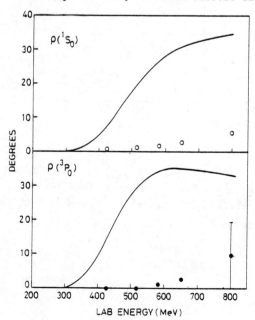

Fig. 2 Values of $\rho(^1S_0)$ and $\rho(^3P_0)$ from Arndt and VerWest, Ref. 8 (full curves) and from Dubois et al., Ref. 10 (circles).

Arndt and VerWest fit large inelasticities in $^1S_0$ and $^3P_0$ from 500 to
800 MeV (Fig 2), whereas I find satisfactory agreement with data using
$\eta\left(^3P_0\right) = \eta\left(^1S_0\right) = 1$. I must declare a prejudice against inelasticities
in these waves as large as those of Arndt and VerWest, for the following
reasons. Close to the inelastic threshold, inelasticity is largely in
$pp{\to}d\pi^+$ and $pp{\to}n\Delta^{++}$. Data on the former demand dominant contributions
from $^1D_2$, $^3P_1$, $^3P_2$ and $^3F_3$, and pp elastic data are compatible with
this; $^3P_0$ is forbidden by parity and angular momentum conservation.
In the second reaction, $pp{\to}n\Delta$, $^1S_0$ can contribute only with L=2 in the
final state, and this will be suppressed by the centrifugal barrier.
Finally, the strong repulsive core is likely to shield inelasticity in
$^3P_0$ and $^1S_0$ (DWBA approximation).

The discrepancies in $\eta\left(^3P_0\right)$ and $\eta\left(^1S_0\right)$ originate at least partly from
a discrepancy of 4 mb in measurements of $\Delta\sigma_L$ (a) at Argonne [17] and
LAMPF [18] and SIN [19] and (b) at TRIUMF [20]. This is shown as Fig 3.
Since $\Delta\sigma_L$ receives contributions directly from inelastic channels, it
is a delicate source of information on the spin dependence of η para-

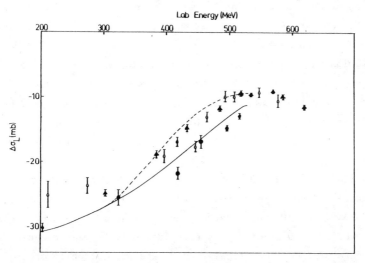

Fig. 3. Values of $\Delta\sigma_L$. TRIUMF results are denoted by full circles,
LAMPF results, Ref. 18 by triangles, and preliminary SIN results, Ref.
19, by open circles. Phase shift fits of Dubois et al., Ref. 10, are
shown by the full line. The dashed line indicates the upper limit
allowed by phase shift analysis if elasticities for J⩾4 are taken from
OPE and remaining inelasticity is all in $^4D_2$.

meters. Unfortunately the discrepancy is comparable to the total
inelastic cross section (5 mb at 500 MeV). Argonne-LAMPF-SIN data pull
$\eta\left(^3P_o\right)$ away from 1, TRIUMF data pulls it towards 1. The only other clue
is that the very complete data set of the Geneva group at 580 MeV pulls
the solution strongly towards $\eta\left(^3P_o\right) = 1$. Hopefully this discrepancy
will be resolved soon by new measurements at Saclay.

The analysis of np data is greatly simplified by two facts. Firstly
there is now general agreement [21] that the I=0 inelastic cross section
is < 1 mb up to 800 MeV, so that one can set all $\eta(I=0) = 1$, except for
minor inelasticities attributed to $^3S_1$, $^1P_1$ and $^3D_1$. Secondly, np data
are sensitive to interferences between I=1 and I=0 amplitudes, and
furthermore these change sign about 90°. Extensive and accurate data
on $d\sigma/d\Omega$, P and Wolfenstein parameters from Harvard and Harwell
near 140 MeV and from TRIUMF [22-24] at 220, 325,425 and 495 MeV lead
to well determined I=0 phase shifts from 140 to 500 MeV.

At lower energies, there is a well known instability in the phase shift
solution due to strong correlations between $\bar\epsilon_1$ and $^1P_1$. This could be
resolved [25] by accurate (±0.02) measurements of A_{LL} near 180°. However,
there is the interesting possibility that potential models are already
capable of eliminating the instability by predicting $\bar\epsilon_1$ accurately over
the energy range as follows. At threshold, Ericson et al [26] have
recently demonstrated that the asymptotic ratio of D and S waves in the
deuteron is determined more accurately by theory than experiment. The
ingredients of the argument are (i) the deuteron binding energy determines
the asymptotic shape of the wave functions, (ii) the long-range tensor
interaction which mixes S and D states is given by OPE, of known strength
and range; these are sufficient, but in addition (iii) the accurately
known quadrupole moment of the deuteron checks the long range contributions
to integrals over S and D waves. This argument could be used above
threshold to determine $\bar\epsilon_1$ theoretically up to energies at which other
exchanges (ρ and ω) affect the tensor force, and this energy is of order
$m_\rho^2/2M = 300$ MeV. Furthermore one has a powerful constraint on the
short-range tensor force from experimental values of $\bar\epsilon_1$ from 142 to
500 MeV. These ingredients are built into the Paris potential, which is
therefore probably the most reliable source of information on $\bar\epsilon_1$ from
25 to 140 MeV. Once $\bar\epsilon_1$ is fixed, $^1P_1$ follows from experimental data.

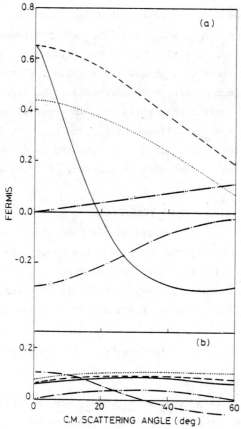

Fig. 4   (a) real and (b) imaginary parts of np charge exchange
          amplitudes at 425 MeV: full line (g–h), dashed m, dash–dot a,
          dotted (g + h), dash–double dot ic.

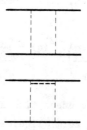

Fig. 5 2π exchange contributions to the long-range part of the Paris
          potential.

Above 500 MeV, there are huge uncertainties in I=0 phase shifts on the basis of published data. However, data have been taken at the HRS at LAMPF at 800 MeV on $D\equiv D_{NN}$, $R\equiv D_{SS}$ and $A\equiv D_{LS}$ and preliminary results are anticipated shortly. Together with P and $d\sigma/d\Omega$ data, it seems quite likely that they will lead to a unique I=0 solution. Unfortunately, they are not the optimum choice of measurements. The data which do most to fix I=0 phase shifts accurately are, in order of priority, A_{NN} and $D_T\equiv K_{NN}$, then $R_T\equiv K_{SS}$, followed by A_{LL} and A_{SS}. It is unlikely that all these need measuring, but an accurate I=0 solution will need some of them at 650 and 800 MeV with an accuracy of ± 0.03 at $\sim 10^{\circ}$ steps from 50 to 165°.

III. (p,n) REACTIONS

A point of interest to nuclear physicists is that phase shift analysis reconstructs very accurately the amplitudes for np→pn from 140 to 500 MeV. These amplitudes are useful in deducing nuclear matrix elements from (p,n) reactions in nuclei. They vary little with energy from 140 to 500 MeV. Fig 4 shows them at 425 MeV in the conventional notation

$$\mathfrak{M} = a + c(\sigma_{N1}+\sigma_{N2}) + m\sigma_{N1}\sigma_{N2} + (g+h)\sigma_{P1}\sigma_{P2} + (g-h)\sigma_{Q1}\sigma_{Q2},$$

where $\underline{Q}$ denotes a unit vector parallel to $\underline{q} = \underline{p}_f - \underline{f}_i$, $\underline{P}$ is parallel to $\underline{p}_f + \underline{p}_i$ and $\underline{N}$ parallel to $\underline{p}_i \times \underline{p}_f$; $\underline{p}_i$ and $\underline{p}_f$ are initial and final neutron momenta. The amplitudes are almost real since

$$f(np\to pn) = \tfrac{1}{2} \left\{ f_{I=1} - f_{I=0} \right\}$$

and the diffraction amplitudes largely cancel.

In discussion of pion condensation, it is common to adopt a NN interaction,[27] dropping a and c,

$$V(q) = \frac{\underline{\tau}_1 \cdot \underline{\tau}_2}{4M^2} \left[g^2_{\pi}\left(g'\underline{\sigma}_1 \cdot \underline{\sigma}_2 - \frac{\underline{\sigma}_1 \cdot \underline{q}}{q^2+m^2_{\pi}}\right)\nu_{\pi}^2(q) - \frac{f_{\rho}^2\left(\underline{\sigma}_1 \times \underline{q}\right)\cdot\left(\underline{\sigma}_2 \times \underline{q}\right)\,\nu_{\rho}^2(q)}{q^2 + m^2_{\rho}} \right].$$

Theoretical discussions of pion condensation are presently inconclusive because of uncertainty about the Fermi liquid parameter g', which is variously taken to lie between 0.35 and 0.7. For g' $\gtrsim$ 0.5, pion condensation does not occur. For free np scattering, the value of the amplitude (g-h) at t=0 is very stable from 140 to 500 MeV, and gives

$$g' = 0.51 \pm 0.02 \qquad\qquad (1)$$

Of course, g' may be different in nuclei or nuclear matter because of off-mass-shell and many-body effects. However, the parameters presently used in models of pion condensation ought to be constrained to give the value of equn (1) for free np scattering.

Experimentally, it should not be difficult [28] to determine g' in nuclei. If one selects a Gamow-Teller transition, so as to eliminate the spin independent term a in $\mathfrak{M}$, and measures the $R = D_{SS}$ parameter (initial and final nucleon polarisations transverse and in the scattering plane), then $R\, d\sigma/d\Omega \simeq |g-h|^2 - |m|^2 - |g+h|^2$ for small q, isolating the term

$$\left(g' - \frac{q^2}{q^2 + m_\pi^2} \right) \sigma_{Q1}\, \sigma_{Q2} \qquad\qquad (2)$$

in $V(q)$. The amplitude $(g-h)$, shown by the full line in Fig 4, has the steep angular dependence characteristic of π exchange, and changes sign at $q^2 \simeq m_\pi^2$ due to cancellation between the two terms in (2). The zero shows up experimentally as a deep dip in the R parameter; in free np scattering it is a very conspicuous feature of the data [23], and is likely to be equally conspicuous in nuclei. This zero determines g' very precisely in terms of m_π^2. A snag, requiring theoretical input, is that the effective value of m_π^2 may be different in nuclei to its free space value, but this is related by theory back to g' itself.

IV. POTENTIAL MODELS

The Paris potential [4,29] is the best potential model fit to NN data up to 425 MeV. Its long range part comes from single and two pion (fig 5) exchange. The latter is calculable from experimental data on πN elastic scattering, so apart from g_π^2, which is experimentally determined [30] to $\pm4\%$, there appears superficially to be no free parameter in the long-range interaction. In reality, a subtraction constant is required in a dispersion relation using the πN data, with the result that the Paris group essentially treats the S-wave $\pi\pi$ scattering length as an adjustable parameter, (and hence the long range central potential); they find $a_o (\pi\pi)=0.59 m_\pi^{-1}$, compared to current algebra predictions $a_o \simeq 0.2 m_\pi^{-1}$. For impact parameters <0.8 f, the theoretical potential is replaced by

10

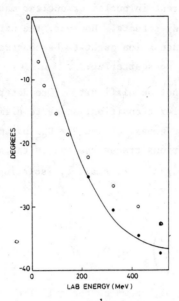

Fig. 6   Phase shift values of δ ($^1P_1$) from Arndt and VerWest, ref. 8
          (full line) compared with original predictions of the Paris
          potential, ref. 29 (open circles) and the revised values of
          Vinh Mau, ref 31 (full circles).

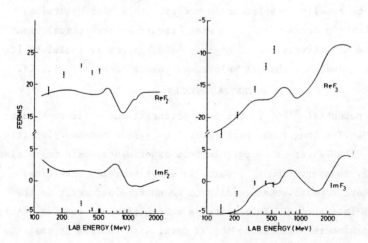

Fig. 7   Np spin-dependent forward amplitudes F_2 and F_3 of Grein and
          Kroll, ref. 33 (full curves), compared with phase shift
          predictions of Dubois et al., ref 10 (crosses).

a phenomenological potential which is constant or linear in lab energy. Since a Yukawa potential with a range of 0.8 f corresponds to an exchanged mass of only 1.75 m_π, one should remember that 2π contributions to the theoretical part of the Paris potential come from very low masses close to the 2π threshold; thus the ρ meson itself makes a rather small contribution to the theoretical Paris potential.

Dubois et al,[4] compared NN high partial waves critically with the Paris potential. They found good agreement for tensor amplitudes (dominated by OPE), fair agreement for spin-orbit, and only qualitative agreement for central (spin-averaged) amplitudes. This comparison showed that there was also room for improvement in some phenomenological low partial waves, particularly $^1P_1$. Since then, Vinh Mau[31] has shown that some fine tuning of the $^1P_1$ prediction is possible without significantly affecting $^1F_3$ and $^1H_5$, with the improved results shown on Fig 6, and correspondingly better fits to np data, particularly $R_T \equiv K_{SS}$.

The overall goodness of fit of the Paris potential to experimental data from 10 to 350 MeV is a χ^2 per point of 2.23, whereas phase shift analysis achieves about 1.5. The conclusion is that the Paris potential is a valuable parametrisation of experimental data for extrapolation to nuclear physics and nuclear matter; however, one achieves a significantly better result by reconstructing amplitudes from the phase shift solutions, where this is practicable.

V. FORWARD DISPERSION RELATIONS

In outline, forward dispersion relations take the form:

$$\text{Re } f(s) = \text{Deuteron pole} + \text{meson-exchange poles} + \frac{1}{\pi} \int_{4M^2}^{\infty} \frac{\text{Im } f(s')ds'}{s' - s} ,$$

The deuteron pole is known from the asymptotic wave function of the deuteron, and Re f(s) and Im f(s) are available from phase shift analysis. Hence, to the extent that one can cut off the dispersion integrals at the top energies of the phase shifts, one has an unambiguous way of evaluating the meson-exchange poles. By choosing appropriate spin combinations, one can isolate meson exchanges with particular quantum numbers, e.g. the unnatural parity sequence 0^-, 1^+, 2^- ...

Grein and Kroll[32] have carried out such an analysis, putting in OPE and 2π exchange calculated (as in the Paris potential) from experimental

12

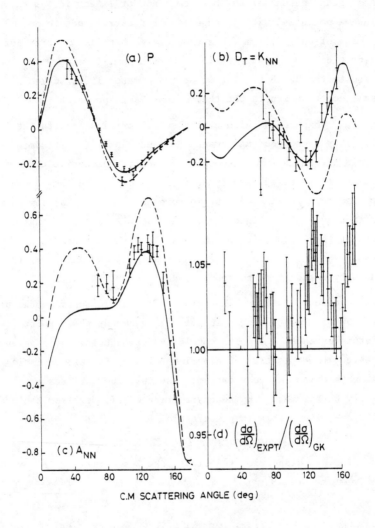

Fig. 8   (a)-(c) Values of P, $D_T \equiv K_{NN}$ and A_{NN} compared with the fit of
          Dubois et al., ref 10 (full lines) and the fit constrained to
          the forward amplitudes of Grein and Kroll, ref 33 (dashed lines).
          (d) A small selection of experimental values of $d\sigma/d\Omega$ compared
          to the fitted values when the fit is constrained to the forward
          amplitudes of Grein and Kroll.

πN data. The remaining discrepancies with the data are interpreted very plausibly in terms of 3π, A_1 and ω exchange. They are even able to relate the 3π contributions sensibly to $\pi\sigma$ and $\pi\rho$. My opinion is that this is the best available dispersion relation analysis at present.

However, Grein and Kroll have since done another analysis [33] with different results, which in my view are unsatisfactory. It is this set of amplitudes which is tabulated in the review of Kroll [3]. Their objective in this second analysis was different to that of the first. They were trying to fit Argonne results for $\Delta\sigma_L$, both pp and np (derived from pd) [34], and to make a case for an I=0 $^1F_3$ dibaryon resonance. They chose to ignore np spin dependent amplitudes from phase shift analysis except at threshold and 140 MeV, and calculated $\Delta\sigma_T$ (np) from theoretical input. The np forward amplitudes which emerged are in serious disagreement with phase shift analysis (Fig 7). If one forces the phase shift analysis to fit these forward amplitudes, χ^2 rises enormously (by $\sim$ 1000 at 425 MeV), and the fits to much of the np data deteriorate by unreasonably large amounts; (examples are illustrated in Fig 8). Agreement cannot be secured by rejecting any one experimental data set (or indeed two). It is easy to locate the origin of the clash. The dominant contributions to I=0 forward amplitudes come from the large partial waves $^1P_1$, $^3D_1$ and $^3D_2$, and these are over-determined by the shape of $d\sigma/d\Omega$, $R_T \equiv K_{SS}$, $D_T \equiv K_{NN}$ and A_{NN}. My conclusion is that this second set of forward amplitudes is unacceptable and there must be something wrong with either the theoretical derivation of $\Delta\sigma_T$(np) or Argonne values of $\Delta\sigma_L$ (np). Grein and Kroll predicted in the np inelastic cross section structure which is ruled out by the data of Dakhno et al [35] (Fig 9). Also, recent LAMPF measurements of the np total cross section [36] from 40 to 770 MeV fail to find any peak corresponding to the predicted I=0 $^1F_3$ dibaryon (Fig 10).

VI. DIBARYONS

Excitement about the possible existence of dibaryon resonances was aroused [37] by the discovery at Argonne [17] of structure in $\Delta\sigma_L$. There followed a phase when wiggles in all varieties of data were labelled as potential dibaryons, but now that the dust has settled there are two clear pp candidates, $^1D_2$ (2175) and $^3F_3$ (2220), corresponding to the peak and dip in $\Delta\sigma_L$ at 550 and 800 MeV in Fig 11. As explained in the previous

14

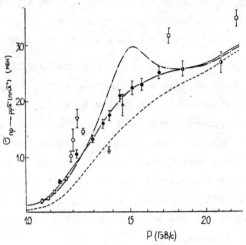

Fig. 9   Total cross sections of the reactions np → ppπ⁻ and np → nnπ⁺,
ref 35.   The dot-dashed curve takes into account dibaryon
resonances predicted by Grein and Kroll, ref 33.

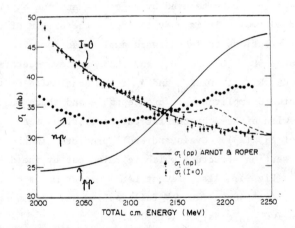

Fig. 10   NN total cross sections, ref. 36. The dashed curve shows
the magnitude of the dibaryon resonance predicted by Grein and
Kroll, ref 33, if added to a smooth background. Note, however,
that the forward amplitude of Grein and Kroll shows smaller
structure than this because of compensating structure in the
background.

section , present evidence is against I=0 dibaryons. If I=1 dibaryon
resonances exist, it is clear that they are associated with the onset
of strong inelasticity and are highly inelastic. For example, after
subtracting off empirical smooth backgrounds, Hoshizaki [11] finds
branching ratios to the NN channel of 0.1 for $^1D_2$ and 0.2 for $^3F_3$, i.e.
they appear at best as small pimples on strong inelastic thresholds.
It is therefore hardly surprising that from elastic data alone some
authors claim the existence of resonant poles, e.g. Edwards [38] and
Bhundari et al [39], whereas others [40-42] argue that conventional
models of inelasticity, correctly unitarised, reproduce well the half-
loops observed in the Argand diagram of each partial wave. It is
clear that coupling to πd is also small [43], from the fact that the πd
total cross section and elastic angular distribution agree accurately
with calculations based on πN amplitudes using the Glauber model plus
a spreading width for the $\Delta(1230)$ allowing for the extra channel
$\pi d \rightarrow \Delta N \rightarrow NN$. Hence the place to look for dibaryons must be in the
$N\Delta$ channel; I have discussed this at the Versailles conference [6].

Fig 11 demonstrates that there are peaks in both $\Delta\sigma_L$ and $\Delta\sigma_T$ at
550 MeV, just below the very large peak in the πd total cross section.
With hindsight, I feel that the peaks in $\Delta\sigma_L$ and $\Delta\sigma_T$ might have been
predicted as follows. The 230 mb peak in the πd total cross section is
due to

$$\pi p \ (+n) \rightarrow \Delta \ (+n)$$
$$\pi n \ (+p) \rightarrow \Delta \ (+p) \ ,$$

In 5% of interactions, the Δ de-excites to a nucleon by interaction
with the spectator:

$$\Delta N \rightarrow NN.$$

Since the spectator has low momentum, it generally carries no angular
momentum, so one expects the final NN state to have quantum numbers
I=1, $J^P = \vec{\frac{3}{2}}^+ + \vec{\frac{1}{2}}^+ = 1^+$ or 2^+; of these, only $J^P = 2^+$ is allowed by
the Pauli principle. By detailed balance, one expects a corresponding
peak in pp $\rightarrow \pi d$ in the $^1D_2$ state. Clebsch-Gordan coefficients are such
that one expects equal contributions of $\sim$8 mb in both $\Delta\sigma_L$ and $\Delta\sigma_T$.
To this one should add some contribution from the same argument
applied to pp $\rightarrow$ NNπ, so the magnitudes of the peaks in Fig 11 are
readily explicable. The downward shift in energy compared to the peak
in the πd total cross section is simply due to the fact that, as the

energy rises, the reaction pp → NΔ acquires one or more units of
angular momentum in the final states, and $^3F_3$ and $^3P_2$ contribute
negatively to $\Delta\sigma_L$ and $\Delta\sigma_T$ respectively. Niskanen [44] has given a semi-

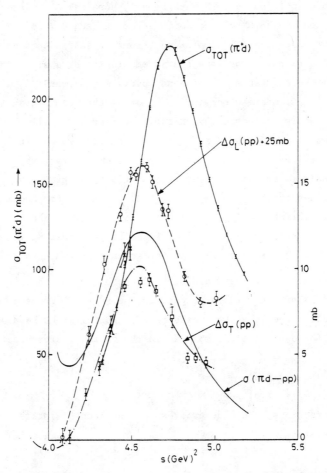

Fig. 11 Values of the πd total cross section, σ(πd→pp) full line,
$\Delta\sigma_L$ + 25 mb from ref. 18 (open circles), and $\Delta\sigma_T$ from
TRIUMF (crosses) and Saclay (open squares) plotted against
s (centre of mass energy squared). The left hand scale
refers to the πd total cross section and the right hand
scale to the remainder.

quantitative version of this argument, and has demonstrated that
one expects strong contributions to inelastic thresholds in successively
$^{1}D_{2}$, $^{3}P_{2}$, $^{3}F_{3}$, $^{1}G_{4}$, $^{3}H_{5}$, $^{1}I_{6}$, etc. This agrees with the known
qualitative features of pp $\to$ dπ^{+} and pp inelasticities from phase
shift analysis. It is plausible (but not yet proven) that this sequence
of inelastic thresholds can account for all the phenomena ascribed to
dibaryons.

REFERENCES

1.   D. V. Bugg, in Progress in Particle and Nuclear Physics, Vol. 7,
     ed. Sir D. H. Wilkinson (Pergamon Press, Oxford, 1981), 47.

2.   P. Signell, in Proceedings of the Telluride Conference on the
     (p,n) reaction and the Nucleon-Nuclear Force, eds. C. D. Goodman
     et al. (Plenum, New York, 1980), 1.

3.   P. Kroll, Physics Data (to be published).

4.   R. Vinh Mau, in Mesons in Nuclei, Vol. 1, eds. M. Rho and
     Sir D. H. Wilkinson (North-Holland, Amsterdam, 1979), 151.

5.   A. Yokosawa, in Reports of the Meeting on Nucleon-Nucleon
     Interactions and Dibaryon Resonances, Hiroshima, HUPD-8001 (1980).

6.   D. V. Bugg, in High Energy Physics and Nuclear Structure, eds.
     P. Catillon, P. Radvanyi and M. Porneuf (North-Holland, Amsterdam,
     1981), 95.

7.   R. Hausammann, Ph.D. thesis, University of Geneva, 1982.

8.   R. A. Arndt and B. J. VerWest, Texas A. & M. preprint
     DOE/ER/05223-29, (1980).

9.   J. Bystricky, C. Lechanoine and F. Lehar, Saclay preprint
     D.Ph.P.E. 79-01 (1979).

10. R. Dubois et al., Nucl. Phys. A377 (1982) 554.

11. N Hoshizaki in High Energy Physics with Polarized Beams and
     Polarized Targets, ed. G. H. Thomas, AIP Conf. Proc. 51 (Amer.
     Inst. Phys, New York, 1979) 399.

12. J. Côté, M. Lacombe, B. Loiseau and W. N. Cottingham, IPNO/TH
     81-41 (1981).

13. J. H. Gruben and B. J. VerWest, Texas A. & M. preprint,
     DOE/ER/05223-44 (1981).

18

14. R. R. Silbar and W. M. Kloet, Nucl. Phys. A338 (1980) 317.

15. A. M. Green and M. E. Sainio, J. Phys. G5 (1979) 503.

16. A. M. Green and M. E. Sainio, J. Phys. G8 (1982) 1337.

17. I. P. Auer et al, Phys. Lett. 67B (1977) 113; ibid 70B
     (1977) 475; Phys. Rev. Lett. 41 (1978) 354.

18. I. P. Auer et al, Argonne preprint ANL-HEP-PR-81-13 (1981).

19. E. Aprile et al., in High Energy Physics with Polarized Beams
     and Polarized Targets, ed. C. Joseph and J. Soffer, (Birkhauser
     Verlag, Basel, 1981), 516.

20. D Axen et al., J. Phys. G7 (1981) L225.

21. B. J. VerWest and R. A. Arndt, Phys. Rev. C25 (1982) 1979.

22. A. S. Clough et al., Phys. Rev. C21 (1980) 988.

23. D. Axen et al., Phys. Rev. C21 (1980) 998.

24. R. Keeler et al., Nucl. Phys. A377 (1982) 529.

25. D. V. Bugg, J. Phys. G 6 (1980) 1329.

26. T.E.O. Ericson and M. Rosa-Clot, Phys. Lett. 110B (1982) 193.

27. J. Delorme, M. Ericson, A. Figureau and N. Giraud, Phys. Lett.
     89B (1980) 327.

28. D. V. Bugg, J. Phys. G 7 (1981) L141.

29. M. Lacombe et al., Phys. Rev. C21 (1980) 861.

30. D. V. Bugg et al., J. Phys. G 4 (1978) 1025.

31. R. Vinh Mau, private communication.

32. W. Grein and P. Kroll, Nucl. Phys. A338 (1980) 332.

33. W. Grein and P. Kroll, Phys. Lett. 96B (1980) 176.

34. H. M. Spinka in AIP Conf. Proc. 51 (382) (High Energy Physics with
     Polarised Beams and Polarised Targets, Argonne, 1978), ed.
     G. H. Thomas (Amer. Inst. Phys., New York, 1979).

35. L. G. Dakhno et al., Phys. Lett. 114B (1982) 409.

36. P. W. Lisowski et al., Phys. Rev. Lett. 49 (1982) 255.

37. K. Hidaka et al., Phys. Lett. 70B (1977) 479.

38. B. J. Edwards, Phys. Rev. D23 (1981) 1978.

39. R. Bhundari, R. A. Arndt, L. D. Roper and B. J VerWest,
     Phys. Rev. Lett. 46 (1981) 1111.

40. W. M. Kloet and R. R. Silbar, Phys. Rev. Lett. $\underline{45}$ (1980) 970.

41. B. J. VerWest, Phys. Rev. $\underline{C25}$ (1982) 482.

42. W. M. Kloet, J. A. Tjon and R. R. Silbar, Phys. Lett. $\underline{99B}$ (1981) 80.

43. Yu. A Simonov and M. Van der Velde, Phys. Lett. $\underline{76B}$ (1978) 277.

44. J. A. Niskanen, Phys. Lett. $\underline{112B}$ (1982) 17.

EFFECTIVE INTERACTIONS AT LOW AND INTERMEDIATE ENERGY

Claude Mahaux[*]

Max-Planck-Institut für Kernphysik, D-6900 Heidelberg, W.-Germany

ABSTRACT

Brueckner's reaction matrix g is an attractive candidate for
the effective nucleon-nucleon interaction at positive as well as at
negative energies. At positive energies, numerical calculations have
been limited to nuclear matter; the results are then applied to finite nuclei with the help of local density approximations. Even in
the case of nuclear matter, g is a complicated operator since it
is nonlocal, state-dependent, nonhermitean and density-dependent;
practical applications request that it be approximated by sets of
local operators. Further work is needed concerning the definition
and the calculation of g in nuclear matter, its replacement by local operators, the local density approximation, the comparison with
experimental data and with empirical effective interactions, and the
relationship with the multiple scattering approach.

1. INTRODUCTION

Elastic and inelastic nucleon-nucleus scattering data can be
accurately fitted in the framework of the optical model and of the
distorted wave Born approximation. These phenomenological fits involve adjustable parameters whose number is finite only because one
implicitly or explicitly uses theoretical models. The meaningfulness
of the adjusted parameter values thus hinges upon the validity of
these models. Models are involved in the description of the wave
function of the target or of the residual nucleus. This aspect of
the analysis mainly belongs to the field of nuclear spectroscopy and
is not discussed here. Rather, we focus upon the microscopic description of the effective interaction between a positive energy nucleon
and a target nucleon.

Theoretical progress on this topic has mainly taken place during
the decade which extends from 1968 (see e.g. ref. [1]) to 1978. A rather exhaustive description can be found in the proceedings of the
1978 Hamburg Conference.[2] Since then, most efforts have been devoted
to the comparison between the theoretical models and the experimental
data. At low[3] as well as at intermediate[4] energies, the agreement is
often sufficiently encouraging to stimulate further theoretical work.
However, disagreements exist, see e.g. refs.[4,5], which raise the nag-
ging worry that the agreement might be partly accidental. This points
to the necessity of critically examining the accuracy of the available models and numerical calculations. This is especially true at
intermediate energy where the high quality data now available provide
stringent tests of the effective interaction. Relatedly, the reaction
mechanism at intermediate energy is hopefully sufficiently simple for

[*] Permanent address : Institut de Physique, Université de Liège,
Bâtiment B5, B-4000 Liège 1, Belgium

enabling one to extract spectroscopic information from nucleon inelastic scattering provided that a reliable effective nucleon-nucleon interaction be available.[6,7]

Since we aim at a unified description of scattering processes at low and intermediate energy, we shall mainly survey the approaches which are based on a <u>realistic</u> nucleon-nucleon interaction. We shall assume that this interaction may be adequately described by a nonrelativistic two-body potential. This restricts our discussion to energies lower or at most only slightly larger than the pion production threshold. At larger energies, one should probably use the multiple scattering approach. Even below the pion production threshold, the possibility exists that a relativistic pseudopotential need be used and that isobar excitations are important. These topics are discussed elsewhere in these proceedings.

Here, we shall mainly discuss a method based on a extension[8] of Brueckner's theory of nuclear matter. As a matter of fact, Brueckner's theory was originally derived from the multiple scattering expansion.[9] The two approaches have been briefly compared in ref.[8]; further work in this direction would be quite instructive.

Section 2 is devoted to the effective nucleon-nucleon interaction in the limiting case of nuclear matter, which constitutes the starting point of all the available calculations. Finite nuclei are considered in Sect. 3 . In keeping with our assigned framework, we shall attempt to provide a critical examination of the accuracy of the approximation adopted by various groups. We shall <u>not</u> discuss the comparison with experimental data, partly because this is the task of the next speaker, and partly because one must beware of mistaking a successful comparison for a justification.

2. EFFECTIVE INTERACTIONS IN NUCLEAR MATTER

2.1. DEFINITION OF THE REACTION MATRIX

Two main methods exist in nuclear matter theory. Until a few years ago, most works adopted the framework of the Brueckner-Bethe low-density expansion which we shall discuss below. Variational approaches have however made much progress in recent years. They have recently been extended to the calculation of the optical-model potential.[10-12] The comparison between these two approaches will certainly prove fruitful.

Brueckner's theory of nuclear matter is based on the hope that most physical properties are dominated by two-nucleon correlations. If this holds, these properties can be described in terms of two-body correlated wave functions (Eq. (2.1.23)) or equivalently in terms of a two-body reaction matrix. This hope also lies at the basis of the applications of Brueckner's theory to nuclear ground states; in this case the calculations are performed either directly,[13] or else proceed via the construction of the reaction matrix in nuclear matter followed by the use of a local density approximation.[14-16] These two approaches are in good agreement[17] provided that some higher-order corrections are taken into account, either in the form of second- and third-order corrections to the Brueckner-Hartree-Fock field, or in the form of a phenomenological readjustment of the interaction and of the inclusion of a rearrangement potential in the local density Hartree-Fock approximation. Thus, Brueckner-Hartree-Fock calculations in nuclear matter probably provide a reasonable starting point for the theoretical description of nuclear scattering at low energy. This should remain true at intermediate energy where the Brueckner-Hartree-Fock field becomes close to the one which would be derived from the multiple scattering expansion. Note, however, that three-body correlations are not negligible at intermediate energy. For instance, it has recently been claimed that unitarity constraints naturally lead to a formulation of the optical-model potential in terms of three-body rather than two-body amplitudes.[18]

Originally, the Bethe-Brueckner expansion was worked out for the average binding energy per nucleon of nuclear matter.[19] It was later extended to the optical-model potential.[8] In a finite nucleus, the optical-model potential for a nucleon with energy E is nonlocal and energy-dependent. We omit reference to the spin and isospin degrees of freedom and write it in the form

$$M(\vec{r},\vec{r}';E) = V(\vec{r},\vec{r}';E) + i\ W(\vec{r},\vec{r}';E) \quad . \qquad (2.1.1)$$

In many-body physics, M is called the "mass operator" or the "self-energy". In symmetric nuclear matter, the nonlocality simplifies since translational invariance entails that M is a function of only the quantity $|\vec{r}-\vec{r}'|$. It also depends on the medium density ρ which is related to the Fermi momentum by

$$\rho = \frac{2}{3\pi^2} k_F^3 \quad . \qquad (2.1.2)$$

It is often convenient to perform a Fourier transform over $|\vec{r}-\vec{r}'|$; for simplicity, we keep the same notation for the Fourier transformed function :

$$M_\rho(|\vec{r}-\vec{r}'|;E) \leftrightarrow M_\rho(k;E) \quad . \qquad (2.1.3)$$

The variable k can be identified with the nucleon momentum.[20]

The leading term of the low-density expansion for M is the Brueckner-Hartree-Fock approximation. It reads

$$M_\rho^{(1)}(k;E) = \sum_{j<k_F} \langle \vec{k},\vec{j}|g[E+e(j)]|\vec{k},\vec{j}\rangle_A \quad , \qquad (2.1.4)$$

where A stands for antisymmetrization :

$$|\vec{k},\vec{j}\rangle_A = |\vec{k},\vec{j} - \vec{j},\vec{k}\rangle \quad . \qquad (2.1.5)$$

Here, $|\vec{k}\rangle$ denotes a plane wave with momentum $\vec{k}$. The upper index (1) refers to "first order". The operator g is the reaction matrix; it is the solution of the integral equation

$$g[w] = v + v \sum_{a,b>k_F} \frac{|\vec{a},\vec{b}\rangle \langle \vec{a},\vec{b}|}{w-e(a)-e(b)+i\delta} g[w] \quad , \qquad (2.1.6)$$

where v is the nucleon-nucleon potential.

The energies e(q) are given by ($\hbar$ = 1)

$$e(q) = \frac{q^2}{2m} + U(q) \quad , \tag{2.1.7}$$

where U(q) is an adjustable "auxiliary" potential. In practice, one tries to choose U(q) in such a way as to optimize the rate of convergence of the low-density expansion. The proper choice of U(q) is a long debated topic in Brueckner's theory. In binding energy calculations, the "standard choice" consists in taking

$$U(q) = 0 \quad \text{for} \quad q > k_F \quad , \tag{2.1.8a}$$

$$U(q) = V^{(1)}(q,e(q)) =: V^{(1)}(q) \quad \text{for} \quad q < k_F \quad . \tag{2.1.8b}$$

This choice is numerically convenient but gives rise to a large discontinuity at $q = k_F$. For instance, one has typically[20]

$$U(k_F-0) \approx -60 \text{ MeV} \tag{2.1.9}$$

for $k_F = 1.35 \text{ fm}^{-1}$ ($\rho = 0.17 \text{ fm}^{-3}$) . Convincing arguments exist for taking a "continuous choice" for U(q) at $q = k_F$, namely for taking

$$U(k_F \pm 0) = M(k_F;e(k_F)) =: M(k_F) \quad , \tag{2.1.10}$$

where M is the exact value of the mass operator. Less convincing arguments exist[8] for extending (2.1.10) to all momenta, i.e. for taking

$$U(q) = M(q;e(q)) =: M(q) \tag{2.1.11}$$

for all values of q . Note that U would then be a complex quantity. Since M is not known, one in practice approximates the right-hand side of Eq. (2.1.11) by the real part of the Brueckner-Hartree-Fock field (2.1.4). One thus takes[21,22]

$$U(q) = V^{(1)}(q;e(q)) = V^{(1)}(q) \tag{2.1.12}$$

for all values of q . At the Fermi surface, i.e. for $q = k_F$, the difference between (2.1.12) and the better prescription (2.1.11) can be as large as ten MeV for $k_F > 1.1 \text{ fm}^{-1}$:

$$V(k_F) - V^{(1)}(k_F) \approx 10 \text{ MeV} \quad . \tag{2.1.13}$$

However, the real part of the Brueckner-Hartree-Fock field is not very sensitive to the choice of U(q) . It changes by typically 2 MeV when U(q) is changed by 10 MeV for $q < k_F$. This is not true for the imaginary part, at least at low energy. Indeed, one has

$$W^{(1)}(k;E) \sim (E-e(k_F))^2 \tag{2.1.14}$$

for E close to e(k_F) . Hence, replacing (2.1.10) by (2.1.12) shifts the zero of the imaginary part by as much as 10 MeV towards negative energies. The threshold behaviour (2.1.14) and the estimate (2.1.13) indicate that the choice (2.1.12) probably leads to a sizeable over-estimate of the imaginary part W of the average field at low energy and at densities which correspond to the target interior.

In summary, the definition (2.1.6) of the reaction matrix invol-ves an auxiliary potential $U(q)$ whose optimal choice is somewhat uncertain. The real part $V^{(1)}$ of the Brueckner-Hartree-Fock field remains fairly stable when $U(q)$ is changed. In contrast, $W^{(1)}(k;E)$ is a sensitive function of $U(q)$ for E smaller than several tens MeV. The quantity

$$W^{(1)}(k;E-e(k_F))\qquad (2.1.15)$$

is probably more stable against changes of $U(q)$;[23] it would thus be desirable to formulate practical applications in terms of the quan-tity (2.1.15). The problem of "the best" choice of $U(q)$ for $q \neq k_F$ could be investigated by estimating the higher-order corrections to the Brueckner-Hartree-Fock approximation. A comparison with other formalisms e.g. the variational and the multiple scattering approa-ches would also be quite instructive.

Let us write Eq. (2.1.6) in the abbreviated form

$$g[w] = v + v\,\frac{Q}{w-e+i\delta}\,g[w]\quad,\qquad (2.1.16)$$

where

$$Q = \sum_{a,b>k_F} |\vec{a},\vec{b}> <\vec{a},\vec{b}|\quad,\qquad (2.1.17)$$

is the Pauli operator which restricts intermediate particles to lie outside the Fermi sea. Equation (2.1.6) is formally quite similar to the equation fulfilled by the nucleon-nucleon transition matrix t in free space, namely

$$t[w] = v + v\,\frac{1}{w-e_o+i\delta}\,t[w]\quad,\qquad (2.1.18a)$$

where e_o is obtained by setting $U = 0$ in Eq. (2.1.7) :

$$e_o|\vec{a},\vec{b}> = \left(\frac{a^2}{2m}+\frac{b^2}{2m}\right)|\vec{a},\vec{b}>\quad.\qquad (2.1.18b)$$

The physical difference between $t[w]$ and $g[w]$ lies in the appea-rance in the latter of medium corrections in the form of Pauli bloc-king effects (approximated by Q) and of binding energy effects (approximated by the auxiliary potential U in e) .

In free space, the wave function of two nucleons with momentum $\vec{k}$ and $\vec{j}$ is the solution of the Lippmann-Schwinger equation

$$\psi_w^0(\vec{k},\vec{j}) = |\vec{k},\vec{j}> + v\,\frac{1}{w-e_o+i\delta}\,\psi_w^0(\vec{k},\vec{j})\quad;\qquad (2.1.19)$$

one has

$$v\,\psi_w^0(\vec{k},\vec{j}) = t[w]\,|\vec{k},\vec{j}>\quad.\qquad (2.1.20)$$

The nucleon-nucleon scattering cross section is determined by the asymptotic value of ψ_w^0 for large r and for the "on-shell" value

$$w_{on} = \frac{k^2}{2m} + \frac{j^2}{2m} \qquad (2.1.21)$$

of the parameter w. In the center-of-mass frame, the on-shell two-body wave function in free space only depends upon the relative momentum $|\vec{k}-\vec{j}|$; we recall that for simplicity we have omitted the spin and isospin degrees of freedom.

Similarly, one can define the wave function of two correlated nucleons embedded in an infinite medium by the relation

$$v \, \psi_w \, |\vec{k},\vec{j}\rangle = g[w] \, |\vec{k},\vec{j}\rangle \quad . \qquad (2.1.22)$$

One has

$$\psi_w(\vec{k},\vec{j}) = |\vec{k},\vec{j}\rangle + v \frac{Q}{w-e+i\delta} \, \psi_w(\vec{k},\vec{j}) \quad . \qquad (2.1.23)$$

Equation (2.1.22) suggests to identify the effective interaction between a nucleon with momenta $\vec{k}$ and $\vec{j}$ with the "on-shell" value $g[w_{on}]$ of the reaction matrix, with now

$$w_{on} = e(k) + e(j) \quad . \qquad (2.1.24)$$

This yields to the following value for the Brueckner-Hartree-Fock approximation to the optical-model potential

$$M_\rho^{(1)}(k) =: M_\rho^{(1)}(k;e(k)) = \sum_{j<k_F} \langle \vec{k},\vec{j}|g[e(k)+e(j)]|\vec{k},\vec{j}\rangle_A \quad . \qquad (2.1.25)$$

Some explanation is in order concerning the on-shell value (2.1.24). Let us first note that (2.1.25) only depends upon one variable (k), while (2.1.4) involves two variables $(k$ and $E)$. In effect, one obtains (2.1.25) from (2.1.4) by using the energy-momentum relation

$$E = \frac{k^2}{2m} + V^{(1)}(k;E) \quad . \qquad (2.1.26)$$

When writing (2.1.25), we have expressed E in terms of k. Instead, we could have used (2.1.26) to express k as a function $k(E)$ of E; one then gets the energy-dependent local potential

$$M^{(1)}(E) =: M^{(1)}(k(E);E) \quad . \qquad (2.1.27)$$

In the present case, the energy-momentum relation (2.1.26) is equivalent to the Perey-Saxon prescription[24] for constructing an energy-dependent potential which is equivalent to a nonlocal one.[20] Georgiev and Mackintosh[25] have demonstrated numerically that this prescription works quite well in the case of a finite nucleus provided that the exchange contribution is not too large; they also found that it is more accurate to replace in Eq. (2.1.26) $V^{(1)}$ by the complex potential $M^{(1)}$. The corresponding $k(E)$ is then complex and $V^{(1)}(k(E);E)$ is a complex quantity since it is a real function of a complex variable. Its imaginary part is proportional to $W^{(1)}(E)$,

One can show quite generally[26-28] that this amounts to writing the local equivalent potential in the form

$$V^{(1)}(E) + i \frac{\widetilde{m}(E)}{m} W^{(1)}(E) \quad , \tag{2.1.28}$$

where $\widetilde{m}(E)$ is the "k-mass" which is defined as follows[20]

$$\frac{\widetilde{m}(E)}{m} = [1 + \frac{m}{k} \frac{\partial V^{(1)}(k;E)}{\partial k}]^{-1}_{k=k(E)} \quad . \tag{2.1.29}$$

The quantity $\widetilde{m}(E)/m$ raises from about 0.6 at densities which correspond to the nuclear interior to unity at zero density. Hence, the nonlocality factor $\widetilde{m}/m$ is quite sizeable in the nuclear interior.

2.2. CALCULATIONAL PROCEDURES

Three main procedures have been used for calculating the matrix elements of $g[w]$ or the correlated wave function $\psi_w(\vec{k},\vec{j})$ in nuclear matter. In practice, these methods all use "angle-average" approximations for the Pauli operator Q and for the quantity $e(a) + e(b)$ which appear in Eq. (2.1.16). Then, one can perform partial wave expansions so that the integral equations (2.1.16) and (2.1.23) reduce to one-dimensional coupled integral equations.

(a) In the method of Brueckner and Gammel[29] the wave function $\psi_w(\vec{k},\vec{j})$ is expanded into partial waves. Its radial components u are solutions of integral equations of the type

$$u^{JSP}_{\ell'\ell k_F}(qr;E) = j_\ell(qr) \, \delta_{\ell\ell'} + \sum_{\ell''} \int v^{JS}_{\ell'\ell''}(r') \, F^P_{\ell''k_F}(E,r,r') \times$$

$$u^{JSP}_{\ell''\ell k_F}(qr';E) \, dr' \quad ; \tag{2.2.1}$$

here, q and P denote relative and total momenta

$$q = \frac{1}{2} |\vec{k}-\vec{j}| \quad , \quad P = |\vec{k}+\vec{j}| \quad , \tag{2.2.2a}$$

and E is related to w by

$$w = E + e(j) \quad . \tag{2.2.2b}$$

The kernel F is calculated by performing an integration over $t = |\vec{a}-\vec{b}|$. Equation (2.2.1) is solved by discretization over r' followed by a matrix inversion. This method has been used by the Liège group.[21] It is economical for hard core potentials provided that the nucleon-nucleon potential $v(r)$ is not too deep near the hard core radius; otherwise the discretization over r' must involve too many bins. For instance, the method is more accurate in the case of Reid's hard core potential than in the case of the Hamada-Johnston potential. It can also be used for soft core potentials provided that the "soft" core is not too strong; it has for instance recently been applied by Lejeune[30] to the Paris nucleon-nucleon potential;[31] in that

case, due caution must be exercised because the Paris interaction is
nonlocal; derivatives of F then appear; this is annoying because
these derivatives are discontinuous at $r = r'$.

   (b) In the method of Haftel and Tabakin[32] one performs a partial
wave decomposition of the matrix elements $\langle \vec{k}',\vec{j}'|g[w]|\vec{k},\vec{j}\rangle$. The
coefficients fulfill an integral equation of the type

$$g_{\ell'\ell k_F}^{JSP} (q',q;E) = v_{\ell'\ell}^{JS}(q',q) + \sum_{\ell''} \int v_{\ell'\ell''}^{JS}(q',q'') H_{k_F}^{P}(q'',q;E) \times$$

$$g_{\ell''\ell k_F}^{JSP} (q'',q;E) \, dq'' \quad , \qquad (2.2.3)$$

where as above q and q' refer to relative momenta and P to the
total momentum. The quantity $v(q,q')$ is the momentum representation
of the nucleon-nucleon interaction. The kernel H has a simple algebraic form. Equations (2.2.3) are solved by discretization of q'' ,
followed by a matrix inversion. The method is accurate and convenient
if $v(q,q')$ is available in algebraic form, which is usually the
case. It does not apply to hard core potentials but can on the other
hand easily cope with velocity-dependent potentials. It has been used
by Grangé and Lejeune[33] and by Faessler et al.[34] in the case of Reid's
soft core potential, and by Grangé and Lejeune[35] in the case of the
Paris interaction.

   (c) In the reference spectrum method one approximates the kernel
of the integral equation (2.1.23) as follows

$$K = \frac{Q}{w-e(a)-e(b)+i\delta} \approx \sum_{n=1}^{N} \frac{c_n}{t^2+d_n[w]+i\delta} = K^R \quad , \qquad (2.2.4)$$

where as above $t = |\vec{a}-\vec{b}|$. The interest of the reference spectrum
parametrization (2.2.4) is that the integral equation (2.1.16) is
then equivalent to a system of coupled differential equations. In the
original version of the reference spectrum approximation,[36] only one
term was included on the right-hand side of (2.2.4) because the authors were focusing on the calculation of the binding energy; in that
case, $d_1[w_{on}]$ is positive $(w_{on} < 2 \, e(k_F))$. Brieva and Rook[22] extended the method to the case $E > e(k_F)$ appropriate for the description of scattering by including two terms in (2.2.4), whose
right-hand side then reads

$$K^R = \frac{c_R}{t^2+\gamma^2} + \frac{c_F}{t^2-t_o^2+i\delta} \quad . \qquad (2.2.5)$$

Here, t_o^2 denotes the pole of the kernel at $w = e(a)+e(b)$, and c_F
the corresponding residue. The extended reference spectrum approximation (2.2.5) only involves two parameters, namely c_R and γ^2 .
These are adjusted for each value of k_F , w and P in such a way
that (2.2.5) approximates the exact kernel as well as possible. The
authors claim a 5 %,[22] occasionally a 2 %[37] accuracy on the kernel.
The accuracy of the fit to the kernel must depend upon P , and mainly upon w and k_F . Von Geramb[38] mentions that the accuracy becomes
very good if one takes three terms on the right-hand side of (2.2.4).

It is not clear to what extent the accuracy of a fit to the kernel of the integral equation (2.1.23) quantitatively reflects the accuracy of the approximate solution of the integral equation. The reference spectrum approximation has been carefully tested in the simpler case of binding energy calculations and was found to be fairly inaccurate in that case. For instance, it yields a 25 % error on the diagonal element of g in the ^{3}S channel, which is the dominant one and which depends most upon density.[39] In particular, the reference spectrum approximation greatly overestimates the defect function u-j in the ^{3}S channel for distances r > 0.7 fm ;[19] it may thereby yield an effective interaction which is inaccurate in the most important intermediate range. Hence, we believe that this method as it stands should be improved. In the case of binding energy calculations, the difference $K - K^R$ between the exact kernel and its reference spectrum approximation is treated as a perturbation which can be included either by one of the iteration procedures reviewed in ref.[40], or else by matrix inversion.[41] We believe that similar improvements should be used in the case E > 0 . Finally, we recall[42] that the angle-average expression used in ref.[22] for the Pauli operator Q is incorrect when the total momentum P is larger than 2 k_F . This domain is important for the scattering of intermediate energy nucleons.

In summary, three main methods exist for calculating the reaction matrix g . We believe that the complexity of these calculations and their practical interest makes it important that they be checked by independent codes. The most flexible and convenient method appears to be that of Haftel and Tabakin, except for hard core nucleon-nucleon potentials. The method of Brueckner and Gammel is reliable for hard core interactions, and also for soft core potentials in which the short range repulsion is not too strong. The accuracy of the reference spectrum as it stands is difficult to assess. Unstabilities in the original code have been acknowledged by Brieva,[43] but we indicated that the method could be improved into a reliable one.

2.3. PARAMETRIZATIONS OF THE REACTION MATRIX

Equation (2.2.3) shows that in momentum representation, the reaction matrix g is a function of the relative momenta q and q' , of the quantum numbers ℓ , ℓ' , J , S , of the "starting energy" w = E+e(j) and of the total momentum P . The latter can be replaced by an average value; this is probably accurate except for the imaginary part of g[E+e(j)] for E close to the Fermi energy e(k_F) . The reaction matrix is thus a complicated operator. In order to use it as an effective interaction in finite nuclei, one must in practice approximate g by simple operators. More precisely, most authors attempt to construct an effective interaction M_{12} between nucleons 1 and 2 which would have the following simple form

$$M_{12} = M_C(r_{12}) + M_{LS}(r_{12}) \, \vec{L}.\vec{S} + M_T(r_{12}) \, S_{12} \quad , \quad (2.3.1)$$

where the functions $M_C(r)$, $M_{LS}(r)$ and $M_T(r)$ are local, i.e. only involve the internucleon distance r . They are different for

each set of values for the total spin S and total isospin T of
the interacting pair; these functions may moreover depend upon the
energy E of nucleon 1 and upon the "local" density ρ .

For simplicity, let us first consider spinless particles. The
only quantum number of interest is then the orbital angular momentum
ℓ . The radial part of the correlated wave function is of the form
$u_{\ell k_F}$ (qr;E) , and the matrix elements of g are given by

$$g_{\ell \ell k_F}(q',q;E) = 4\pi \int_0^\infty r\, j_\ell(q'r)\, v_\ell(r)\, u_{\ell k_F}(qr;E)\, dr . \quad (2.3.2)$$

Heretofore, practically all authors agree in defining the effective
interaction M by the requirement that it should reproduce the same
matrix element as g when used in the <u>plane wave Born</u> approximation,
namely

$$g_{\ell \ell k_F}(q';q;E) = 4\pi \int_0^\infty r^2\, j_\ell(q'r)\, M_{\ell k_F}(r;E)\, j_\ell(qr) . \quad (2.3.3)$$

We emphasize that other definitions are possible. For instance,
the authors of ref.[44] require that M should have the same matrix
elements as g in a harmonic oscillator basis.

One possibility of fulfilling Eq. (2.3.3) consists in taking[45]

$$M_{\ell k_F}(r;E) = \frac{v_\ell(r)\, u_{\ell k_F}(qr;E)}{r\, j_\ell(qr)} . \quad (2.3.4)$$

In general the right-hand side of Eq. (2.3.4) depends upon q unless
$u_{\ell k_F}$ can be factorized as follows

$$u_{\ell k_F}(qr;E) \approx f_{\ell k_F}(r;E)\, j_\ell(qr) . \quad (2.3.5)$$

Negele[14] argued that (2.3.5) approximately holds for two on-shell
nucleons (E = e(k)) with momenta k and j smaller than k_F . It
has not been checked whether (2.3.5) is valid when the energy E of
nucleon 1 is positive; in that case, $u_{\ell k_F}$ is a <u>complex</u> function.

In practice one gets rid of the q dependence of the right-
hand side of (2.3.4) by performing an average over the momentum $\vec{j}$
of the "target" nucleons[46]

$$M_{\ell k_F}(r;k;E) = \frac{\sum\limits_{j<k_F} v_\ell(r)\, j_\ell(qr)\, u_{\ell k_F}(qr;E)}{r \sum\limits_{j<k_F} j_\ell^2(qr)} , \quad (2.3.6)$$

where we recall that $q = \frac{1}{2}|\vec{k}-\vec{j}|$. The prescription (2.3.6) is a
plausible one only if (2.3.5) holds for the relevant range of values
for q . This has <u>not</u> yet been checked in the case $k > k_F$, E > 0 .
It appears somewhat doubtful that (2.3.5) is accurate for the imaginary part of $u_{\ell k_F}$(r;E) when the energy E is smaller than several
MeV.

The a priori requirement that M be <u>local</u> is a stringent one.
Indeed, the off-diagonal elements are then all determined by the dia-

gonal ones.[47] For s-waves, for instance, one has $(q' > q)$

$$<q'|M_o|q> = <\tfrac{1}{2}(q'+q)|M_o|\tfrac{1}{2}(q'+q)> - <\tfrac{1}{2}(q'-q)|M_o|\tfrac{1}{2}(q'-q)> .$$

In this example, the reaction matrix cannot be replaced by a local interaction unless this relation is fulfilled. Within the requirement that the effective interaction be local, one could thus content oneself of requesting it to fit the diagonal elements of the g-matrix, as done for instance by Elliott and collaborators.[48,49] These authors have shown that the requirement

$$g_{\ell\ell k_F}(q,q;E) = 4\pi \int_o^\infty r^2 \, j_\ell^2(qr) \, M_{\ell k_F}(r;E) \, dr \qquad (2.3.7a)$$

fully determines $M_{\ell k_F}(r;E)$. One has indeed

$$M_{\ell k_F}(r;E) = \int_o^\infty f_\ell(qr) \, g_{\ell\ell k_F}(q,q;E) \, dq \quad , \qquad (2.3.7b)$$

where f_ℓ is a known function. In the present case, however, this way of defining $M_{\ell k_F}(r;E)$ may be too demanding, since it involves the diagonal elements of g for all values of q . In effect, the prescription (2.3.6) amounts to taking into account that in the present context one is mainly interested in reproducing the diagonal elements of g for a limited range of values for the wave number q , once the energy E is fixed.

Further complications and to some extent further arbitrariness arise when the spin degree of freedom is taken into account. One could define an effective interaction as follows[50]

$$M^{JS}_{\ell'\ell k}(r;k_F;E) = \frac{\sum\limits_{j<k_F} j_{\ell'}(qr) \sum\limits_{\ell''} v^{JS}_{\ell'\ell''}(r) \, u^{JS}_{\ell''\ell k_F}(qr;E)}{r \sum\limits_{j<k_F} j_{\ell'}(qr) \, j_\ell(qr)} . \qquad (2.3.8)$$

If $M(r)$ could be exactly written in the form (2.3.1), one could find the coefficients M_C , M_{LS} and M_T from Eq. (2.3.8) without ambiguity.[51,52] In practice, this is not true, since for instance one may have to insert terms proportional to $(\vec{L}_{12}\cdot\vec{S}_{12})^2$, $(\vec{L}_{12}\cdot\vec{L}_{12})^2$, ... on the right-hand side of Eq. (2.3.1). In addition to the average over j , one must therefore perform averages over orbital and total angular momenta in order to construct an "average" local interaction $\bar{M}$ of the form (2.3.1). The particular averaging prescription adopted by Brieva and Rook[46] is similar in spirit to the ones proposed by Negele,[14] Sprung[53] and others : it is chosen in such a way as to exactly reproduce the Brueckner-Hartree-Fock approximation $M^{(1)}$ of the mass operator in nuclear matter when used in a Born approximation, i.e. when sandwiched between Bessel functions. Note, however, that only diagonal elements of the reaction matrix enter in the expression of $M^{(1)}$, and that furthermore $M^{(1)}$ involves specific weighting factors. It should thus be investigated whether these particular averaging prescriptions (over $\vec{\jmath}$, ℓ' , ℓ , J) are still appropriate if one wants to use the local effective interaction for describing elastic and inelastic nucleon scattering by nuclei.

Brieva and Rook[46] introduced direct $(\overline{M}_d)$ and exchange $(\overline{M}_{ex})$ components of their central interaction in the usual way :

$$\overline{M}_d(r;k_F;E) = (16)^{-1} (\overline{M}^{00} + 3\,\overline{M}^{01} + 3\,\overline{M}^{10} + 9\,\overline{M}^{11}) , \quad (2.3.9a)$$

$$\overline{M}_{ex}(r;k_F;E) = (16)^{-1} (3\,\overline{M}^{01} + 3\,\overline{M}^{10} - \overline{M}^{00} - 9\,\overline{M}^{11}) , \quad (2.3.9b)$$

where $\overline{M}^{ST}$ denotes the angular momentum average of $M_{(1)}$ in the subspace of total spin S and total isospin T . Then, $M_\rho^{(1)}(k;E)$ is exactly given by

$$M_\rho^{(1)}(k;E) = \rho \int \overline{M}_d(r;k_F;E)\, d^3r + \int \rho\, \rho_{ex}(s;k_F)\, \overline{M}_{ex}(s;\rho;E)\, j_0(ks)\, d^3s ,$$

$$(2.3.10)$$

where

$$\rho\, \rho_{ex}(s;k_F) = \rho\, \frac{3}{(k_F s)^3} [\sin(k_F s) - (k_F s)\cos(k_F s)] \quad (2.3.11)$$

is the off-diagonal part of the single-particle density matrix.

In the original work of Brieva and Rook,[46] some complications are introduced by their use of the reference spectrum method for calculating the reaction matrix. This led them to introduce a large discontinuity in $M(r)$ at the hard core radius $r = r_c$. If the Brueckner-Gammel method had been used, the effect of the hard core would have been to introduce a delta function contribution $\delta(r-r_c)$ in the effective interaction, which would then have been set to zero for $r < r_c$.

It is not convenient to use an effective interaction whose dependence upon r is not given by an algebraic expression and may even be a discontinuous function of r .[46] Thus, Brieva, von Geramb and Rook[38,43,54] computed the Fourier transform of each component of $M(r)$ and fitted these transforms with a sum of Gaussians form factors[55] or of Yukawa form factors.[56] These fits can be transformed back to spatial coordinate representations if one wishes. The components $\overline{M}_C^{ST}$, $\overline{M}_{LS}^{ST}$ and $\overline{M}_T^{ST}$ of the effective interaction have been tabulated for energies E which extend from 0 up to about 180 MeV , and for the Fermi momenta $k_F = 0.6$, $0.8 - (0.1) - 1.4$ fm^{-1}.[55,56]

2.4. HIGHER-ORDER CONTRIBUTIONS

The Brueckner-Hartree-Fock approximation (2.1.4) to the optical-model potential in nuclear matter is the leading term of a low-density expansion.[8] The expressions of the higher-order terms can formally be written down, but little work has been done towards their evaluation. The available estimates[42] are admittedly quite crude. The size of the higher-order terms depends upon the choice of the auxiliary potential $U(q)$. It is not possible to choose $U(q)$ in such a way that the Brueckner-Hartree-Fock approximation $M^{(1)}(k;E)$ be equal to the exact mass operator $M(k;E)$. Accordingly, it would be quite surprizing if detailed agreement with the experimental data could be obtained without introducing adjustable parameters in the effective interaction. This had not yet been done partly because the need has not yet been exhibited and partly because one does not yet how to perform this "renormalization" in a plausible way.

Hints about how to "renormalize" the effective interaction could probably be obtained from the two extreme limits of negative energies on the one hand and of intermediate energies on the other hand. It seems physically reasonable to require that at low energy the effective interaction should be close to those which have been successfully used in the Hartree-Fock calculations of nuclear ground state properties. These calculations are surveyed in refs.[19,40,57]. They all have to include modifications of the effective interactions in order to reproduce the empirical saturation properties of nuclear matter or of ^{208}Pb. The various effective forces yield similar results for the ground state properties, but may have rather different direct and exchange components, and also rather different density-dependences of the short and long range parts of the effective interaction. Negele[14] found it exceedingly difficult to solve the Hartree-Fock equation when a repulsive exchange force is present and therefore constructed a purely attractive effective exchange interaction. Remnants of this difficulty probably persist in the case of scattering and may lead to inaccuracies when strong short range repulsion exists in the exchange part of the effective interaction. This is the case in the interaction of Brieva and Rook.[46]

The other extreme limit at which one would like to impose constraints on the renormalization of the effective interaction is the case of high energy nucleons. There, one believes that the impulse approximation becomes fairly accurate. Hence, one would probably want the effective interaction $\tilde{M}$ to yield in that limit the same matrix elements as the free nucleon-nucleon transition matrix. In ref.[4], it was found that the effective interaction of Bauhoff, Brieva, von Geramb, Rook and Palla still has a sizeable density dependence at $E = 140$ MeV. Hence, it is not clear yet what should be considered as "intermediate" energies.

2.5. DISCUSSION

Independent codes should be developed for calculating the matrix elements $<q',P|g[w]|q,P>$ of the complex reaction matrix in the case of nuclear matter, for Fermi momenta between 0.5 fm^{-1} and 1.4 fm^{-1}. Several groups appear to have that capability within fairly easy access.

Only one prescription[46] has been applied for constructing a local effective interaction from the reaction matrix. The plausibility and the practical usefulness of this prescription should be further investigated; other prescriptions can and should be formulated and confronted with the existing one. In particular, these prescriptions should be required to provide a smooth transition from the effective interaction which describes nuclear ground states to the interaction which is used at intermediate energy.[6]

3. EFFECTIVE INTERACTIONS IN NUCLEI

3.1. INTRODUCTION

Two main approaches have been used to construct the optical-model potential in nuclei from the numerical results computed in the case of nuclear matter. In the first method (Sect. 3.2) it is assumed that the optical-model potential felt by a nucleon at the distance r from the target center is the same as the potential that it would feel in an infinite medium with uniform density $\rho(r)$, where $\rho(r)$ is taken from experiment. In the second method (Sect. 3.3), it is assumed that the effective interaction between two nucleons at the average location r is the same as in an infinite medium with uniform density $\rho(r)$. These two methods are discussed in Sect. 3.4. The second method can be extended to the description of inelastic nucleon scattering (Sect. 3.5).

3.2. LOCAL DENSITY APPROXIMATION FOR THE OPTICAL-MODEL POTENTIAL

In symmetric nuclear matter, the optical-model potential is a function of the medium density ρ, of the nucleon momentum k and of the energy E. In refs.[58-60], the optical-model potential for neutrons was first approximated by the expression

$$M_{LDA}(r;E) = M_{\rho(r)}^{(1)}(k(r;E);E) \quad , \qquad (3.2.1)$$

where $\rho(r)$ is the experimental matter distribution, while $k(r;E)$ is determined by the energy-momentum relation (see Eq. (2.1.26))

$$\frac{k^2(r;E)}{2m} + V_{\rho(r)}^{(1)}(k(r;E);E) = E \quad . \qquad (3.2.2)$$

This local density approximation can identically be written in the form (2.3.10), with k determined by (3.2.2), while k_F is replaced by the r-dependent local Fermi momentum $k_F(r)$, where

$$\rho(r) = \frac{2}{3\pi^2} k_F^3(r) \quad . \qquad (3.2.3)$$

Despite its crude character, this approximation leads to good agreement with the volume integrals of the empirical real and imaginary parts of the optical-model potential.[3,59,60] However, the calculated root mean square radii are significantly smaller than the empirical ones.[60] The main origin of this failure is best exhibited by comparing the real part of the first term on the right-hand side of Eq. (2.3.10), namely

$$M_d(r) = \rho(r) \int \overline{M}_d(s) \, d^3s \qquad (3.2.4)$$

with the Hartree approximation, namely

$$M_H(r) = \int \rho(r') \, \overline{M}_d(|\vec{r}-\vec{r}'|) \, d^3r' \quad , \qquad (3.2.5)$$

we have dropped the variables k , k_F , E which are irrelevant for the present discussion. Expressions (3.2.4) and (3.2.5) are equal in the domain where ρ is independent of r (i.e. in the nuclear interior), but differ at the nuclear surface unless $\overline{M}_d$ is a delta function. Hence, the crude local density approximation (3.2.1) is in error at the nuclear surface because it does not take into account the effect of the finite range of the effective interaction $\overline{M}_d$. This suggests the following "improved" local density approximation[60]

$$V_{ILDA}(r;E) = (t_V\sqrt{\pi})^{-3} \int V_{LDA}(r';E) \, \exp(-|\vec{r}-\vec{r}'|^2/t_V^2) \, d^3r' \,, \quad (3.2.6)$$

and similarly for the imaginary part W . Here, t_V is a range parameter adjusted in such a way as to reproduce the root mean square radius of the real part of the empirical potential well. One finds

$$t_V \approx 1.2 \text{ fm} \quad . \qquad (3.2.7)$$

The imaginary part requires a larger range t_W . It can be verified analytically and numerically that the volume integrals of the potential wells V_{LDA} and V_{ILDA} are about the same for Woods-Saxon-like potentials; this explains why the crude local density approximation yields good agreement with the empirical value of the volume integrals.

3.3. LOCAL DENSITY APPROXIMATION FOR THE EFFECTIVE INTERACTION

Brieva and Rook[46] proposed to use the following local density approximation, based on Eq. (2.3.10) :

$$M_{LDA}^{BR}(r;E) = \int \rho(r') \, \overline{M}_d(|\vec{r}-\vec{r}'|;k_F(r);E) \, d^3r' + \int \rho\left(\frac{\vec{r}+\vec{r}'}{2}\right) \rho_{ex}(|\vec{r}-\vec{r}'|;\times$$

$$k_F(r)) \, \overline{M}_{ex}(|\vec{r}-\vec{r}'|;k_F(r);E) \, j_0(k(r)|\vec{r}-\vec{r}'|) \, d^3r' \,, \quad (3.3.1)$$

where $k_F(r)$ is defined by Eq. (3.2.3) and $k(r)$ by Eq. (3.2.2). This approximation has the merit of automatically including the range of the effective interaction.

The volume integrals derived from the approximations (3.2.1), (3.2.6) and (3.3.1) only differ by less than five per cent. The root mean square radii deduced from (3.2.6) and (3.3.1) are larger than those associated with (3.2.1) and are in fair agreement with the empirical values. We recall that in the case of (3.2.6) this agreement is achieved at the expense of introducing two adjustable parameters.

Little detailed comparison exists between the shapes of the potentials derived from approximations (3.2.1), (3.2.6) and (3.3.1), respectively. Reference[46] contains a comparison between (3.2.1) and

(3.3.1), but is marred by the existence of spurious wiggles[43] due to numerical inaccuracies in the calculation of the reaction matrix. It would be of interest to compare (3.2.6) and (3.3.1). Indeed, (3.3.1) is a priori expected to be a better founded way of improving (3.2.1), while on the other hand (3.2.6) has the merit of being simple and therefore convenient for practical applications.

3.4. DISCUSSION

Let us now try to evaluate the reliability of the theoretical constructions of the optical-model potential described above. We start with limitations or inaccuracies which already exist in the case of nuclear matter, and then turn to the case of nuclei.

(a) The Brueckner-Hartree-Fock approximation to the mass operator is only the first-order term of a low-density expansion. Higher-order terms are not negligible.[61] They have been estimated to yield a repulsive correction to the real part of the potential; this correction appears to be equal to about ten per cent of the Brueckner-Hartree-Fock field.[42] The magnitude of the higher-order corrections to the imaginary part has barely been investigated.[62]

(b) Even within the Brueckner-Hartree-Fock approximation, the quantity which should be identified with the imaginary part of the optical-model potential is not $W^{(1)}(E)$ but rather $(\tilde{m}/m) W^{(1)}(E)$ (see Eq. (2.1.28)), where the coefficient $\tilde{m}/m$ rises from about 0.6 at densities which correspond to the nuclear interior to unity at zero density. This factor $\tilde{m}/m$ has not been applied in the available calculations, in particular in refs.[46,52,59,60,63]

(c) The same correction factor $\tilde{m}/m$ should multiply the symmetry potential. This was taken into account in ref.[59] but not in refs.[46,52,63]. It appears plausible that it should also be applied to the spin-orbit component[51] but this has not yet been studied.

(d) The second-order (so-called "rearrangement") term yields a significant contribution to the symmetry potential. This was included in ref.[59] but neglected in refs.[46,52,63].

(e) The energy-momentum relation (2.1.26) should include the effect of the Coulomb potential and of the symmetry potential. Both of these were included in ref.[52], but the symmetry potential was omitted in ref.[60].

(f) One should investigate to what extent the mass operator in nuclear matter depends upon the nucleon-nucleon interaction in free space. It appears that Reid's soft core potential (supplemented by suitable components acting in high partial waves) and the Paris potential are realistic candidates in the present context.

(g) The local effective interaction $\overline{M}$ exactly reproduces the Brueckner-Hartree-Fock approximation when introduced in Eq. (2.3.10). However, its individual matrix elements $<q'|\overline{M}|q>$ are different from those of the reaction matrix. The difference between these matrix elements should be evaluated in order to investigate to what extent the use of a local effective interaction is justified, especially for the imaginary part of the interaction.

I now turn to the local density approximations used for constructing the optical-model potential in nuclei. The evaluation of their accuracy is difficult, and I can only make a few general remarks.

(h) The dynamical properties of nuclear matter are quite different from those of nuclei. In particular, the Brueckner-Hartree-Fock approximation does not include any collectivity effects. In nuclear matter these could be taken into account by summing ring diagrams. However, this would not be of much use for the application to finite nuclei in which vibrational and rotational collective excitations play an important role, in particular for the imaginary part of the optical-model potential at low energy.[26,64,65] One possible procedure would be to include explicitly the collective nuclear excitations up to about 20 MeV . In order to avoid double counting, the reaction matrix should then be calculated from an auxiliary potential $U(q)$ which has a small gap $(\approx 10$ MeV$)$ at $q = k_F$.

(i) The center-of-mass motion is neglected in the local density approximations outlined above. This is not only a kinematical approximation. Indeed, a spurious inelastic channel is implicitly included in which the center-of-mass of the target is excited. According to Hughes, Fallieros and Goulard,[66] this may lead to a fifty per cent overestimate of $|W|$ in the internal region of light nuclei and at low energy.

(j) The effective interaction $\overline{M}$ depends upon the Fermi momentum k_F (see Eq. (2.3.8)), and so does the mass operator M in nuclear matter. In the local density approximation, one introduces a radial dependent Fermi momentum $k_F(r)$ either by the relation (3.2.3) or by closely related definitions like[67]

$$\frac{1}{2} \left| \rho(\vec{r}_1) + \rho(\vec{r}_2) \right| = \frac{2}{3\pi^2} k_F^3(r_{12}) \quad , \qquad (3.4.1)$$

$$\left| \rho(\vec{r}_1) \ \rho(\vec{r}_2) \right|^{\frac{1}{2}} = \frac{2}{3\pi^2} k_F^3(r_{12}) \quad . \qquad (3.4.2)$$

These two choices only lead to 5 % changes of the real part of the calculated optical-model potential. However, the imaginary part is more significantly affected.[52]

Another possible prescription for $k_F(r)$ is suggested by Eq. (3.2.2) and reads

$$\frac{1}{2m} k_F^2(r) + V_{k_F(r)}(k_F(r);\mu) = \mu \quad , \qquad (3.4.3)$$

where μ is the Fermi energy of the nucleus $(\approx - 8$ MeV$)$. The problem with the definition (3.4.3) is that k_F becomes imaginary when the potential V becomes larger than μ , i.e. in the classically forbidden region. Hence, it appears preferable to use another method for taking into account the difference between the Fermi energy in nuclear matter on the one hand and in nuclei on the other hand. A related difficulty is that the local wave number $k(r)$ is imaginary when a proton tunnels through the Coulomb barrier. A novel method for constructing a local equivalent optical-model potential has been proposed by Horiuchi[85] and recently applied by Kohno and Sprung.[71] It leads to a striking angular momentum dependence of the local potential.[71]

(k) There exists a "starting energy correction" because of the fact that the Fermi energy is approximately equal to $\mu \approx -8$ MeV in nuclei, while in nuclear matter $e(k_F)$ rises from about -20 MeV at $k_F \approx 1.35$ fm^{-1} to zero at $k_F = 0$. In nuclear ground state calculations, this is taken into account by modifying the effective interaction.[14,16,50] A similar correction has not yet been introduced in the scattering case. Its need is exhibited by the fact that $W^{(1)}(k;E)$ vanishes at $E = e(k_F)$, while it should vanish at $E = \mu$. The situation is further complicated by the existence of a Coulomb field and of a symmetry potential. Lejeune[68] has proposed to modify some parametric expressions contained in ref.[60]; however, his prescription lacks theoretical justification and only aims at avoiding the explicit occurrence of some inconsistencies at low energy; it is not supported by recent analyses of experimental data.[69]

(ℓ) The density-dependent Hartree-Fock equations contain a so-called "rearrangement" potential which involves the derivative of the effective interaction with respect to the density. This rearrangement potential arises when one applies a variational principle to the ground state energy. Since no variational procedure is used in the scattering case, rearrangement contributions do not seem to occur. This would not be satisfactory since one expects a smooth transition from the bound to the scattering case. It can be checked that in nuclear matter the rearrangement potential in effect exists and is represented by the second-order term in the low-density expansion of the mass operator.[8,70] The rearrangement potential appears to decrease with increasing energy and to become negligible above several tens MeV.[71] This probably reflects the fact that the density dependence of the reaction matrix decreases when the energy increases.

(m) The local density approximation of Sect. 3.2 involves the sum of the direct and exchange contributions. It has been proposed to identify the "strength" (volume integral) of the effective interaction with the ratio $V(\rho;E)/\rho$.[60] The dependence of this "strength" upon density and energy is larger than the dependence of the effective interaction (2.3.9a) and (2.3.9b) because it contains the contribution of the space exchange term.[72]

(n) The first and the second terms on the right-hand side of Eqs. (2.3.10) and (3.3.1) nearly have the same magnitude but have opposite signs. A delicate cancellation is thus involved between the direct and exchange contributions. This is somewhat troublesome because the large exchange contribution involves approximations, in the form of the Slater approximation for the mixed density matrix (Eq. (2.3.11)) and of the Slater energy approximation for the introduction of a local wave number. It may thus be useful to try to construct another effective interaction which would lead to a smaller exchange contribution. This was performed by Negele[14] in the case of ground states and should be attempted in the case of scattering states.

(o) Instead of using in Eq. (2.3.11) the value of k_F given in (3.2.3), one may take the modified form proposed by Campi and Bouyssy.[73] This mainly modifies the surface properties of the real part of the calculated optical-model potential.[52]

(p) From its very definition, an effective interaction is meant to be used in some prescribed basis. In Sect. 2.3, the effective in-

teraction was required to have (on the average) the same diagonal matrix elements as the reaction matrix in a <u>plane wave basis</u>, see Eq.
(2.3.3). The Michigan group[44] rather requested that the effective interaction should have about the same matrix elements as the reaction matrix in a truncated harmonic oscillator basis. This "M3Y" effective interaction is real and independent of energy and of density. It should be viewed as corresponding to an average over a domain of densities and of energies; the size of this domain is determined by the number of oscillator shells. This renders the M3Y interaction inadequate for energies larger than about 50 MeV. Nevertheless, one might take advantage of the possibility of using different bases. Indeed, one may wonder whether it would not be appropriate to request the effective interaction to have the same matrix elements as the reaction matrix between other radial functions than the Bessel functions, e.g. between the radial parts of distorted waves in the vicinity of the energy of interest.

(q) The validity of the local density approximation for ground state calculations has been discussed by Negele.[74] At first sight, the errors involved would be expected to become "appreciable in the surface where the density varies significantly over a distance comparable with the range of the effective interaction; the local density approximation was believed to break down at low density because the de Broglie wave length is large".[40]

Actually, the success of the ground state calculations makes one suspect that the local density approximations are more reliable than one might expect. The "local" approximations essentially lie in the definition (3.2.3) of a local Fermi momentum and in the definition (3.2.2) of a local nucleon momentum. Both of these approximations can be improved, as we briefly indicated. In ref.[60], it was proposed that the basic requirement for using local density approximations is that the effective interaction varies "smoothly" with density, more precisely, that the strength S of the effective interaction varies little over the range t of the interaction, i.e.

$$ t \, \frac{d\rho}{dr} \, \frac{d|S|}{d\rho} \ll |S| \quad . \tag{3.4.4} $$

(r) Another characteristic of the local density approximation is that it is assumed that the Pauli exclusion operator in nuclei can be approximated by its value in nuclear matter. The comparison carried out in ref.[17] supports the accuracy of this approximation in the case of nuclear ground states. This accuracy should improve with increasing energy since the importance of the Pauli operator then decreases.

(s) One of the striking outcomes of the local density approximations is that the real part of the calculated optical-model potential has a "wine-bottle bottom shape" at intermediate energies. This expression refers to the fact that the calculated potential still has an attractive pocket at the nuclear surface at the energy at which it changes sign in the nuclear interior. This was first pointed out in ref.[58] and has since been repeatedly verified. The physical origin of this property is discussed in ref.[75]; it mainly derives from the Pauli operator Q in the definition (2.1.16) of the reaction matrix. This wine-bottle bottom shape is compatible, if not required, by the recent experimental data, see e.g. refs.[5,76-78].

There exist, however, two puzzling differences between the numerical results obtained in the framework of the local density approximation of Sect. 3.2 on the one hand, and in the framework of the local density approximation of Sect. 3.3 on the other hand. The first
difference is that the former predicts the phenomenon to occur above
180-200 MeV ,[58,75,79] while the latter already exhibits it at   135
MeV .[4,52] The second difference is that the local density approximation of Sect. 3.3 also predicts that at intermediate energy the real
potential well has a narrow attractive dip at the nuclear center,
while that of Sect. 3.2 does not. It should be investigated whether
these differences reflect discrepancies in the numerical values of
the reaction matrix in nuclear matter, or whether they are due to the
detailed nature of the two local density approximations. It appears
unlikely that they might arise from the difference in the free nucleon-nucleon interaction used as input.

Two remarks should still be made concerning this wine-bottle
bottom shape. Firstly, good fits to the elastic scattering data can
probably be obtained with potential wells which do not have a wine-
bottle bottom shape; analyses of inelastic scattering data may be
useful in better pinning down the shape of the optical-model potential at intermediate energy.[80] Secondly, a wine-bottle bottom shape
is also predicted by semi-phenomenological relativistic Dirac-Hartree
models.[81] This again warns that we must beware of mistaking agreement
for justification.[40]

3.5. NUCLEON INELASTIC SCATTERING

The main interest of a realistic effective interaction like the
one discussed in Sect. 2.3 is that it enables one to make definite
predictions on inelastic scattering cross sections, provided that
the distorted wave Born approximation is accurate. The Born approximation appears particularly justified above 100 MeV . By suitably
selecting the residual states,[6] one can test the reliability of the
effective interaction, in particular its density dependence, its
energy dependence and the strength of its central, spin-orbit and
tensor components.[4,82]

Further thought should be devoted to the physical interpretation
of the imaginary part of the effective interaction when it is used
as a transition operator for describing inelastic scattering,[83] to
the possibility of using distorted rather than plane waves for the
basis in which the effective interaction is defined and to the fact
that the channel coupling matrix elements of the effective interaction may have to be multiplied by the correction factor $\tilde{m}/m$ of
paragraph (b) above.

Effective interactions are also used for constructing ion-ion
potentials. Here, however, the Pauli operator which appears in the
defining equation (2.1.16) of the reaction matrix takes a more complicated form, and the imaginary part should include explicitly surface vibrational modes.[84]

4. CONCLUSIONS

It appears that little progress has been accomplished during the last four years concerning the proper definition and the accurate calculations of an effective interaction based on a realistic nucleon-nucleon interaction. However, the semi-quantitative success encountered by previously constructed effective interactions and the availability of accurate elastic and inelastic scattering data at intermediate energy call for a renewed theoretical effort. This effort should mainly aim at a better calculation of the reaction matrix in nuclear matter, at an evaluation of the size and of the density dependence of higher-order corrections, at the accuracy of the parametrization of the reaction matrix in terms of local potentials, at the improvement of the local density approximation and at a detailed comparison with the multiple scattering approach. Simultaneously, one should perform systematic analyses of the experimental data.

I am grateful to the members of the Max-Planck-Institut für Kernphysik of Heidelberg for their kind hospitality.

REFERENCES

1. D. Slanina and H. McManus, Nucl.Phys. A116, 271 (1968).
2. H.V. von Geramb, ed., "Microscopic Optical Potentials" (Springer Verlag, 1979).
3. J. Rapaport, Physics Reports 87C, 25 (1982).
4. J. Kelly, W. Bertozzi, T.N. Buti, F.W. Hersman, C. Hyde, M.V. Hynes, B. Norum, F.N. Rad, A.D. Bacher, G.T. Emery, C.C. Foster, W.P. Jones, D.W. Miller, B.L. Berman, W.G. Love and F. Petrovich, Phys.Rev.Lett. 45, 2012 (1980).
5. H.O. Meyer, J. Hall, W.W. Jacobs, P. Schwandt and P.P. Singh, Phys.Rev. C24, 1782 (1981).
6. W.G. Love and M.A. Franey, Phys.Rev. C24, 1073 (1981).
7. W.G. Love, M.A. Franey and F. Petrovich, International Conference on Spin Excitations in Nuclei (Plenum Press, 1982).
8. J. Hüfner and C. Mahaux, Ann.Phys. (N.Y.) 73, 525 (1972).
9. K.A. Brueckner, C.A. Levinson and H.M. Mahmoud, Phys.Rev. 95, 217 (1954).
10. B. Friedman and V.R. Pandharipande, Phys.Lett. 100B, 205 (1981).
11. S. Fantoni, B.L. Friman and V.R. Pandharipande, Nucl.Phys. A386, 1 (1982).
12. E. Krotscheck, R.A. Smith and A.D. Jackson, Phys.Lett. 104B, 421 (1981).
13. K.T.R. Davies, R.J. McCarthy and P.U. Sauer, Phys.Rev. C6, 1461 (1972).
14. J.W. Negele, Phys.Rev. C1, 1260 (1970).
15. J. Nemeth and D. Vautherin, Phys.Lett. 32B, 561 (1970).
16. X. Campi and D.W. Sprung, Nucl.Phys. A194, 401 (1972).
17. K.T.R. Davies, R.J. McCarthy, J.W. Negele and P.U. Sauer, Phys. Rev. C10, 2607 (1974).

18. F. Cannata, J.P. Dedonder and F. Lenz, Ann.Phys. (N.Y.) in press.
19. H.A. Bethe, Ann.Rev.Nucl.Sci. $\underline{21}$, 93 (1971).
20. J.P. Jeukenne, A. Lejeune and C. Mahaux, Physics Reports $\underline{25C}$, 83 (1976).
21. J.P. Jeukenne, A. Lejeune and C. Mahaux, Phys.Rev. $\underline{C10}$, 1391 (1974).
22. F.A. Brieva and J.R. Rook, Nucl.Phys. $\underline{A291}$, 299 (1977).
23. H. Orland and R. Schaeffer, Nucl.Phys. $\underline{A299}$, 442 (1978).
24. F. Perey and D.S. Saxon, Phys.Rev. $\underline{10}$, 107 (1964).
25. B.Z. Georgiev and R.S. Mackintosh, Phys.Lett. $\underline{73B}$, 250 (1978).
26. V. Bernard and Nguyen Van Giai, Nucl.Phys. $\underline{A327}$, 397 (1979).
27. S. Fantoni, B.L. Friman and V.R. Pandharipande, Phys.Lett. $\underline{104B}$, 89 (1981).
28. J.W. Negele and K. Yazaki, Phys.Rev.Lett. $\underline{47}$, 71 (1981).
29. K.A. Brueckner and J.L. Gammel, Phys.Rev. $\underline{109}$, 1023 (1958).
30. A. Lejeune, private communication
31. M. Lacombe, B. Loiseau, J.M. Richard, R. Vinh Mau, J. Côté, P. Pirès and R. de Tourreil, Phys.Rev. $\underline{C21}$, 861 (1980).
32. M.I. Haftel and F. Tabakin, Nucl.Phys. $\underline{A158}$, 1 (1970).
33. P. Grangé and A. Lejeune, Nucl.Phys. $\underline{A327}$, 335 (1979).
34. A. Faessler, I. Izumoto, S. Krewald and R. Sartor, Nucl.Phys. $\underline{A359}$, 509 (1981).
35. P. Grangé and A. Lejeune, private communication.
36. H.A. Bethe, B.H. Brandow and A.G. Petscheck, Phys.Rev. $\underline{129}$, 225 (1963).
37. F.A. Brieva and J.R. Rook, J.Phys.Soc.Japan $\underline{44}$, 539 (1978).
38. H.V. von Geramb, in Proceedings of the XVI Bormio Workshop (1978).
39. P.J. Siemens, Nucl.Phys. $\underline{A141}$, 225 (1970).
40. D.W.L. Sprung, Advances in Nuclear Physics $\underline{5}$, 225 (1972).
41. W. Legindgaard, Nucl.Phys. $\underline{A297}$, 429 (1978).
42. C. Mahaux, in Microscopic Optical Potentials, edited by H.V. von Geramb (Springer Verlag, 1979), p. 1.
43. F.A. Brieva, in Microscopic Optical Potentials, edited by H.V. von Geramb (Springer Verlag, 1979), p. 84.
44. G.F. Bertsch, J. Borysowicz, H. McManus and W.G. Love, Nucl. Phys. $\underline{A284}$, 399 (1977).
45. Y. Eisen, B.D. Day and E. Friedman, Phys.Lett. $\underline{56B}$, 313 (1975); Y. Eisen and B.D. Day, Phys.Lett. $\underline{63B}$, 253 (1976).
46. F.A. Brieva and J.R. Rook, Nucl.Phys. $\underline{A291}$, 317 (1977).
47. M.K. Srivastava, Nucl.Phys. $\underline{A144}$, 236 (1970).
48. J.P. Elliott, H.A. Mavromatis and E.A. Sanderson, Phys.Lett. $\underline{24B}$, 358 (1967).
49. J.P. Elliott, A.D. Jackson, H.A. Mavromatis, E.A. Sanderson and B. Singh, Nucl.Phys. $\underline{A121}$, 241 (1968).
50. D.W.L. Sprung and P.K. Banerjee, Nucl.Phys. $\underline{A168}$, 273 (1971).
51. F.A. Brieva and J.R. Rook, Nucl.Phys. $\underline{A297}$, 206 (1978).
52. H.V. von Geramb, in Microscopic Optical Potentials, edited by H.V. von Geramb (Springer Verlag, 1979), p. 104.
53. D.W.L. Sprung, Nucl.Phys. $\underline{A182}$, 97 (1972).
54. F.A. Brieva, H.V. von Geramb and J.R. Rook, Phys.Lett. $\underline{79B}$, 177 (1978).
55. F.A. Brieva, H.V. von Geramb and J.R. Rook, University of Hamburg, internal report (1978).

56. W. Bauhoff, H.V. von Geramb and G. Palla, University of Hamburg, internal report (1980).
57. J. Nemeth, in The Structure of Nuclei (International Atomic Energy Agency, Vienna, 1972), p. 287.
58. J.P. Jeukenne, A. Lejeune and C. Mahaux, in Nuclear Self-Consistent Fields, edited by G. Ripka and M. Porneuf (North-Holland Publ. Comp., Amsterdam, 1975), p. 155.
59. J.P. Jeukenne, A. Lejeune and C. Mahaux, Phys.Rev. C15, 10 (1977).
60. J.P. Jeukenne, A. Lejeune and C. Mahaux, Phys.Rev. C16, 80 (1977).
61. B.D. Day, Phys.Rev.Lett. 47, 226 (1981); Phys.Rev. C24, 1203 (1981).
62. J.P. Jeukenne, A. Lejeune and C. Mahaux, Nukleonika 20, 181 (1975).
63. F.A. Brieva and J.R. Rook, Nucl.Phys. A307, 493 (1978).
64. N. Vinh Mau and A. Bouyssy, Nucl.Phys. A257, 189 (1976); A. Bouyssy, H. Ngô and N. Vinh Mau, Nucl.Phys. A371, 173 (1981).
65. F. Osterfeld, J. Wambach and V.A. Madsen, Phys.Rev. C23, 179 (1981).
66. T.A. Hughes, S. Fallieros and B. Goulard, Nuclei and Particles 1, 93 (1971).
67. C.W. Wong and S.A. Moszkowski, Nucl.Phys. A239, 209 (1975).
68. A. Lejeune, Phys.Rev. C21, 1107 (1980).
69. S. Kailas, S.K. Gupta, M.K. Mehta and G. Singh, Phys. Rev. C26, 830 (1982).
70. C.A. Engelbrecht and H.A. Weidenmüller, Nucl.Phys. A184, 385 (1972).
71. R. Sartor, Nucl.Phys. A289, 329 (1977); M. Kohno and D.W.L. Sprung, McMasters University preprint (1982).
72. J.P. Jeukenne and C. Mahaux, Z. Physik A302, 233 (1981).
73. X. Campi and A. Bouyssy, Phys.Lett. 73B, 263 (1978).
74. J.W. Negele, in Effective Interactions and Operators in Nuclei, edited by B.R. Barrett (Springer Verlag, 1975), p. 270.
75. C. Mahaux, in Common Problems in Low- and Medium-Energy Nuclear Physics, edited by B. Castel, B. Goulard and F.C. Khanna (Plenum Publ. Corp., 1979), p. 265.
76. L.G. Arnold, B.C. Clark, R.L. Mercer and P. Schwandt, Phys.Rev. C23, 1949 (1981).
77. L.G. Arnold, B.C. Clark, E.D. Cooper, H.S. Sherif, D.A. Hutcheon, P. Kitching, J.M. Cameron, R.P. Liljestrand, R.N. MacDonald, W.J. MacDonald, C.A. Miller, G.C. Neilson, W.C. Olsen, D.M. Sheppard, G.M. Stinson, D.K. McDaniels, J.R. Tinsley, R.L. Mercer, L.W. Swensen, P. Schwandt and C.E. Stronach, Phys.Rev. C25, 936 (1982).
78. H.O. Meyer, P. Schwandt, G.L. Moake and P.P. Singh, Phys.Rev. C23, 616 (1981).
79. J.P. Jeukenne, A. Lejeune and C. Mahaux, in Proceedings of the International Conference on the Interactions of Neutrons with Nuclei, Lowell, edited by E. Sheldon (National Technical Information Service, Springfield, 1976), vol. 1, p. 451.
80. G.R. Satchler, preprint (1982).
81. M. Jaminon, C. Mahaux and P. Rochus, Phys.Rev.Lett. 43, 1097 (1979).
82. S. Kosugi and T. Kosugi, Phys.Lett. 113B, 437 (1982).

83. F. Todd Baker, A. Scott, W.G. Love, J.A. Mourey, W.P. Jones and J.D. Wiggins, Nucl.Phys. <u>A386</u>, 45 (1982).
84. S.B. Khadkikar, L. Rikus, A. Faessler and R. Sartor, Nucl.Phys. <u>A369</u>, 495 (1981).
85. H. Horiuchi, Progr.Theor.Phys. <u>63</u>, 725 (1980); ibid. <u>64</u>, 184 (1980).

MICROSCOPIC OPTICAL POTENTIALS*

H.V. von Geramb
Theoretische Kernphýsik
Universität Hamburg
Luruper Chaussee 149, 2oo0 Hamburg 5o, W. Germany

ABSTRACT

Various versions of infinite nuclear matter reaction matrix elements (t-matrix elements) were generated from free NN potentials. Most emphasis is put on calculations with the momentum dependent Paris potential as input. Thereupon is parametrized a complex energy and density dependent interaction (25-5oo MeV with convenient Yukawa-form factors to serve as *effective interaction*. It is the driving two-body transition operator in applications to finite nuclei using a *local density approximation*. Studies concentrate on elastic channel microscopic optical model potential and its general structure as compared to phenomenological potentials. Further detailed checks on our effective interactions were performed with exemplary studies of (p,p') transition

*This work represents a combined effort together with K. Nakano.

I. INTRODUCTION

Heuristic direct nucleon nucleus reaction analyses creep
around a microscopic understanding [1]. They aim to start with
an elementary, assumed known NN-interaction and obtain quantitative reproductive/predictive results for nucleon-nucleus scattering
situations. The present conference sets a boundary with projectile
energies ranging from, say, 1oo to 4oo MeV. As expected from the
pronounced minimum in the NN reaction cross section in this energy
interval we find the nucleus most transparent. Nevertheless, the
nucleus is not as transparent as for electrons, where lowest order
Born approximation yields already satisfactory results. In contrary
the coupling potential between projectile and target nucleon was
and remains a tough problem. Because of the nucleon's spin and
isospin, and indistinguishability from the target nucleons we may
invent an endless row of consistency checks. Many experimental data
are now available for these checks and the IUCF staff has achieved
great merits with its frontier experiments.

The theoretical study of elastic and inelastic scattering processes is a fundamental step in understanding the nuclear many
body problems and we have today good reasons to pursue such studies
with low order approximations [2,3]. The nuclear matter approach,
which we follow here, has itself established as a qualitative and
quantitative method to reconcile the sucess of the phenomenological
local optical model potential with a purely microscopic model.

The essential quantity in this approach is the *effective interaction* which is identified with an operator that reproduces certain reaction matrix elements. We distinguish and study four
different levels. They are schematically summarized in Fig. 1 for
an initial condition in infinite
nuclear matter. The simplest approximation (1) neglects the *Fermi motion*
(frozen nucleus approximation) of the
target nucleons and *Pauli blocking*
effects (Q=1).

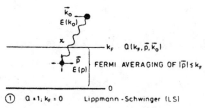

Fig. 1

The t-matrix elements are identical with the free NN scattering amplitudes, which are solutions of a Lippmann-Schwinger equation

$$t = V + VG_o t .\tag{1}$$

To take the Fermi motion into account, version ② averages the free scattering amplitudes within a rectangular Fermi sea for the target nucleon in symmetric (N = Z) nuclear matter

$$|\vec{p}| \leq k_F , \qquad k_F = (\frac{3\pi^2}{2} \rho_{NM})^{1/3} .\tag{2}$$

An account of the Pauli-blocking is obtained in version ③ where we replace the Lippmann-Schwinger equation by the Bethe-Goldstone equation. The Pauli operator Q is approximated in its spherical averaged standard form [4,5]. With version ④ we account additionally for dispersive effects in the single particle energies [6]. It is our most sophisticated version, so far computed and applied. Some more details to this calculation we present in sections 2 and 4.

II. COMPARISON OF THE EFFECTIVE NN INTERACTION BY LOVE-FRANEY AND THE HAMBURG POTENTIAL

In a recent publication by Love and Franey [7] experimental NN-data have been used between 1oo and 8oo MeV to tailor a local, complex and energy dependent effective interaction

$$V_{12}(r) = V^C(r) + V^{LS}(r)(\vec{L}\cdot\vec{S}) + V^T(r) S_{12}\tag{3}$$

Since our version ① and ② t-matrix ($k_F \to o$) elements based on the Paris potential [8] are supposed to reproduce the same experimental data (at least up to 35o MeV), a comparison should be quite illustrative.

We follow the idea of Siemens [9] and generate the local energy and density dependent potential from the correlated free two body wave function

$$\psi^{(+)} = \phi + G_o^{(+)} V\psi^{(+)} \quad . \tag{4}$$

As two body potential we used the momentum dependent Paris potential [8] and solved the integral equation with standard partial wave techniques. We use the same classification in central spin orbit and tensor components as used in Ref. 7; Eq. (3).

$$V(r) = \sum_{ST} v_o^{ST}(r) \, P^S P^T + \sum_T v_1^T(r) (\vec{L} \cdot \vec{S})$$
$$\tag{5}$$

$$+ v_2^T(r) S_{12} = \sum_k (R_k \cdot S_k)$$

The tensor amplitudes are obtained with

$$<L \| R_k \| L'><ST \| S_k \| ST> = \sum_J \hat{k} \, \hat{J} (-)^{J-L'-S} (\hat{T})^{\frac{1}{2}} \begin{Bmatrix} L & S & J \\ S & L' & k \end{Bmatrix} \tag{6}$$

$$(LSJT \mid t \mid L'SJT)$$

with t-matrix elements computed with the correlated two body wave functions

$$(LSJT \mid t(\vec{k},\vec{k}) \mid L'SJT)$$

$$= \frac{\sum_{L''} \int_{|p| \le k_F} d^3p \; j_L(\frac{1}{2}|\vec{k}-\vec{p}|r) \; v_{LL''}^{J,ST}(r) \; u_{L'L''}^{J,ST}(r,k_F,\vec{k},\vec{p})}{\int_{|p| \le k_F} d^3p \; j_L(\frac{1}{2}|\vec{k}-\vec{p}|r) \; j_{L'}(\frac{1}{2}|\vec{k}-\vec{p}|r)} \tag{7}$$

The central component $S_0 = P^S P^T$ and

$$R_0 = R_0^{(LL,ST)} = \frac{\sum\limits_{J} \hat{J}(LSJT|t|LSJT)}{S \cdot L} \tag{8}$$

The spin orbit component $S_1 = \frac{1}{2}(\vec{\sigma}_1 + \vec{\sigma}_2) \cdot P^T$ and

$$R_1 = R_1^{(LL,T)} \cdot \vec{L} = \frac{\sum\limits_{J} \hat{J}<\vec{L}\cdot\vec{S}>(L1JT|t|L1JT)}{2L \cdot L(L+1)} \vec{L} \tag{9}$$

The tensor components, $S_2 = [\vec{\sigma}_1 \times \vec{\sigma}_2]_2 \cdot P^T$

$$R_2 = 3 \left(\frac{8\pi}{15}\right)^{1/2} R_2^{(LL',T)} V_{12}(\hat{r}) \tag{10}$$

The ℓ-dependence is eliminated with averaging between plane wave states. It yields for the central interaction

$$V_0^{ST}(r) = \frac{\sum\limits_{L\geq o} \epsilon(L+S+T+1)\hat{L} \, R_0^{(LL,ST)} \cdot W_{LL}(r)}{\sum\limits_{L\geq o} \epsilon(L+S+T+1)L \, W_{LL}(r)} \tag{11}$$

the spin orbit interaction

$$V_1^T(r) = \frac{1}{3} \frac{\sum\limits_{L\geq 1} \epsilon(L+T)L(2L+3)R_1^{(LL,T)}(r)W_{LL}(r)}{\sum\limits_{L>1} \epsilon(L+T)L(2L+3)W_{LL}(r)}$$

$$+ \frac{1}{3} \frac{\sum\limits_{L>1} \epsilon(L+T)\hat{L} \, R_1^{(LL,T)}(r)W_{LL}(r)}{\sum\limits_{L>1} \epsilon(L+T)L \, W_{LL}(r)} \tag{12}$$

$$+ \frac{1}{3} \frac{\sum\limits_{L>1} \epsilon(L+T)(L+1)(2L-1)R_1^{(LL,T)}(r)W_{LL}(r)}{\sum\limits_{L>1} \epsilon(L+T)(L+1)(2L-1)W_{LL}(r)}$$

the tensor interaction

$$V_2^T(r) = \frac{\sum\limits_{L\neq L'} \varepsilon(L+T)\sqrt{\hat{L}\hat{L}'}\,\hat{J}<LL'oo|2o>\{^{L\,1\,J}_{1\,L'\,2}\}R_2^{(LL';T)}(r)W_{LL'}(r)}{\sum\limits_{L\neq L'} \varepsilon(L+T)\sqrt{\hat{L}\hat{L}'}\hat{J}<LL'oo|2o>\{^{L\,1\,J}_{1\,L'\,2}\}W_{LL'}(r)}$$
(13)

with the notation

$$\varepsilon(n) = \begin{cases} o & n \text{ odd} \\ 1 & n \text{ even} \end{cases}$$
(14)

$$W_{LL'} = \int\limits_{|\vec{p}|\leq k_F} d^3p \; j_L(\tfrac{1}{2}|\vec{k}-\vec{p}\,|r)j_{L'}(\tfrac{1}{2}|\vec{k}-\vec{p}|r)$$
(15)

The above expressions are not handy to use and we parametrize an analytical expression in coordinate space in terms of Yukawa form factors. However, the direct parametrization of V(r) in co-ordinate space puts undue emphasis on small r values. We first Fourier transform V(r) into a momentum space operator V(k) and then apply a χ^2 fitting. This procedure permits to put all emphasis on the physical significant momentum (transfer) region and cut off poorly understood high momentum regions.
The momentum space Yukawa expressions are for

central
$$V_o^{ST}(k) = \sum_{i=1}^{4} \frac{4\pi\cdot V_i^{ST}}{\mu_i^2+k^2}$$
(16)

spin orbit
$$V_1^T(k) = \sum_{i=1}^{4} 4\pi\cdot V_i^{LS,T}\{\tan^{-1}(\frac{k}{\mu_i})\,\frac{1}{k^2} - \frac{\mu_i}{k(k^2+\mu_i^2)}\}$$
(17)

tensor
$$V_2^T(k) = \sum_{i=1}^{4} \frac{32\pi k^2 V_i^{TN,T}}{(k^2+\mu_i^2)^3}$$
(18)

The mass parameters μ_i (inverse ranges) are a priori fixed
to cover the range contained in the free NN-interaction. The
potential strength parameters V_i are uniquely obtained by
minimizing

$$\chi^2 = \int_0^{5\,fm^{-1}} \{V(k) - V_\alpha(k)\}^2 dk \,. \tag{19}$$

The r-space representation of the effective
interaction is easily obtained by backtransformation and is
of the form

$$central \quad V_0^{ST}(r) = \sum_{i=1}^{4} V_i^{ST} \frac{e^{-\mu_i r}}{r} \tag{20}$$

$$spin\ orbit \quad V_1^{T}(r)\,(\vec{L}\cdot\vec{S}) = \sum_{i=1}^{4} V_i^{T} \frac{e^{-\mu_i r}}{r} \ (\vec{L}\cdot\tfrac{1}{2}(\vec{\sigma}_1+\vec{\sigma}_2)) \tag{21}$$

$$tensor \quad V_2^{T}(r)\,S_{12} = \sum_{i=1}^{4} V_i^{T}\,r^2 \frac{e^{-\mu_i r}}{r}(\frac{3(\vec{\sigma}_1\cdot\vec{r})\,(\vec{\sigma}_2\vec{r})}{r^2} - (\vec{\sigma}_1\vec{\sigma}_2)) \tag{22}$$

The enclosed tables contain a selection of this interaction parameters and they are also available on magnetic tape/cards
on request. The volume integral refers to

$$4\pi \int_0^{\infty} V_k(r)\,r^2 dr \tag{23}$$

In the following comparison of the effective interaction by
Love and Franey [7] (FL) with the Hamburg potential (HH) we use
our version ① respectively ② potential with $k_F \to 0$.
We compare all potentials in k-space as defined with Eqs. 16-18.
The Hamburg potentials refer to the form before fitting with
Yukawa form factors. Figs. 2-5 represent a selection in energy
325 MeV and 1oo MeV. Despite fine details we observe as most

51

prominent,

- central SE compares well
- central TE compares well
- central SO is not unique

with respect to a deltafunction contribution in r-space
(constant in k-space). This deltafunction ambiguity shows a
random character with respect to real, imaginary and energy
dependence; Figs. 2a - 3b, 4a - 4d;

- the non central components (spin orbit and tensor) agree
  reasonable for T = 1, but have little resemblance for T = o;
  Figs. 4e - 4h.

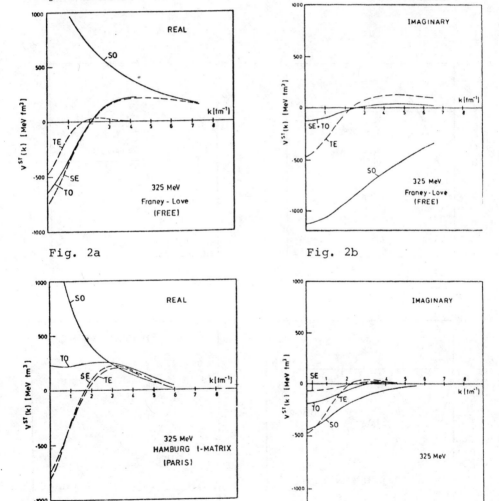

Fig. 2a

Fig. 2b

Fig. 3a

Fig. 3b

52

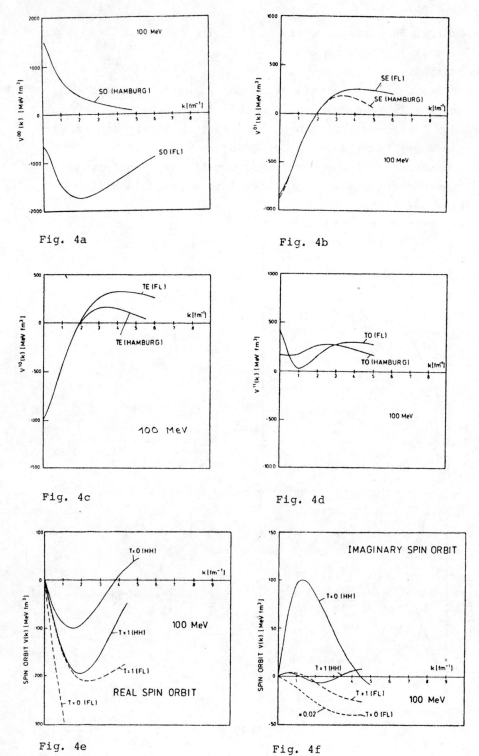

Fig. 4a

Fig. 4b

Fig. 4c

Fig. 4d

Fig. 4e

Fig. 4f

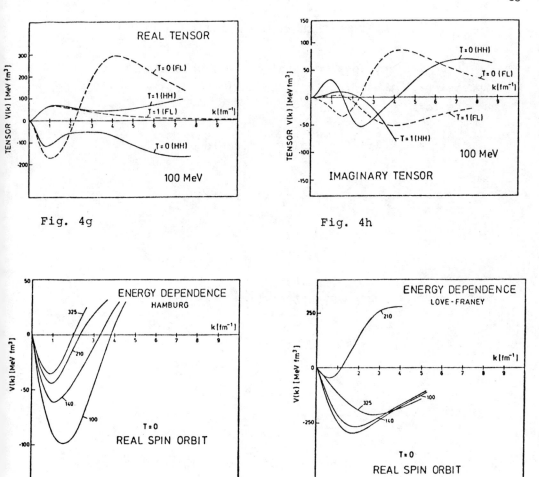

Fig. 4g

Fig. 4h

Fig. 5a

Fig. 5b

- The energy dependence for Love-Franey components appear   not
smooth for central SO TO; spin orbit and tensor T = o;
Figs. 5a-b.
For all components in common is a   different high
momentum behaviour (k > 1.5 fm^{-1}).

54

III. FEATURES OF THE MICROSCOPIC OPTICAL POTENTIAL

The equivalent local central optical potential in a folding expression of the effective interaction, Ref. 1, page 1o9 ff, has the form

$$U(\vec{r},E) = \int \delta \ (\vec{r}-\vec{r}') \{ \ \int \rho(s) t^D(\vec{r},\vec{s};E) d^3 s$$

$$+ \ \rho(r,r') t^{EX}(r,r') j_o (k|\vec{r}-\vec{r}'|) \} d^3 r' \tag{24}$$

with direct and exchange interaction mixtures

$$t^{D,EX}_{p(p)} = t^{D,EX}_{n(n)} = \frac{1}{4}(SE \pm 3TO) \tag{25}$$

and

$$t^{D,EX}_{p(n)} = t^{D,EX}_{n(p)} = \frac{1}{8}(3TE + SE \pm SO \pm 3TO) \tag{26}$$

Recently, strong evidence has been put forward that medium energy real optical model potentials have a non Woods-Saxon shape. The empirical evidence concentrates on ^{12}C elastic scattering data by H.O. Meyer et al. [1o]. Theoretical evidence comes from relativistic formulations [11] as well as from non relativistic, microscopic optical model potentials. How this aspect is related to the two nucleon effective interaction is shown next.

As can be seen from Eqs. 24-26, the even state interaction SE, TE have equal sign matrix elements in direct and exchange. Their individual contributionsare shown as a function of energy in Fig. 6.

The underlying effective interaction is based on Hamada-
Johnston's free interaction [12] of the kind (4). The
attractive direct potential in this figure reflects the attractive low momentum behaviour as known from Figs. 2-3.

The exchange contribution is *repulsive*. This reflects predominant high momentum contributions (k > 1.5 fm^{-1}) where the
interaction   is repulsive   . For 6o MeV projectiles
the sampled region is around the incident momentum k = 1.7 fm
and yields   an almost zero exchange potential. In Fig. 7 the
direct and exchange contributions are summed up and we distinguish
even and odd state contributions. The even state contributions
show already a surface structure which is emphasized when combined with the odd state contribution, Fig. 8. The imaginary
potential contributions are shown in Fig. 9. This decomposition
has also been made for the Hamburg interaction (4) based on the
Paris potential. The essential diffrence in the t-matrix interaction between Hamada-Johnston and Paris is a significant weaker
TO interaction for Paris (factor 5) and a weaker repulsive potential for SE, TE, for k ~ 3fm^{-1}.

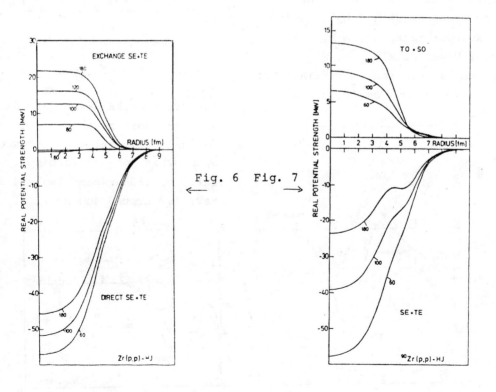

Fig. 6   Fig. 7
←   →

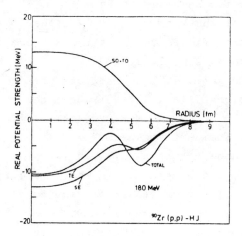

The net result is a real optical model potential, Fig. 1o. The spin-orbit potential is quite comparable with Hamada-Johnston or Paris potential as input. Fig. 12 compares its features for ^{12}C at 2oo and 4oo MeV.

Fig. 8

The above discussion should serve as an introduction for a comparison between our microscopic optical model and the phenomenological fits with double Woods-Saxon parametrization by H.O. Meyer et al.[13]. Fig. 12 shows various real potentials for 2oo MeV protons on ^{12}C.

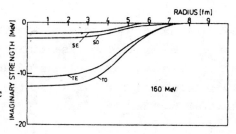

Fig. 9

The differences are significant and more work is required to ha definite answers. The microscop spin orbit potentials display a ratio of 1:3 between 1oo and 2o MeV; 1:2 around 4oo MeV.

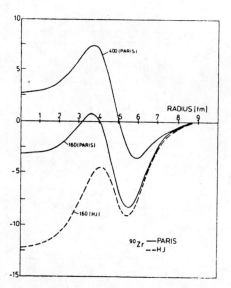

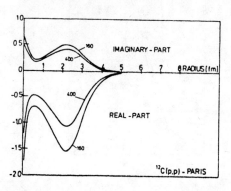

Fig. 11

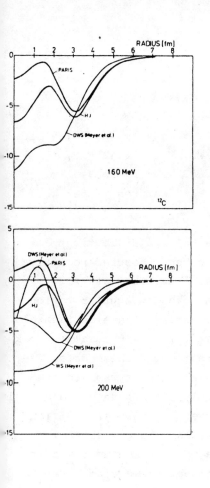

Fig. 12a

Fig. 12b

Fig. 13

The phenomenological spin orbit potentials favour a ratio 1:1 in the 1oo to 2oo MeV region, respectively strong fluctuations are observed. At 4oo MeV the situation is similar even Bauhoff's potential seems in agreement with the microscopic potentials. Other sets of parameters yield not worse fits.

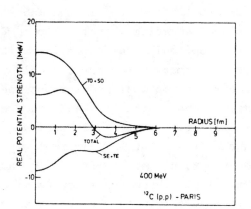

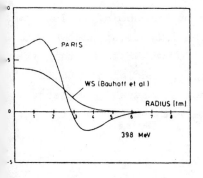

Fig. 14

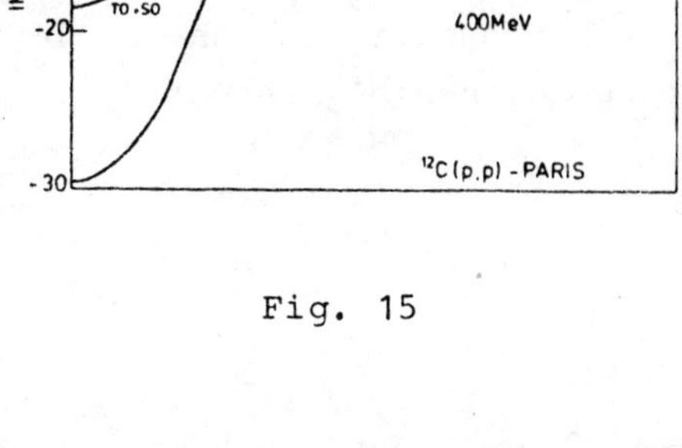

Fig. 15

For completeness we add results of 4oo MeV proton scattering
on ^{12}C, Figs. 13-15. The underlying data for the phenomenological analyses have been measured by Ch. Glashausser et al. [14].
The phenomenological Woods-Saxon OMP was fitted by W. Bauhoff [15)
and he found parameters V = -4.3 MeV, r = 1.o84 fm, a = o.591 fm;
W = 27.6 MeV, r = 1.129 fm, a = o.629 fm; V_{SO} = 4.19 MeV,
r = o.933 fm, a = o.611 fm; W_{SO} = - 3.683 MeV, r = o.999 fm,
a = o.53 fm. This potential agrees in all features rather well
with the microscopic OMP.

Cross section and analyzing power angular distributions for
proton elastic scattering from ^{12}C have been measured recently
at the Indiana University Cyclotron Facility. The data with
16o and 2oo MeV polarized proton beams cover an angular region
from about 5° to $16o^{\circ}$ in the lab and constitute an experimental
study of momentum transfers up to 6 fm^{-1}. It was natural to look
for an optical model potential which reproduced these data. Various attempts were made but only partial success was achieved
because the optical model reproduced only the region of relatively
small momentum transfer and required the inclusion of modified
radial form factors as discussed above. The changes alter the
angular distributions particularly the large momentum transfer
behaviour, but not sufficiently. Additionally, we discovered
that simple three parameter Fermi distributions for the nuclear
matter density of ^{12}C yield an exaggerated backward cross section
due to a non-physical behaviour of the spin orbit potential near
the origin. When experimental density distributions were used,
the backward-angle enhancement disappeared.

With projectile energies above the pion threshold, it is known
that the nucleus should not be considered only as an assembly of
nucleons interacting through a two body force. It is certain that
baryon resonances as well as mesons exist in nuclei in addition
to the nucleons. In particular one cannot understand the large
momentum properties of any nuclear state without introducing
mesonic degrees of freedom [16).

This point is equivalent to proclaim the failure of the phenomenological or microscopic optical potential which *knows only
of nucleons*. We add to the OMP a doorway isobar channel where
a nucleon is transformed into a $\Delta(1236)$ and follow a coupled
channel approach. The first channel labelled with the subcript
zero consists of the projectile and the target nucleons in its
ground state. The associated channel Hamiltonian contains the potential as we know it from the microscopic or the phenomenological optical potentials. In our analyses we use the double
Woods-Saxon by Meyer et al. [13]. The second channel distinguishes
a $J^{\pi} = 3/2^{+}$ particle (Δ) with a mass equal to 1.3o the nucleon
rest mass, and the target nucleus in a low excited 1^{+}, T=1 (15.1 MeV)
state.

The coupled equations for the two channel situation read

$$Eu_{o} - (T_{o}+V_{oo})u_{o} = V_{o1}u_{1}$$

$$(E+Q)u_{1} - (T_{1}+V_{11})u_{1} = V_{1o}u_{o}. \tag{27}$$

The value for Q was taken as -3oo MeV.

We treated the coupling potentials V_{o1}, V_{1o} microscopically
and the diagonal potential V_{11} phenomenologically. The parameters
for V_{11} were V = 3o MeV, W = 3o MeV, r = 1.o9 fm, a = o.54 fm,
V_{LS} = 9 MeV, r = 1.o9 fm, a = o.54 fm. The transition potentials
were computed microscopically with a folding model and the transition potential N + N -> N + Δ of Sugawara and v. Hippel [17].
The short range correlations in this potential were treated with
a standard correlation function with standard parameters.
The radial form factors for the transition potential is shown in
Fig. 16. Without any further adjustments we have performed numerical
analyses of 16o and 2oo MeV elastic proton scattering data from
^{12}C. The results for the different cross sections are shown in
Figs. 17-18. The solid line refers to an analysis without explicit
Δ coupling. The dashed line refers to the situation with coupling.

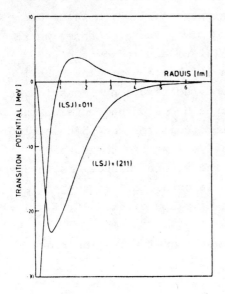

Fig. 16 - Microscopic transition
form factors for the
intermediate Δ excitation
due to central and tensor
components in the Sugawara
and v. Hippel potential.

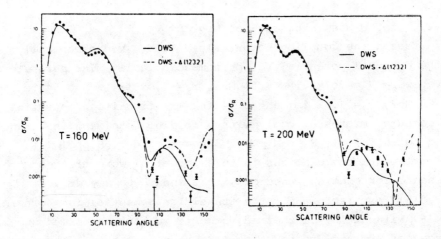

Fig. 17   Fig. 18

IV. SELECTED EXAMPLES OF (p,p') TRANSITIONS

Inelastic nucleon scattering is known to cover many
possibilities to cross check the effective interactions. The
microscopic antisymmetrized transition amplitude

$$T_{fi} = \sum_{j_1 j_2} <J_i \ I \ M_i M | J_f M_f> \ S(j_1 j_2 J_i J_f \ I)$$

$$<\psi^{(-)} \phi_{j_2} \ U(\vec{r}_1,\vec{r}_2; \rho(\frac{r_1+r_2}{2})) | \psi_i^{(f)} \phi_{j_1}>_a$$

(28)

carries the local density approximation (LDA) of the transition
operator and is readily computed with central tensor and spin
orbit forces. In Figs. 19-22 we show some selected test cases
for ^{12}C and ^{90}Zr. The main purpose is to demonstrate the sensitivity of all ingredients entering Eq. (28). The ^{12}C transition
is based on shell model wave functions by Cohen and Kurath [18]
and phenomenological optical model parameters [13]. The underlying
effective interaction is of the kind ③ . The best agreement
with data is obtained with a double Woods-Saxon OMP. This result
however should not mean that we give preference to the double
Woods-Saxon parametrization of the optical model. It is meant
to show the high sensitivity of cross section data in particular
to optical model parameters. In Figs. 2o-21 are shown 6^+ and 8^+
analyzing power results. These transitions are believed to be
dominated from $(g9/2)^2$. Again, we have used two sets of optical
model parameters: HH refers to V = 25.41 MeV, r = 1.21 fm,
a = o.77 fm, W = 1o.31 MeV, r = 1.39 fm, a = o.55 fm, V_{LS} = 3.74 MeV,
W_{LS} = - 1.o3 MeV, r = 1.1 fm, a = o.57 fm. UGA refers to
V = 22.2 MeV, r = 1.23 fm, a = o.8 fm, W = 9.6 MeV, r = 1.323 fm,
a = o.64 fm, V_{LS} = 2.6 MeV, W_{LS} = - 1.92 MeV, r = 1.o7 fm,
a = o.58 fm. The bound state wave functions are computed with a
Woods-Saxon potential, r = 1.38 fm, a = o.65 fm, BE = 1.95 MeV,
potential strength adjusted to match the binding energy BE.

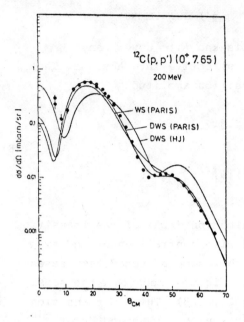

$^{12}C(p,p')(0^+, 7.65)$

200 MeV

WS (PARIS)

DWS (PARIS)

DWS (HJ)

Fig. 19

The potential parameters Po and P4 refers to potentials ① respectively ④ derived from the Paris potential. More detailed discussion of these analyses is presented by Prof. Alan Scott during this conference.

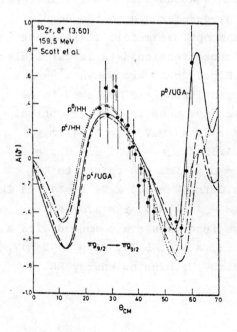

$^{90}Zr, 8^+ (3.60)$
159.5 MeV
Scott et al.

p^0/UGA

p^0/HH

p^4/HH

p^4/UGA

$\pi g_{9/2} \longrightarrow \pi g_{9/2}$

Fig. 2o

$^{90}Zr, 6^+ (3.45)$
159.5 MeV
Scott et al.

$\pi g_{9/2} \longrightarrow \pi g_{9/2}$

p^4/HH

p^0/HH

p^4/UGA

p^0/UGA

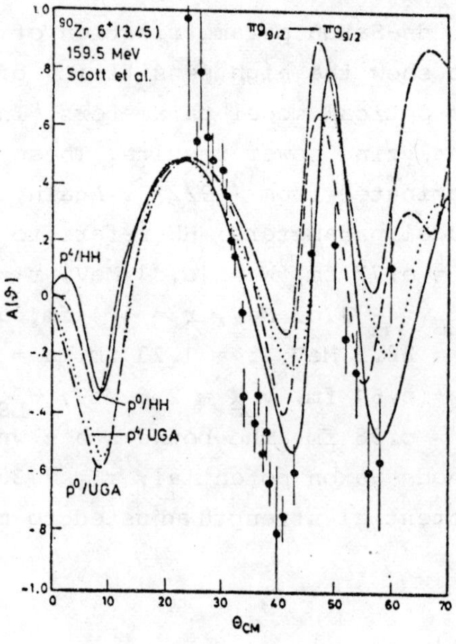

Fig. 21

V. BRIEF EXPLANATION TO THE TABLE OF EFFECTIVE INTERACTIONS

We have tabulated the parameters for the effective NN-interaction using the Paris potential. The following tables list the
mass parameters (inverse ranges) and interaction strengths in
Yukawa representation as defined with the Eqs. (2o-22) in section
2. The first part contains version ① in the energy interval
(the energy refers here to the projectile energy in the lab
system impinging on infinite nuclear matter) between 25 and 5oo
MeV. For 2oo MeV, version ② is added for Fermi momenta o.5,
o.8, 1.o, 1.2, 1.4 fm^{-1}. Additionally is added the 2oo MeV result of a version ③ calculation for the same Fermi momenta
as for the Fermi averaged case. This latter table has a bold face
"q" in the left top corner.

PROJ. LAB. ENERGY = 50.0 MEV. INCIDENT MOMENTUM = 1.55 (FM)**-1.
TARGET FERMI MOMENTUM = 0.00 (FM)**-1

COMPLEX PARAMETERS FOR THE SUM OF YUKAWA POTENTIALS

	<<< T = 0 >>>		<<< T = 1 >>>	
	< REAL >	< IMAG >	< REAL >	< IMAG >

CENTRAL SINGLET INTERACTION

MASS	POTENTIAL STRENGTH, U			
0.69	0.4197694300+02	-0.5544291700+01	-0.1411471900+02	0.7752428400+00
3.10	0.1022116860+03	-0.1297879600+03	-0.3439987800+04	-0.3089944360+04
5.40	0.6701471700+04	0.1284132800+04	0.5437339800+05	0.4795897500+05
5.90	-0.7328974000+04	-0.1173349100+04	-0.5360278110+05	-0.4727570900+05
VOLUME INTEGRAL	1483.8	-251.62	-791.19	-418.18

CENTRAL TRIPLET INTERACTION

MASS	POTENTIAL STRENGTH, U			
0.69	-0.1796843300+02	0.5547709700+01	0.1142746270+01	-0.2392478100+01
3.10	-0.2030026130+04	0.4968234470+04	-0.1419466100+04	-0.2328592500+03
5.40	0.3120168300+05	0.7458572400+05	0.2277200400+05	0.2765540200+04
5.90	-0.3058244020+05	-0.7330358600+05	-0.2148541400+05	-0.2526384100+04
VOLUME INTEGRAL	-723.09	-918.38	230.44	-87.825

L-S INTERACTION

MASS	POTENTIAL STRENGTH, U			
0.76	0.2384194900+01	-0.1315773500+00	-0.1243172200+01	0.1149287900+01
1.60	-0.5280754600+02	0.4551388600+02	0.1803667810+02	0.1631312700+02
3.80	-0.3550917200+04	0.1020443000+04	-0.2906614400+04	0.3672088000+03
5.70	0.7743472980+04	-0.2314210000+04	0.5532099970+04	0.6724368500+03
VOLUME INTEGRAL	-302.53	213.50	-328.30	45.601

TENSOR INTERACTION

MASS	POTENTIAL STRENGTH, U			
0.76	-0.2318827600+00	0.5365227800+00	0.3213320700+00	0.6710961300+00
1.60	-0.6304163400+02	0.2345188900+02	0.1847075800+02	-0.2997292500+02
3.80	0.5055399200+04	-0.3633184400+04	0.1322755100+04	0.3927127800+04
5.70	-0.2653388200+04	0.1349744460+05	-0.5880262230+04	-0.2346534040+05
VOLUME INTEGRAL	-844.90	41.389	343.42	-446.91

PROJ. LAB. ENERGY = 25.0 MEV. INCIDENT MOMENTUM = 1.10 (FM)**-1.
TARGET FERMI MOMENTUM = 0.00 (FM)**-1

COMPLEX PARAMETERS FOR THE SUM OF YUKAWA POTENTIALS

	<<< T = 0 >>>		<<< T = 1 >>>	
	< REAL >	< IMAG >	< REAL >	< IMAG >

CENTRAL SINGLET INTERACTION

MASS	POTENTIAL STRENGTH, U			
0.69	0.4046051800+02	0.4374237600+01	-0.1013634600+02	-0.1634321500+01
3.10	0.1496588400+03	-0.6217838200+03	0.3769965400+04	-0.4774041140+04
5.40	0.6808820650+04	-0.3849925200+02	0.5893704700+05	0.7455415300+05
5.90	-0.6640304600+04	0.1134375500+03	-0.5805079500+05	-0.7354803400+05
VOLUME INTEGRAL	1456.0	-172.40	-754.90	-707.82

CENTRAL TRIPLET INTERACTION

MASS	POTENTIAL STRENGTH, U			
0.69	-0.9917689200+01	-0.5640315700+01	0.1574351200+00	-0.1911348400+01
3.10	-0.6452196300+03	0.6909988700+04	-0.1389016100+04	-0.1089396460+03
5.40	0.9400327900+04	0.1039029000+05	0.2270227690+05	0.1547342750+04
5.90	-0.9075509900+04	-0.1019894100+06	-0.2146066810+05	-0.1464304450+04
VOLUME INTEGRAL	-330.70	-1226.2	224.20	-53.990

L-S INTERACTION

MASS	POTENTIAL STRENGTH, U			
0.76	0.5838009820+01	-0.1463941410+01	-0.5808568500+00	0.1622003650+01
1.60	-0.1267692900+03	0.4382557710+02	0.2901544100+02	0.1158612600+02
3.80	-0.7003641500+04	0.1939888940+04	0.3099446700+04	-0.2613599700+04
5.70	0.1502217900+05	-0.4130170400+04	-0.5802891040+04	0.4474824000+04
VOLUME INTEGRAL	-779.94	274.01	-323.07	37.790

TENSOR INTERACTION

MASS	POTENTIAL STRENGTH, U			
0.76	0.8552791900+00	0.5116702100+00	0.6773729100+00	0.6620801190+00
1.60	0.1002176300+03	0.1340339100+02	0.3024042000+01	-0.2981035500+02
3.80	0.1083299150+05	-0.4899553700+04	0.3243937000+04	-0.3792200200+04
5.70	0.5214274500+05	0.1860725800+05	-0.1733346600+05	-0.2262213400+05
VOLUME INTEGRAL	-766.92	-172.77	122.80	-436.10

PROJ. LAB. ENERGY = 75.0 MEV.   INCIDENT MOMENTUM = 1.90 (FM)**-1.
TARGET FERMI MOMENTUM = 0.00 (FM)**-1

COMPLEX PARAMETERS FOR THE SUM OF YUKAWA POTENTIALS

<<< T = 0 >>>   <<< T = 1 >>>

< REAL >   < IMAG >   < REAL >   < IMAG >

CENTRAL SINGLET INTERACTION

MASS	POTENTIAL STRENGTH, V		POTENTIAL STRENGTH, V	
0.69	0.42796080D+02	-0.56458254D+01	-0.16283330D+02	0.98607100D+00
3.10	0.11219538D+03	-0.30157217D+03	-0.34816655D+04	0.21637169D+04
5.40	0.64736453D+04	-0.28083339D+04	-0.55877404D+05	0.33327015D+05
5.90	-0.70937254D+04	-0.26660496D+04	-0.55221836D+05	-0.32803771D+05
VOLUME INTEGRAL	1505.2	-297.51	-837.44	-283.32

CENTRAL TRIPLET INTERACTION

MASS	POTENTIAL STRENGTH, V		POTENTIAL STRENGTH, V	
0.69	-0.20293549D+02	0.74078498D+01	0.18203070D+01	-0.22693554D+01
3.10	-0.28758222D+04	-0.38792471D+04	-0.14070783D+04	-0.34254940D+03
5.40	0.45337695D+05	0.58661113D+05	0.22553276D+05	0.36970963D+04
5.90	-0.44699776D+05	-0.57608885D+05	-0.21295186D+05	-0.33102732D+04
VOLUME INTEGRAL	-894.65	-785.82	239.80	-109.59

L-S INTERACTION

MASS	POTENTIAL STRENGTH, V		POTENTIAL STRENGTH, V	
0.76	0.38092010D+00	-0.56658465D-01	0.10640870D+01	0.61966768D+00
1.60	-0.28866365D+02	0.57688385D+02	0.80517316D+01	0.14636043D+02
3.80	0.21823142D+04	0.62972348D+03	-0.27322987D+04	-0.36704335D+03
5.70	0.48967366D+04	-0.15972283D+04	0.52663754D+04	0.72433796D+04
VOLUME INTEGRAL	-138.62	212.19	-324.49	46.065

TENSOR INTERACTION

MASS	POTENTIAL STRENGTH, V		POTENTIAL STRENGTH, V	
0.76	-0.64562272D+00	0.77942211D-01	0.12277653D+00	0.56138285D+00
1.60	-0.52060447D+02	0.35776584D+02	0.26536165D+02	-0.25058709D+02
3.80	0.29574865D+04	-0.32459900D+04	0.24623328D+03	0.34526341D+04
5.70	-0.17125307D+05	0.12032466D+05	0.59492865D+03	-0.20665585D+05
VOLUME INTEGRAL	-898.64	114.92	464.57	-389.04

PROJ. LAB. ENERGY = 100.0 MEV.   INCIDENT MOMENTUM = 2.20 (FM)**-1.
TARGET FERMI MOMENTUM = 0.00 (FM)**-1

COMPLEX PARAMETERS FOR THE SUM OF YUKAWA POTENTIALS

<<< T = 0 >>>   <<< T = 1 >>>

< REAL >   < IMAG >   < REAL >   < IMAG >

CENTRAL SINGLET INTERACTION

MASS	POTENTIAL STRENGTH, V		POTENTIAL STRENGTH, V	
0.69	0.43155527D+02	-0.53965997D+01	-0.17227239D+02	0.57834547D+00
3.10	0.15176400D+03	-0.40918565D+03	-0.36497200D+04	-0.15453225D+04
5.40	0.57764655D+04	0.41289201D+05	0.41289201D+05	-0.23679362D+05
5.90	-0.63884399D+04	-0.39643369D+04	-0.58842433D+05	-0.23277436D+05
VOLUME INTEGRAL	1520.6	-329.29	-873.59	-204.33

CENTRAL TRIPLET INTERACTION

MASS	POTENTIAL STRENGTH, V		POTENTIAL STRENGTH, V	
0.69	-0.19750261D+02	0.89906490D+01	0.22296642D+01	-0.19448416D+01
3.10	-0.35225378D+04	-0.32620755D+04	-0.13706660D+04	-0.42995725D+03
5.40	0.56242939D+05	0.50117548D+05	0.22172171D+05	0.43795769D+04
5.90	-0.55227829D+05	-0.49341555D+05	-0.20976551D+05	-0.38729463D+04
VOLUME INTEGRAL	-971.43	-717.22	249.08	-124.33

L-S INTERACTION

MASS	POTENTIAL STRENGTH, V		POTENTIAL STRENGTH, V	
0.76	-0.72109381D+00	-0.58007339D+00	0.68229037D+00	0.39761875D+00
1.60	-0.23673166D+02	0.69643228D+02	0.12927584D+01	0.10244343D+02
3.80	-0.14104038D+04	0.39754446D+03	0.25974733D+04	-0.32011153D+03
5.70	0.33359425D+04	0.11865569D+04	0.50501247D+04	0.69457246D+03
VOLUME INTEGRAL	-69.029	216.27	-315.67	47.352

TENSOR INTERACTION

MASS	POTENTIAL STRENGTH, V		POTENTIAL STRENGTH, V	
0.76	-0.79961399D+00	-0.27239890D+00	0.28686384D-01	0.43936092D+00
1.60	-0.50351295D+02	0.40040984D+02	0.30195642D+02	0.20121838D+02
3.80	0.19882195D+04	-0.28332561D+04	-0.32763393D+03	0.29017722D+04
5.70	-0.12576658D+05	0.10563819D+05	0.40887168D+04	-0.17361473D+05
VOLUME INTEGRAL	-943.57	129.14	527.43	-323.00

PROJ. LAB ENERGY = 125.0 MEV. INCIDENT MOMENTUM = 2.45 (FM)**-1.
TARGET FERMI MOMENTUM = 0.00 (FM)**-1

COMPLEX PARAMETERS FOR THE SUM OF YUKAWA POTENTIALS

	<<< T = 0 >>>		<<< T = 1 >>>	
	< REAL >	< IMAG >	< REAL >	< IMAG >

CENTRAL SINGLET INTERACTION

MASS	POTENTIAL STRENGTH, U		POTENTIAL STRENGTH, U	
0.69	0.432281640+02	-0.500215382D+02	0.174613560+02	0.504307490-01
3.10	0.206292260+03	-0.498171620+03	-0.384209714D+04	-0.111947250+04
5.40	0.485758270+04	0.516117570+04	0.632054530+05	0.172089210+05
5.90	0.546314710+04	-0.496776280+04	-0.627441230+05	-0.169190180+05
VOLUME INTEGRAL	1531.9	-353.14	-897.33	-154.17

CENTRAL TRIPLET INTERACTION

MASS	POTENTIAL STRENGTH, U		POTENTIAL STRENGTH, U	
0.69	-0.181757170+02	0.977346300+01	0.244707120-01	0.157000237D+01
3.10	-0.396573090+04	-0.293621380+04	-0.132225190+04	-0.493465550+03
5.40	-0.636172030+05	0.460112560+05	0.217080420+05	0.479970420+04
5.90	-0.630103000+05	-0.454772030+05	-0.205909510+05	-0.420038880+04
VOLUME INTEGRAL	-996.56	-673.26	257.24	-134.64

L-S INTERACTION

MASS	POTENTIAL STRENGTH, U		POTENTIAL STRENGTH, U	
0.76	-0.114501250+01	-0.128370860+00	0.345456020+00	0.380215710+00
1.60	-0.274425850+02	0.784471770+02	-0.271903410+01	0.533352350+01
3.80	-0.885732010+03	0.251596720+03	0.294487760+04	0.260034860+03
5.70	0.230919180+04	-0.927367290+03	0.487836050+04	0.627461090+03
VOLUME INTEGRAL	-37.284	217.42	-305.19	50.846

TENSOR INTERACTION

MASS	POTENTIAL STRENGTH, U		POTENTIAL STRENGTH, U	
0.76	-0.812349990+00	-0.429129250+00	0.422706640+02	0.339651850+00
1.60	-0.520655750+02	0.375956530+02	0.315067190+02	-0.161218540+02
3.80	-0.140539840+04	-0.235417230+04	0.622887570+03	0.240987060+04
5.70	-0.991128810+04	0.893389940+04	0.590862220+04	-0.143658830+05
VOLUME INTEGRAL	-982.34	122.40	558.32	-263.42

PROJ. LAB ENERGY = 150.0 MEV. INCIDENT MOMENTUM = 2.69 (FM)**-1.
TARGET FERMI MOMENTUM = 0.00 (FM)**-1

COMPLEX PARAMETERS FOR THE SUM OF YUKAWA POTENTIALS

	<<< T = 0 >>>		<<< T = 1 >>>	
	< REAL >	< IMAG >	< REAL >	< IMAG >

CENTRAL SINGLET INTERACTION

MASS	POTENTIAL STRENGTH, U		POTENTIAL STRENGTH, U	
0.69	0.431254670+02	-0.461715260+01	0.173437020+02	-0.408271620+00
3.10	0.267305120+03	-0.569145900+03	-0.400663150+04	-0.830525490+03
5.40	0.385799810+04	0.590455290+04	0.664502630+05	0.130038630+05
5.90	-0.445989820+04	-0.568033380+04	-0.660631190+05	-0.128219370+05
VOLUME INTEGRAL	1540.4	-372.16	-909.33	-121.55

CENTRAL TRIPLET INTERACTION

MASS	POTENTIAL STRENGTH, U		POTENTIAL STRENGTH, U	
0.69	-0.165276660+02	-0.982341080+01	0.253335630+01	-0.121457980+01
3.10	-0.421347650+04	0.274671350+04	0.120024880+04	-0.535231670+03
5.40	-0.676283260+05	0.442454150+05	0.212312220+05	0.457429610+04
5.90	-0.670097430+05	-0.439176180+05	-0.201978760+05	-0.430465900+04
VOLUME INTEGRAL	-992.25	-639.15	263.93	-142.28

L-S INTERACTION

MASS	POTENTIAL STRENGTH, U		POTENTIAL STRENGTH, U	
0.76	-0.109849680+01	-0.194368220+01	0.102078840+00	0.458144130+00
1.60	-0.353174310+02	0.835884960+02	-0.499964870+01	0.834621440+00
3.80	0.496512430+03	-0.165316270+03	0.241462210+04	-0.192851930+03
5.70	0.158845570+04	-0.767617140+03	0.473920930+04	0.542984730+03
VOLUME INTEGRAL	-22.710	215.00	-294.77	56.249

TENSOR INTERACTION

MASS	POTENTIAL STRENGTH, U		POTENTIAL STRENGTH, U	
0.76	-0.769648650+00	-0.448244920+00	0.762746770-02	0.261197060+00
1.60	-0.553994160+02	0.314810830+02	0.316699810+02	0.130567760+02
3.80	-0.104139230+04	-0.187340050+04	0.187340050+04	0.199982330+04
5.70	-0.814660030+04	0.735404040+04	0.674728320+04	-0.118414810+05
VOLUME INTEGRAL	-1016.6	108.74	570.14	-213.85

PROJ. LAB ENERGY = 175.0 MEV, INCIDENT MOMENTUM = 2.90 (FM)**-1.
TARGET FERMI MOMENTUM = 0.00 (FM)**-1

COMPLEX PARAMETERS FOR THE SUM OF YUKAWA POTENTIALS

<<< T = 0 >>>   <<< T = 1 >>>
< REAL >   < IMAG >   < REAL >   < IMAG >

CENTRAL SINGLET INTERACTION

MASS	POTENTIAL STRENGTH, V		POTENTIAL STRENGTH, V	
0.69	0.42904528D+02	-0.42147248D+01	-0.17090047D+02	-0.74363124D+00
3.10	0.33158226D+03	-0.62416222D+03	0.41211683D+04	-0.64042896D+03
5.40	0.28281680D+04	0.63921381D+04	0.68793214D+05	0.10397263D+05
5.90	-0.34289949D+04	-0.61359904D+04	-0.68467956D+05	-0.10316375D+05
VOLUME INTEGRAL	1546.9	-387.70	-910.80	-100.62

CENTRAL TRIPLET INTERACTION

MASS	POTENTIAL STRENGTH, V		POTENTIAL STRENGTH, V	
0.69	-0.15177712D+02	-0.95513455D+01	0.25355199D+01	0.90682448D+00
3.10	0.43014720D+04	-0.26434106D+04	0.21939420D+04	-0.55899055D+03
5.40	0.68976215D+05	0.43482797D+05	-0.20785213D+05	0.49419563D+04
5.90	-0.68338652D+05	-0.43323115D+05	-0.19835031D+05	-0.42195129D+04
VOLUME INTEGRAL	-970.55	-609.60	269.27	-148.42

L-S INTERACTION

MASS	POTENTIAL STRENGTH, V		POTENTIAL STRENGTH, V	
0.76	-0.76635832D+00	-0.24615465D+00	0.61899919D+01	0.56734554D+00
1.60	0.44609381D+02	0.85494990D+02	-0.61407898D+01	0.29904318D+01
3.80	-0.19583527D+03	0.12047251D+03	0.23494407D+04	0.12491617D+03
5.70	0.10077561D+04	-0.67498330D+03	0.46234115D+04	0.44979125D+03
VOLUME INTEGRAL	-16.297	209.87	-285.16	62.925

TENSOR INTERACTION

MASS	POTENTIAL STRENGTH, V		POTENTIAL STRENGTH, V	
0.76	-0.71372310D+00	-0.36844328D+00	0.36016850D-02	0.20483094D+00
1.60	0.58958100D+02	0.23578895D+02	0.31316072D+03	-0.10818353D+02
3.80	0.76144895D+03	-0.14223849D+04	-0.80721983D+03	0.16697403D+04
5.70	0.67787980D+04	0.59060106D+04	0.70308444D+04	-0.97919844D+04
VOLUME INTEGRAL	-1049.1	95.519	571.40	-173.81

PROJ. LAB ENERGY = 200.0 MEV, INCIDENT MOMENTUM = 3.11 (FM)**-1.
TARGET FERMI MOMENTUM = 0.00 (FM)**-1

COMPLEX PARAMETERS FOR THE SUM OF YUKAWA POTENTIALS

<<< T = 0 >>>   <<< T = 1 >>>
< REAL >   < IMAG >   < REAL >   < IMAG >

CENTRAL SINGLET INTERACTION

MASS	POTENTIAL STRENGTH, V		POTENTIAL STRENGTH, V	
0.69	0.42597312D+02	-0.38397282D+01	-0.16819363D+02	-0.95361189D+00
3.10	0.39657771D+03	-0.66400442D+03	-0.41816194D+04	-0.51954654D+03
5.40	0.17995924D+04	0.66333984D+04	-0.70216561D+05	0.88644248D+03
5.90	-0.24018605D+04	-0.63427545D+04	-0.69946912D+05	-0.88734073D+04
VOLUME INTEGRAL	1551.9	-400.72	-903.22	-87.756

CENTRAL TRIPLET INTERACTION

MASS	POTENTIAL STRENGTH, V		POTENTIAL STRENGTH, V	
0.69	-0.14209229D+02	-0.90126625D+01	-0.24866246D+01	-0.65564624D+00
3.10	0.42747100D+04	-0.25713687D+04	-0.11722193D+04	-0.56844296D+03
5.40	0.68510640D+05	0.42994995D+05	0.20939070D+05	0.47419953D+04
5.90	-0.67861836D+05	-0.42967299D+05	-0.19522381D+05	-0.39797569D+04
VOLUME INTEGRAL	-938.50	-582.73	273.47	-153.77

L-S INTERACTION

MASS	POTENTIAL STRENGTH, V		POTENTIAL STRENGTH, V	
0.76	-0.28546200D+00	0.28184487D+01	0.17261115D+00	0.67837902D+00
1.60	-0.53845472D+02	0.84872118D+02	-0.68325878D+01	0.61628669D+01
3.80	0.40552082D+02	-0.10444404D+03	-0.22946012D+04	-0.57791126D+02
5.70	0.57243022D+03	-0.62872277D+03	0.45247978D+04	0.35187697D+03
VOLUME INTEGRAL	-13.831	203.01	-276.57	70.312

TENSOR INTERACTION

MASS	POTENTIAL STRENGTH, V		POTENTIAL STRENGTH, V	
0.76	-0.75581913D+00	-0.21421826D+00	0.25032283D-01	0.17595519D+00
1.60	-0.60576717D+02	0.15142211D+02	0.30711281D+02	-0.93305972D+01
3.80	0.46600424D+03	-0.10288566D+04	0.80572601D+03	0.14087067D+04
5.70	-0.54588467D+04	0.46874581D+04	0.69960758D+04	-0.81526657D+04
VOLUME INTEGRAL	-1089.1	88.573	567.34	-140.53

PROJ. LAB. ENERGY = 250.0 MEV, INCIDENT MOMENTUM = 3.47 (FM)**-1,
TARGET FERMI MOMENTUM = 0.00 (FM)**-1

COMPLEX PARAMETERS FOR THE SUM OF YUKAWA POTENTIALS

| | <<< T = 0 >>> | | <<< T = 1 >>> | |
| | < REAL > | < IMAG > | < REAL > | < IMAG > |

CENTRAL SINGLET INTERACTION

MASS	POTENTIAL STRENGTH, U			
0.69	0.418477730+02	-0.321587840+02	0.164304300+02	-0.108985090+01
3.10	0.522889090+03	-0.702763220+04	-0.416697220+04	0.400614720+03
5.40	-0.109949172D+03	0.64372581D+03	0.759410440+05	-0.754268840+04
5.90	-0.505122340+03	0.607196870+04	-0.707689250+05	-0.768792600+04
VOLUME INTEGRAL	1558.8	-421.71	-867.21	-77.469

CENTRAL TRIPLET INTERACTION

MASS	POTENTIAL STRENGTH, U			
0.69	-0.132547720+02	0.777670100+01	0.232272260+01	0.314591280+00
3.10	-0.402852050+04	-0.244449840+04	0.109300570+04	-0.557516230+03
5.40	0.648840600+05	0.417257810+05	0.198092260+05	0.398274200+04
5.90	-0.642733370+05	-0.418657080+05	-0.190833870+05	-0.317179330+04
VOLUME INTEGRAL	-858.79	-535.81	279.61	-163.85

L-S INTERACTION

MASS	POTENTIAL STRENGTH, U			
0.76	0.799914760+00	0.310352990+01	0.313127420+00	0.871629250+00
1.60	0.696088780+02	0.787154250+01	0.774574560+01	0.110420910+02
3.80	0.373847320+03	0.124114230+03	-0.220533230+04	0.728230090+02
5.70	-0.376597710+02	-0.620642270+03	0.436475520+04	0.148947210+03
VOLUME INTEGRAL	-13.559	186.83	-262.21	85.744

TENSOR INTERACTION

MASS	POTENTIAL STRENGTH, U			
0.76	-0.621798670+00	-0.157003300+01	0.617494120+01	0.148041554D+00
1.60	0.653386180+02	0.323557740+01	0.295891900+02	-0.730282580+01
3.80	0.489004880+02	-0.578081860+03	-0.759410790+03	0.102690770+03
5.70	-0.314196440+04	0.359309150+04	0.649082400+04	-0.577065470+04
VOLUME INTEGRAL	-1134.3	77.429	550.62	-93.136

PROJ. LAB. ENERGY = 300.0 MEV, INCIDENT MOMENTUM = 3.80 (FM)**-1,
TARGET FERMI MOMENTUM = 0.00 (FM)**-1

COMPLEX PARAMETERS FOR THE SUM OF YUKAWA POTENTIALS

| | <<< T = 0 >>> | | <<< T = 1 >>> | |
| | < REAL > | < IMAG > | < REAL > | < IMAG > |

CENTRAL SINGLET INTERACTION

MASS	POTENTIAL STRENGTH, U			
0.69	0.410132830+02	-0.162996840+02	0.274832260+01	-0.101395170+01
3.10	0.635999510+03	-0.408337480+04	-0.696387470+03	0.357711090+03
5.40	-0.170662290+04	0.700295290+04	0.548993370+04	0.707413770+04
5.90	0.106436140+04	-0.700293790+05	-0.504242350+04	-0.726626900+04
VOLUME INTEGRAL	1562.9	-438.35	-812.89	-79.096

CENTRAL TRIPLET INTERACTION

MASS	POTENTIAL STRENGTH, U			
0.69	-0.131342010+02	0.667247450+01	0.216325970+01	-0.133621030+00
3.10	-0.369187600+04	-0.230677620+04	0.103286970+04	-0.524378970+03
5.40	0.604844980+05	0.396998740+05	0.194678410+05	0.298060980+04
5.90	-0.600080400+05	-0.395283000+05	-0.188691110+05	-0.213177640+04
VOLUME INTEGRAL	-771.60	-498.13	284.33	-174.31

L-S INTERACTION

MASS	POTENTIAL STRENGTH, U			
0.76	0.175564870+01	0.297777370+01	0.417553730+00	0.102988340+01
1.60	0.804782100+02	0.691764740+02	-0.865922520+01	0.147726660+02
3.80	0.575886680+03	0.176074640+03	0.213466020+04	0.196658750+03
5.70	-0.411054050+03	0.667489060+03	0.424170170+04	0.57081441D+02
VOLUME INTEGRAL	-14.847	169.84	-250.67	100.70

TENSOR INTERACTION

MASS	POTENTIAL STRENGTH, U			
0.76	0.615114720+00	0.176152230+00	0.921771610-01	0.133482780+00
1.60	0.662110150+02	0.369700220+01	0.287344000+02	0.605782840+01
3.80	0.650290770+03	0.450883710+03	-0.649557140+03	0.772420670+03
5.70	-0.616651540+03	0.362635160+04	0.581628680+04	0.421339070+04
VOLUME INTEGRAL	-1180.0	93.274	531.98	-61.171

PROJ. LAB. ENERGY = 350.0 MEV. INCIDENT MOMENTUM = 4.11 (FM)**-1,
TARGET FERMI MOMENTUM = 0.00 (FM)**-1

COMPLEX PARAMETERS FOR THE SUM OF YUKAWA POTENTIALS

	<<< T = 0 >>>		<<< T = 1 >>>	
	< REAL >	< IMAG >	< REAL >	< IMAG >

CENTRAL SINGLET INTERACTION

MASS	POTENTIAL STRENGTH, U			
0.69	0.4017508D+02	-0.2501287D+02	-0.1636024D+02	-0.8711822D+00
3.10	0.7323144D+03	-0.6574141D+04	-0.3865553D+04	-0.3328268D+03
5.40	-0.2922514D+04	0.4006811D+04	0.6871797D+04	0.6597998D+04
5.90	0.2235963D+04	-0.3472765D+04	-0.6891096D+05	-0.6847749D+04
VOLUME INTEGRAL	1565.8	-452.62	-749.63	-86.863

CENTRAL TRIPLET INTERACTION

MASS	POTENTIAL STRENGTH, U			
0.69	-0.1340790D+02	0.5804203D+01	0.2026893D+02	-0.4942517D-01
3.10	-0.3369073D+04	-0.2156130D+04	-0.9859507D+03	-0.4833633D+03
5.40	-0.5688770D+05	0.3709804D+05	0.1928011D+05	0.1902579D+04
5.90	-0.5662470D+05	-0.3735281D+05	-0.1879221D+05	-0.1037183D+04
VOLUME INTEGRAL	-685.37	-469.80	288.94	-185.73

L-S INTERACTION

MASS	POTENTIAL STRENGTH, U			
0.76	0.2457588D+01	-0.2623179D+02	0.5129638D+00	0.1173223D+01
1.60	-0.8661381D+02	0.5868034D+01	-0.9756194D+01	-0.1797634D+02
3.80	0.6873207D+03	0.2354635D+03	0.2077861D+04	0.3192888D+03
5.70	0.6260956D+03	0.7301765D+03	0.4147481D+04	-0.2601402D+03
VOLUME INTEGRAL	-15.718	153.47	-240.84	114.53

TENSOR INTERACTION

MASS	POTENTIAL STRENGTH, U			
0.76	-0.6946559D+00	0.2042572D+00	0.9521198D-01	0.9188444D-01
1.60	-0.6488117D+02	0.5181035D+01	0.3865698D+02	-0.4539769D+01
3.80	-0.1270215D+04	0.5301929D+03	0.5620863D+04	0.5808638D+03
5.70	0.1852592D+04	0.4530704D+04	0.5172150D+04	0.3120541D+04
VOLUME INTEGRAL	-1230.4	109.53	512.25	-44.315

PROJ. LAB. ENERGY = 400.0 MEV. INCIDENT MOMENTUM = 4.39 (FM)**-1,
TARGET FERMI MOMENTUM = 0.00 (FM)**-1

COMPLEX PARAMETERS FOR THE SUM OF YUKAWA POTENTIALS

	<<< T = 0 >>>		<<< T = 1 >>>	
	< REAL >	< IMAG >	< REAL >	< IMAG >

CENTRAL SINGLET INTERACTION

MASS	POTENTIAL STRENGTH, U			
0.69	0.3936104D+02	-0.2349273D+02	-0.1650764D+02	-0.7417509D+00
3.10	0.8131385D+03	-0.5986322D+04	-0.3693998D+04	-0.3053390D+03
5.40	-0.3781749D+04	0.2214927D+04	0.6765975D+04	0.5836986D+04
5.90	0.3034663D+04	-0.1593535D+04	-0.6807629D+05	-0.6078877D+04
VOLUME INTEGRAL	1568.0	-465.55	-663.89	-97.895

CENTRAL TRIPLET INTERACTION

MASS	POTENTIAL STRENGTH, U			
0.69	-0.1379862D+02	0.5190678D+01	0.1915738D+01	-0.1302246D-01
3.10	-0.3102730D+04	-0.2001403D+04	-0.9455165D+03	-0.4429962D+03
5.40	-0.5458201D+05	0.3417238D+05	0.1915004D+05	0.8809365D+03
5.90	-0.5458346D+05	-0.3441405D+05	-0.1876069D+05	0.4963049D+01
VOLUME INTEGRAL	-604.16	-450.06	294.21	-198.19

L-S INTERACTION

MASS	POTENTIAL STRENGTH, U			
0.76	0.2882100D+01	-0.2173192D+02	0.5001889D+00	0.1312818D+01
1.60	-0.8879419D+02	0.4860390D+01	0.1080329D+02	-0.2091903D+02
3.80	0.7359211D+03	0.2883983D+03	0.2032123D+04	0.4333792D+03
5.70	0.7317436D+03	0.7856893D+03	0.4876068D+04	0.4547120D+03
VOLUME INTEGRAL	-15.752	138.39	-231.89	127.14

TENSOR INTERACTION

MASS	POTENTIAL STRENGTH, U			
0.76	-0.6992628D+00	0.4843888D+00	0.1117193D+00	0.8770199D-01
1.60	-0.6380460D+02	0.6744376D+01	0.2960688D+02	-0.3543205D+01
3.80	-0.1825559D+04	0.6922348D+04	0.4858471D+03	0.4481503D+03
5.70	0.3964914D+04	0.5530250D+04	0.4569185D+04	-0.2385363D+04
VOLUME INTEGRAL	-1269.0	176.58	496.92	-29.271

PROJ. LAB. ENERGY = 500.0 MEV, INCIDENT MOMENTUM = 4.91 (FM)**-1.
TARGET FERMI MOMENTUM = 0.00 (FM)**-1

COMPLEX PARAMETERS FOR THE SUM OF YUKAWA POTENTIALS

<<< T = 0 >>>   <<< T = 1 >>>

< REAL >   < IMAG >   < REAL >   < IMAG >

CENTRAL SINGLET INTERACTION

MASS	POTENTIAL STRENGTH, V		POTENTIAL STRENGTH, V	
0.69	0.378031920+02	-0.22632553D+01	-0.16785993D+02	-0.61074207D+00
3.10	0.941395090+03	-0.46266426D+03	0.34329478D+04	0.22477383D+03
5.40	-0.468965260+04	-0.15744702D+04	0.67167558D+05	0.35090016D+04
5.90	0.377980260+04	0.23616095D+04	0.68069933D+05	-0.36756701D+04
VOLUME INTEGRAL	1572.3	-490.71	-559.71	-124.76

CENTRAL TRIPLET INTERACTION

MASS	POTENTIAL STRENGTH, V		POTENTIAL STRENGTH, V	
0.69	-0.14457584D+02	-0.42810331D+01	0.171927650+01	0.15162813D-01
3.10	-0.27676145D+04	-0.17061250D+04	-0.86424581D+03	-0.37945714D+03
5.40	-0.53580057D+05	-0.28087816D+05	0.18799797D+05	0.84992087D+03
5.90	-0.54164770D+05	-0.28232093D+05	-0.18583837D+05	0.17606899D+04
VOLUME INTEGRAL	-463.96	-431.42	308.21	-226.45

L-S INTERACTION

MASS	POTENTIAL STRENGTH, V		POTENTIAL STRENGTH, V	
0.76	0.30818952D+01	0.12989758D+01	0.73124753D+00	0.15830331D+01
1.60	0.85295674D+02	0.32142897D+02	-0.12423493D+02	-0.26238163D+02
3.80	0.72310379D+03	0.35288948D+03	-0.15635323D+04	0.63865399D+03
5.70	-0.75408908D+03	0.83727838D+03	0.39786170D+04	-0.80291950D+03
VOLUME INTEGRAL	-14.029	112.78	-215.00	149.14

TENSOR INTERACTION

MASS	POTENTIAL STRENGTH, V		POTENTIAL STRENGTH, V	
0.76	-0.10274207D+01	-0.26690874D+00	0.15249790D+00	0.94492941D-01
1.60	-0.60140554D+02	-0.38966611D+00	0.26938286D+02	-0.17441134D+01
3.80	-0.26768319D+04	-0.93330437D+03	0.34104792D+03	0.27140556D+03
5.70	0.68282843D+04	0.78109039D+04	0.35442315D+04	-0.14843107D+04
VOLUME INTEGRAL	-1404.3	107.45	474.84	-6.5903

PROJ. LAB. ENERGY = 450.0 MEV. INCIDENT MOMENTUM = 4.66 (FM)**-1.
TARGET FERMI MOMENTUM = 0.00 (FM)**-1

COMPLEX PARAMETERS FOR THE SUM OF YUKAWA POTENTIALS

<<< T = 0 >>>   <<< T = 1 >>>

< REAL >   < IMAG >   < REAL >   < IMAG >

CENTRAL SINGLET INTERACTION

MASS	POTENTIAL STRENGTH, V		POTENTIAL STRENGTH, V	
0.69	0.38573541D+02	-0.22861683D+01	-0.16666184D+02	-0.65393391D+00
3.10	0.88128253D+03	-0.53051498D+04	-0.35469115D+04	-0.26889824D+03
5.40	-0.43416269D+04	0.29951279D+03	0.67128914D+05	0.47793865D+04
5.90	0.35194468D+04	0.40690094D+03	0.67786287D+05	-0.49904128D+04
VOLUME INTEGRAL	1570.0	-478.09	-619.83	-110.76

CENTRAL TRIPLET INTERACTION

MASS	POTENTIAL STRENGTH, V		POTENTIAL STRENGTH, V	
0.69	-0.14170856D+02	0.46595387D+02	0.18170136D+01	0.26198879D-02
3.10	-0.29034072D+04	-0.18498495D+04	-0.90616870D+03	-0.40766542D+03
5.40	-0.55547672D+05	0.31124377D+05	0.19004713D+05	-0.45155823D+03
5.90	-0.53838412D+05	-0.31261860D+05	-0.18704972D+05	0.94375532D+03
VOLUME INTEGRAL	-530.11	-437.72	300.54	-211.77

L-S INTERACTION

MASS	POTENTIAL STRENGTH, V		POTENTIAL STRENGTH, V	
0.76	0.30737741D+01	0.17168471D+01	0.67398391D+00	0.14501982D+00
1.60	0.88058396D+02	0.39673586D+02	-0.11710405D+02	-0.23678867D+02
3.80	0.74273638D+03	0.32816143D+03	-0.19948691D+04	0.53953784D+03
5.70	-0.76548512D+03	0.82255470D+03	0.40216512D+04	-0.63658623D+03
VOLUME INTEGRAL	-15.089	124.83	-223.37	138.63

TENSOR INTERACTION

MASS	POTENTIAL STRENGTH, V		POTENTIAL STRENGTH, V	
0.76	-0.87443350D+00	-0.88212489D-01	0.98209376D-01	0.22850699D-01
1.60	-0.61502970D+02	-0.23050093D+01	-0.27197200D+02	-0.20332093D+01
3.80	-0.23074020D+04	-0.87249420D+03	0.41510237D+03	0.33802576D+03
5.70	0.56523600D+04	0.65150740D+04	0.40402963D+04	-0.18319395D+04
VOLUME INTEGRAL	-1335.8	143.28	479.59	-26.853

PROJ. LAB. ENERGY = 200.0 MEV.   INCIDENT MOMENTUM = 3.11 (FM)**-1.
TARGET FERMI MOMENTUM = 0.80 (FM)**-1

COMPLEX PARAMETERS FOR THE SUM OF YUKAWA POTENTIALS

	<<< T = 0 >>>		<<< T = 1 >>>	
	< REAL >	< IMAG >	< REAL >	< IMAG >

CENTRAL SINGLET INTERACTION

MASS	POTENTIAL STRENGTH, U			
0.69	0.42365293D+02	0.38035438D+01	0.16820952D+02	-0.82261101D+00
3.10	0.41143839D+03	0.64864905D+03	0.41142236D+04	-0.57536140D+03
5.40	0.16783711D+04	0.61332981D+04	0.69206045D+05	0.96703792D+04
5.90	-0.22947500D+04	0.58186517D+04	-0.68946578D+05	-0.96672695D+04
VOLUME INTEGRAL	1551.1	-406.00	-889.48	-96.542

CENTRAL TRIPLET INTERACTION

MASS	POTENTIAL STRENGTH, U			
0.69	0.14494208D+02	0.87407875D+02	0.24080478D+01	-0.67950347D+00
3.10	-0.41276981D+04	0.25742300D+04	-0.11697569D+04	-0.55433271D+03
5.40	0.66252854D+05	0.42758160D+05	0.20584915D+05	0.44230094D+04
5.90	-0.65621557D+05	-0.42693756D+05	-0.19766573D+05	-0.36608254D+04
VOLUME INTEGRAL	-918.03	-582.82	271.38	-157.89

L-S INTERACTION

MASS	POTENTIAL STRENGTH, U			
0.76	0.17361336D+00	-0.26496930D+00	0.18943886D+00	0.72025986D+00
1.60	-0.58701284D+02	0.80969043D+02	-0.70879757D+01	0.67934995D+01
3.80	0.13423006D+03	0.11764363D+03	-0.22768785D+04	-0.23854590D+02
5.70	0.39997529D+03	-0.62865578D+03	0.44947194D+04	0.29565817D+03
VOLUME INTEGRAL	-12.858	199.04	-273.67	75.918

TENSOR INTERACTION

MASS	POTENTIAL STRENGTH, U			
0.76	-0.67121014D+00	0.22985576D+00	0.34109639D-01	0.14136224D+00
1.60	0.60884598D+02	0.14747937D+02	0.30467186D+02	-0.76657396D+02
3.80	0.42130246D+03	0.18793384D+04	-0.75181537D+03	0.12313645D+04
5.70	0.52284841D+04	0.49970814D+04	0.65840550D+04	-0.70849845D+04
VOLUME INTEGRAL	-1073.3	84.148	556.94	-117.04

PROJ. LAB. ENERGY = 200.0 MEV.   INCIDENT MOMENTUM = 3.11 (FM)**-1.
TARGET FERMI MOMENTUM = 0.50 (FM)**-1

COMPLEX PARAMETERS FOR THE SUM OF YUKAWA POTENTIALS

	<<< T = 0 >>>		<<< T = 1 >>>	
	< REAL >	< IMAG >	< REAL >	< IMAG >

CENTRAL SINGLET INTERACTION

MASS	POTENTIAL STRENGTH, U			
0.69	0.42496651D+02	-0.38257787D+01	0.16819945D+02	-0.90162018D+00
3.10	0.40308742D+03	-0.65791838D+03	-0.41541336D+04	-0.54302995D+03
5.40	0.17452042D+04	0.64350408D+04	0.69799898D+05	0.69300402D+04
5.90	-0.23529794D+04	-0.61348818D+04	-0.69533796D+05	-0.92101731D+04
VOLUME INTEGRAL	1551.4	-402.83	-897.71	-91.442

CENTRAL TRIPLET INTERACTION

MASS	POTENTIAL STRENGTH, U			
0.69	-0.14328954D+02	-0.89084439D+01	0.24517445D+01	-0.66561678D+00
3.10	-0.42139395D+04	-0.25729615D+04	-0.11711522D+04	-0.56277183D+03
5.40	0.67570490D+05	0.42905837D+05	0.20468901D+05	0.46148435D+04
5.90	-0.66929988D+05	-0.42862765D+05	-0.19616947D+05	-0.38522728D+04
VOLUME INTEGRAL	-930.23	-582.96	272.56	-155.39

L-S INTERACTION

MASS	POTENTIAL STRENGTH, U			
0.76	-0.91909392D-01	0.27480466D-01	0.17904496D+00	0.69479841D+00
1.60	-0.55888071D+02	0.82394218D+02	0.69342594D+01	0.64030652D+01
3.80	0.79695375D+02	0.10977277D+03	0.22874689D+04	-0.44411488D+02
5.70	0.50060090D+03	-0.62852392D+03	0.45126336D+04	0.32972708D+03
VOLUME INTEGRAL	-13.364	201.49	-275.41	72.567

TENSOR INTERACTION

MASS	POTENTIAL STRENGTH, U			
0.76	-0.66422287D+00	-0.24940938D+00	0.19381517D+00	0.13153736D+00
1.60	-0.61461907D+02	0.15309933D+02	0.30760267D+02	-0.81724209D+01
3.80	0.46958178D+03	-0.10584072D+04	-0.78755329D+04	0.13212344D+04
5.70	0.54411404D+04	0.48432520D+04	0.68419215D+04	-0.76585498D+04
VOLUME INTEGRAL	-1076.1	82.985	562.19	-133.57

PROJ. LAB. ENERGY = 200.0 MEV. INCIDENT MOMENTUM = 3.11 (FM)**-1.
TARGET FERMI MOMENTUM = 1.00 (FM)**-1

COMPLEX PARAMETERS FOR THE SUM OF YUKAWA POTENTIALS

	<<< T = 0 >>>		<<< T = 1 >>>	
	< REAL >	< IMAG >	< REAL >	< IMAG >

CENTRAL SINGLET INTERACTION

MASS	POTENTIAL STRENGTH, V			
0.69	0.4225868800+02	-0.3784694700+01	0.1682342900+02	-0.7548987600+00
3.10	0.4186189700+03	-0.6402170900+03	0.4080230300+04	-0.6007891500+03
5.40	0.1623971200+04	0.5860989800+04	0.6871184900+05	0.1002890600+05
5.90	-0.2242824340+04	-0.5533333610+04	-0.6845957280+05	-0.1001948600+05
VOLUME INTEGRAL	1SS1.0	-408.83	-882.26	-100.65

CENTRAL TRIPLET INTERACTION

MASS	POTENTIAL STRENGTH, V			
0.69	0.1462589940+02	-0.8591577500+01	0.2375439800+01	-0.6902644500+00
3.10	0.4055183100+04	-0.2573545400+04	0.1168576000+04	0.5470067500+03
5.40	0.6517271920+05	0.4260404200+05	0.2068867840+05	0.4256087700+04
5.90	-0.6455747800+05	-0.4252180000+05	-0.1988854100+05	-0.3492534160+04
VOLUME INTEGRAL	-907.51	-582.34	270.48	-160.17

L-S INTERACTION

MASS	POTENTIAL STRENGTH, V			
0.76	0.3870338900+00	-0.2566070200+01	0.1986014700+00	0.7438786600+00
1.60	0.6095528400+02	0.7893626800+02	-0.7232093100+01	-0.7168638200+01
3.80	0.1784886600+03	0.1246345300+04	-0.2265758720+04	-0.5240510200+01
5.70	0.3177376200+03	-0.6292131000+03	0.6299637000+04	0.2647771200+03
VOLUME INTEGRAL	-12.570	196.75	-272.11	78.844

TENSOR INTERACTION

MASS	POTENTIAL STRENGTH, V			
0.76	-0.6841233390+00	-0.2129104400+00	0.3909096700-01	0.1296341000+00
1.60	-0.6038448900+02	0.1441462200+02	0.3031237700+02	0.6991981200+01
3.80	0.3782352100+03	-0.1102811400+04	-0.7225758300+03	0.1151509200+04
5.70	0.5039027220+04	0.5145321300+04	0.6374189200+04	-0.6598974100+04
VOLUME INTEGRAL	-1072.5	86.459	551.58	-106.11

PROJ. LAB. ENERGY = 200.0 MEV. INCIDENT MOMENTUM = 3.11 (FM)**-1.
TARGET FERMI MOMENTUM = 1.20 (FM)**-1

COMPLEX PARAMETERS FOR THE SUM OF YUKAWA POTENTIALS

	<<< T = 0 >>>		<<< T = 1 >>>	
	< REAL >	< IMAG >	< REAL >	< IMAG >

CENTRAL SINGLET INTERACTION

MASS	POTENTIAL STRENGTH, V			
0.69	0.4214145500+02	-0.3763870200+01	0.1682921600+02	-0.6800356400+00
3.10	0.4269465100+03	-0.6300951500+03	0.4041964300+04	-0.6266400900+03
5.40	0.1563356900+04	0.5536287200+04	0.6872014750+05	0.1038847550+05
5.90	-0.2192212120+04	-0.5193189900+04	-0.6792290800+05	-0.1036814600+05
VOLUME INTEGRAL	1SS1.1	-873.89	-412.18	-104.99

CENTRAL TRIPLET INTERACTION

MASS	POTENTIAL STRENGTH, V			
0.69	0.1477917200+02	-0.8421711800+01	0.2342900500+01	-0.7010545500+00
3.10	0.3977917000+04	-0.2570024250+04	0.1167169700+04	-0.5386334200+03
5.40	0.6402092100+05	0.4239018600+05	0.2080361600+05	0.4061512900+04
5.90	-0.6341668100+05	-0.4228838200+05	-0.2003117100+05	-0.3297413000+04
VOLUME INTEGRAL	-895.57	-581.42	269.60	-162.91

L-S INTERACTION

MASS	POTENTIAL STRENGTH, V			
0.76	0.6148994800+00	-0.2469884200+01	0.2102294690+00	0.7226769300+00
1.60	0.6333452200+02	0.7657847600+02	-0.7412580700+01	-0.7641988600+01
3.80	0.2258434810+03	0.1328714100+03	-0.2256796500+04	0.1701261900+02
5.70	0.2290439700+03	-0.6304717690+03	0.4461298200+04	0.2278289500+03
VOLUME INTEGRAL	-12.386	193.95	-270.26	82.222

TENSOR INTERACTION

MASS	POTENTIAL STRENGTH, V			
0.76	-0.7045037600+00	-0.2031739700+00	0.3979499500-01	0.1126440750+00
1.60	-0.5981388800+02	0.1430888000+02	0.3019810900+02	0.6223624700+01
3.80	0.3271862300+03	-0.1135956100+04	-0.6922050900+03	0.1064807900+04
5.70	0.4815452200+04	0.5339265700+04	0.6143655100+04	-0.6877116300+04
VOLUME INTEGRAL	-1073.0	89.216	544.94	-95.182

PROJ. LAB. ENERGY = 200.0 MEV. INCIDENT MOMENTUM = 3.11 (FM)**-1.
TARGET FERMI MOMENTUM = 0.50 (FM)**-1

COMPLEX PARAMETERS FOR THE SUM OF YUKAWA POTENTIALS

q

	<<< T = 0 >>>		<<< T = 1 >>>	
	< REAL >	< IMAG >	< REAL >	< IMAG >

CENTRAL SINGLET INTERACTION

MASS	POTENTIAL STRENGTH, U		POTENTIAL STRENGTH, U	
0.69	0.425895290+02	-0.370011680+01	0.167904410+02	-0.866572860+00
3.10	0.412057610+03	-0.637020470+03	0.415934040+04	-0.512973590+03
5.40	0.197046620+04	0.610422840+04	0.699591340+05	0.871407510+04
5.90	-0.258266050+04	-0.581317660+04	-0.697071300+05	-0.872139830+04
VOLUME INTEGRAL	1579.2	-398.61	-897.65	-86.778

CENTRAL TRIPLET INTERACTION

MASS	POTENTIAL STRENGTH, U		POTENTIAL STRENGTH, U	
0.69	-0.142853600+02	0.847588250+01	0.248623740+01	-0.649685900+00
3.10	-0.420659940+04	0.244436770+04	0.118996380+04	-0.535215600+03
5.40	0.675536600+05	-0.408387050+05	0.214999080+05	0.449633480+04
5.90	-0.669142850+05	0.408677410+05	-0.214293050+05	-0.383328600+04
VOLUME INTEGRAL	-923.19	-553.17	241.46	-163.15

L-S INTERACTION

MASS	POTENTIAL STRENGTH, U		POTENTIAL STRENGTH, U	
0.76	0.841864130-02	-0.267956640+01	0.181019030+00	0.693325020+00
1.60	0.556714700+02	0.799470610+02	0.692892910+01	0.658558090+01
3.80	0.118276700+03	0.137696610+03	-0.266888650+04	-0.845216820+01
5.70	0.362250770+03	-0.631347640+03	0.439344461D+04	0.236323910+03
VOLUME INTEGRAL	-29.985	209.77	-305.28	66.806

TENSOR INTERACTION

MASS	POTENTIAL STRENGTH, U		POTENTIAL STRENGTH, U	
0.76	-0.656985900+00	-0.244274070+00	0.636924610-01	0.942942860-01
1.60	-0.620098750+02	0.149065280+02	0.289992840+02	-0.627931774D+01
3.80	0.362453510+03	-0.104259100+03	-0.528937040+04	0.108328850+04
5.70	0.513491790+04	0.486940710+04	0.507925530+04	0.660205280+04
VOLUME INTEGRAL	-1097.6	85.935	517.26	-89.173

PROJ. LAB. ENERGY = 200.0 MEV. INCIDENT MOMENTUM = 3.11 (FM)**-1.
TARGET FERMI MOMENTUM = 1.40 (FM)**-1

COMPLEX PARAMETERS FOR THE SUM OF YUKAWA POTENTIALS

	<<< T = 0 >>>		<<< T = 1 >>>	
	< REAL >	< IMAG >	< REAL >	< IMAG >

CENTRAL SINGLET INTERACTION

MASS	POTENTIAL STRENGTH, U		POTENTIAL STRENGTH, U	
0.69	0.420168600+02	-0.374152180+01	0.168402220+02	-0.603093030+00
3.10	0.436280560+03	-0.618457660+03	0.400085990+04	-0.650587530+03
5.40	0.149844440+04	0.516460980+04	0.676081650+05	0.107020510+05
5.90	-0.214360400+04	-0.480387480+04	-0.673793430+05	-0.106777230+05
VOLUME INTEGRAL	1551.4	-416.00	-864.58	-109.29

CENTRAL TRIPLET INTERACTION

MASS	POTENTIAL STRENGTH, U		POTENTIAL STRENGTH, U	
0.69	-0.149368100+02	0.824125620+01	0.231231650+01	-0.710926890+00
3.10	-0.389664570+04	0.256257030+04	0.116548250+04	-0.529516870+03
5.40	0.628483370+05	-0.421024960+05	0.209306900+05	0.384569410+04
5.90	-0.622638950+05	0.419804960+05	-0.201889620+05	-0.308094660+04
VOLUME INTEGRAL	-882.63	-579.98	268.80	-166.11

L-S INTERACTION

MASS	POTENTIAL STRENGTH, U		POTENTIAL STRENGTH, U	
0.76	0.845907280+00	-0.236222080+01	0.224835500+00	0.806495800+00
1.60	-0.659945370+02	0.735541780+02	0.763075180+01	0.821654570+01
3.80	0.235781900+03	0.142123430+03	-0.224472850+04	-0.425803150+04
5.70	0.138520870+03	-0.632489730+03	0.444166570+04	0.185350840+03
VOLUME INTEGRAL	-12.329	190.68	-268.15	85.958

TENSOR INTERACTION

MASS	POTENTIAL STRENGTH, U		POTENTIAL STRENGTH, U	
0.76	-0.723033260+00	-0.190003930+00	0.404688390-01	0.894463920-01
1.60	-0.593172770+02	0.142575040+02	0.300641800+02	0.536596310+01
3.80	-0.272212180+03	-0.117342270+03	-0.658948600+04	0.974641170+03
5.70	-0.457300250+04	0.555555740+04	0.589288780+04	-0.554283910+04
VOLUME INTEGRAL	-1074.0	93.588	537.67	-84.999

PROJ. LAB. ENERGY = 200.0 MEV. INCIDENT MOMENTUM = 3.11 (FM)**-1,
TARGET FERMI MOMENTUM = 1.00 (FM)**-1

q

COMPLEX PARAMETERS FOR THE SUM OF YUKAWA POTENTIALS

| | <<< T = 0 >>> | | <<< T = 1 >>> | |
| | < REAL > | < IMAG > | < REAL > | < IMAG > |

CENTRAL SINGLET INTERACTION

MASS	POTENTIAL STRENGTH, U			
0.69	0.425526510+02	-0.328798350+01	-0.166958110+02	-0.639560490+00
3.10	0.447718050+03	-0.553403280+03	-0.411336360+04	-0.478903390+03
5.40	0.271455930+04	0.455148360+04	0.695356100+05	0.803404930+04
5.90	-0.335043910+04	-0.426759430+04	0.693315580+05	-0.803669880+04
VOLUME INTEGRAL	1669.0	-389.59	-882.01	-82.113

CENTRAL TRIPLET INTERACTION

MASS	POTENTIAL STRENGTH, U			
0.69	-0.143881370+02	0.694209040+01	0.250267870+02	-0.615760620+00
3.10	-0.403001640+04	-0.207443560+04	-0.122827890+04	-0.439181430+03
5.40	0.652483800+05	0.347513700+05	0.260169001+05	0.346581180+04
5.90	-0.646642790+05	-0.347597810+05	-0.263051770+05	-0.299235290+04
VOLUME INTEGRAL	-874.80	-468.13	175.64	-177.20

L-S INTERACTION

MASS	POTENTIAL STRENGTH, U			
0.76	0.603464890+00	0.232888630+01	0.112133240+00	0.729075950+00
1.60	0.586523610+02	0.665660570+02	0.659086820+01	0.768226070+01
3.80	0.307406140+03	0.199524460+03	0.223303580+04	0.109342990+03
5.70	0.106545390+03	0.593230380+03	0.410550570+04	-0.262018500+02
VOLUME INTEGRAL	-48.472	220.28	-380.58	63.173

TENSOR INTERACTION

MASS	POTENTIAL STRENGTH, U			
0.76	-0.652040650+00	0.189533240+00	0.160540810+00	0.229599990-01
1.60	-0.627628880+02	0.122584910+02	0.250104410+02	-0.169996630+01
3.80	-0.463419130+02	0.100933260+04	0.466498470+01	0.475463250+03
5.70	-0.383747970+04	0.509750030+04	0.157124070+04	0.200615900+04
VOLUME INTEGRAL	-1160.3	97.431	434.57	14.267

PROJ. LAB. ENERGY = 200.0 MEV. INCIDENT MOMENTUM = 3.11 (FM)**-1,
TARGET FERMI MOMENTUM = 0.80 (FM)**-1

q

COMPLEX PARAMETERS FOR THE SUM OF YUKAWA POTENTIALS

| | <<< T = 0 >>> | | <<< T = 1 >>> | |
| | < REAL > | < IMAG > | < REAL > | < IMAG > |

CENTRAL SINGLET INTERACTION

MASS	POTENTIAL STRENGTH, U			
0.69	0.425526510+02	-0.348416480+00	-0.167419950+02	-0.741445460+00
3.10	0.432188810+03	0.594012950+03	-0.413190890+04	-0.497784700+03
5.40	0.231768910+04	0.528512460+04	0.695496510+04	0.840065910+04
5.90	-0.294343260+04	-0.520022970+04	0.694541000+05	0.840549410+04
VOLUME INTEGRAL	1624.5	-394.47	-889.36	-84.638

CENTRAL TRIPLET INTERACTION

MASS	POTENTIAL STRENGTH, U			
0.69	-0.143614950+02	0.765741300+01	0.249235880+02	-0.634692180+00
3.10	-0.411042750+04	-0.225075040+04	-0.121219350+04	-0.484488630+03
5.40	0.662581840+05	0.376753680+05	0.241829000+05	0.400218510+04
5.90	-0.656460790+05	-0.376704690+05	-0.240999140+05	-0.345455120+04
VOLUME INTEGRAL	-898.50	-508.24	202.15	-172.65

L-S INTERACTION

MASS	POTENTIAL STRENGTH, U			
0.76	0.347165850+00	0.249013040+01	0.160865330+00	0.713183850+00
1.60	0.526750910+02	0.727700220+02	0.620578940+01	0.717973230+01
3.80	0.224704760+03	0.176295290+03	0.578279440+02	-0.582794940+03
5.70	0.916368950+02	0.618575660+03	0.422569420+04	0.860510430+02
VOLUME INTEGRAL	-44.567	217.20	-346.16	63.880

TENSOR INTERACTION

MASS	POTENTIAL STRENGTH, U			
0.76	-0.652069300+00	0.215043850+00	0.126668700+00	0.598398240-01
1.60	-0.623465210+02	0.134086310+02	0.263933520+02	-0.362598440+01
3.80	-0.147896600+03	0.102932500+04	0.206234580+03	0.722623570+03
5.70	-0.445077940+04	0.501443690+04	0.290441210+04	0.360796930+04
VOLUME INTEGRAL	-1129.1	91.628	465.15	-24.599

PROJ. LAB. ENERGY = 200.0 MEV. INCIDENT MOMENTUM = 3.11 (FM)**-1,
TARGET FERMI MOMENTUM = 1.20 (FM)**-1

q

COMPLEX PARAMETERS FOR THE SUM OF YUKAWA POTENTIALS

| | <<< T = 0 >>> | | <<< T = 1 >>> | |
| | < REAL > | < IMAG > | < REAL > | < IMAG > |

CENTRAL SINGLET INTERACTION

MASS	POTENTIAL STRENGTH, U			
0.69	0.425744830+02	-0.305117610+01	-0.166389390+02	-0.532200700+00
3.10	-0.462749030+03	-0.502291030+03	-0.409786610+04	-0.450438500+03
5.40	0.329725530+04	0.367794880+04	0.694803050+05	0.750116590+04
5.90	-0.394170790+04	-0.340404720+04	-0.693015520+05	-0.750131480+04
VOLUME INTEGRAL	1726.8	-382.02	-873.24	-78.429

CENTRAL TRIPLET INTERACTION

MASS	POTENTIAL STRENGTH, U			
0.69	-0.143885740+02	0.619303220+01	0.252034790+01	-0.587783860+00
3.10	-0.393923980+04	-0.186064010+04	-0.124414000+04	-0.386001450+03
5.40	0.641997130+05	0.312031090+05	0.280643600+05	0.275031160+04
5.90	-0.636574260+05	-0.311226951+05	-0.287534170+05	-0.233716520+04
VOLUME INTEGRAL	-844.55	-420.90	153.90	-178.74

L-S INTERACTION

MASS	POTENTIAL STRENGTH, U			
0.76	0.836605785+00	0.213795360+01	0.191030050-01	0.745966160+00
1.60	-0.587747350+03	0.594167870+02	-0.397418810+01	-0.825242270+01
3.80	0.391996290+03	0.219978280+03	-0.223643050+04	0.161789500+03
5.70	-0.294532820+03	-0.549508800+03	0.399666710+04	-0.135721610+03
VOLUME INTEGRAL	-43.105	218.82	-419.52	64.023

TENSOR INTERACTION

MASS	POTENTIAL STRENGTH, U			
0.76	0.651785410+00	-0.116394280+00	0.183893931+00	-0.101242450-01
1.60	0.634088620+02	0.110706141+02	0.240137850+02	-0.632789330+01
3.80	-0.276802500+03	0.974745520+03	0.156800310+03	0.260380670+03
5.70	-0.311586650+04	0.512268950+04	0.540586820+03	0.654058320+03
VOLUME INTEGRAL	-1199.5	103.45	413.13	44.420

PROJ. LAB. ENERGY = 200.0 MEV. INCIDENT MOMENTUM = 3.11 (FM)**-1,
TARGET FERMI MOMENTUM = 1.40 (FM)**-1

q

COMPLEX PARAMETERS FOR THE SUM OF YUKAWA POTENTIALS

| | <<< T = 0 >>> | | <<< T = 1 >>> | |
| | < REAL > | < IMAG > | < REAL > | < IMAG > |

CENTRAL SINGLET INTERACTION

MASS	POTENTIAL STRENGTH, U			
0.69	0.426140910+02	-0.277329100+01	-0.165720700+02	-0.427238860+00
3.10	-0.474748550+03	-0.442579220+03	-0.408766440+04	-0.410669500+03
5.40	0.412441460+04	0.270000020+04	0.695554150+05	0.677661570+04
5.90	-0.477223000+04	-0.244318680+04	-0.694044830+05	-0.677408310+04
VOLUME INTEGRAL	1800.1	-370.37	-862.93	-73.367

CENTRAL TRIPLET INTERACTION

MASS	POTENTIAL STRENGTH, U			
0.69	-0.143644150+02	-0.526222420+01	0.254816990+01	-0.549892970+00
3.10	-0.383832500+04	-0.161480910+04	-0.126002910+04	-0.265382980+03
5.40	0.631521870+05	0.270483540+05	0.303092550+05	0.187046630+04
5.90	-0.626696800+05	-0.270881240+05	-0.314141920+05	-0.149907590+04
VOLUME INTEGRAL	-806.80	-367.66	148.75	-176.40

L-S INTERACTION

MASS	POTENTIAL STRENGTH, U			
0.76	0.101332690+01	0.191212430+01	0.132161540+00	0.761937120+00
1.60	0.575148550+02	0.514855300+02	-0.132238380+01	-0.886208600+01
3.80	0.473104390+03	0.214153710+03	-0.225998170+04	0.210966980+03
5.70	-0.456365030+03	0.485179770+03	0.391295080+04	-0.233321570+03
VOLUME INTEGRAL	-25.069	209.84	-462.67	66.425

TENSOR INTERACTION

MASS	POTENTIAL STRENGTH, U			
0.76	0.645366610+00	-0.141453930+00	0.198724080+00	-0.367251180-01
1.60	0.649822890+02	0.973659390+01	0.234210420+02	0.116008020+01
3.80	-0.535021750+03	-0.914778580+03	0.268645150+03	0.928936220+02
5.70	-0.231259650+04	0.503255020+04	-0.114241900+03	0.344682070+03
VOLUME INTEGRAL	-1245.2	108.73	403.35	63.246

REFERENCES

1) Microscopic Optical Models, ed. H.V. von Geramb,
    Lect. Notes in Physics 89, Springer(1979)

2) J. Hüfner and C. Mahaux, Ann. Phys. (N.Y.) 73, 525 (1972)

3) A.K. Kerman, H. MacManus, and R.M. Thaler, Ann. Phys. (N.Y.)
    8, 551 (1959)

4) J.P. Jeukenne, A. Lejeune and C. Mahaux, Phys. Rep. 25C,
    85 (1976)
    F. Brieva and J.R. Rook, Nucl. Phys. A291, 299 (1977);
    A291, 317 (1977); A297, 2o6 (1978)

5) J.P. Jeukenne, A. Jejeune, and C. Mahaux, Phys.Rev. C1o,
    1391 (1974)

6) C. Mahaux, The Many-Body Problem Jastow Correlations Versus
    Brueckner Theory, ed. R. Guardiola and J. Ros, Lect. Notes
    in Phys. 138, 5o (1981), Springer

7) W.G. Love, M.A. Franey, Phys. Rev. C24, 1o73 (1981)

8) M. Lacombe, B. Loiseau, J.M. Richard, R. Vinh Mau, J. Coté,
    P. Pirès, and R. de Tourreil, Phys. Rev. C21, 861 (198o)

9) P.J. Siemens, Nucl Phys. A141, 225 (197o)

1o) H.O. Meyer, P. Schwandt, G.L. Moake and P.P. Singh,
    Phys. Rev. C23, 616 (1981);
    H.O. Meyer, J.R. Hall, W.W. Jacobs, P. Schwandt and P.P. Sing
    Phys. Rev. C24, 1734 (1981);
    J.R. Comfort, G.L. Moake, C.C. Foster, P. Schwandt, C.D. Good
    J. Rapaport, W.G. Love, Phys. Rev. C24, 1834 (1981)
    J. R. Comfort, Phys. Rev. C24, 1844 (1981)

11) M. Jaminon, C. Mahaux, P. Rochus, Phys. Rev. C22, 2o27 (198o)
    E.D. Cooper, PhD thesis, Edmonton, Alberta (1981)
    L.G. Arnold, B.C. Clark, R.L. Mercer and P. Schwandt,
    Phys. Rev. C23, 1949 (1981)

12) T. Hamada, D. Johnston, Nucl. Phys. 34, 382 (1962)

13) H.O. Meyer, private communication and references cited
    under Ref. 1o).

14) Ch. Glashausser, private communication and progress report
    Rutgers University, New Brunswick (1982)

15) W. Bauhoff, private communication, to be published

16) L. Kisslinger, Mesons in Nuclei, ed. M. Rho and D. Wilkinson,
    Vol. 1, 261 (1979), North Holland
    A.M. Green, Reports on Progress in Phys., 11o9 (1976)
    H.J. Weber and H. Arenhövel, Phys. Rep. 36C, 277 (1978)
    W. Weise, Nucl. Phys. A374, 5o5c (1982)
    E.J. Monitz, Nucl. Phys. A374, 557c (1982)

17) H. Sugawara, F. von Hippel, Phys. Rev. 172, 1764 (1968)

18) S. Cohen, D. Kurath, Nucl. Phys. 73, 1 (1965)

COUPLED CHANNELS AND QUARKS
IN NUCLEON-NUCLEON SCATTERING

Earle L. Lomon
Massachusetts Institute of Technology, Cambridge, MA 02139

ABSTRACT

In q.c.d. hadron interactions are approximated by asymptotically free quarks at short range, and by exchange and production of confined hadrons at long range. The R-matrix formalism is introduced to match the exterior hadronic description to a complete set of internal q^6 states at an appropriate radius expected to be smaller than the equilibrium bag radius. The long range interaction among NN, NΔ, $\Delta\Delta$ and NN*(1440) channels is consistent with meson exchange potentials and scattering data up to 800 MeV laboratory kinetic energy. This predicts a deuteron with 4.0% D state and 2.3% total isobar component. The first $^1S_0$ and $^3S_1$, q^6 states, obtained from the standard MIT bag model parameters, satisfy the R-matrix condition of being at the energies of the $^1S_0$ and $^3S_1$ - $^3D_1$ wave function zeros for a quark boundary radius R = 1.10 fm (corresponding to an interhadron separation r_0 = 1.25 fm). A zero corresponds to a pole in the logarithmic derivative of a partial-wave function, whose residue is determined by the derivative of the q^6 state energy with respect to the boundary radius. These R-matrix residues are substantially smaller than the values which fit the $^1S_0$ and $^3S_1$ - $^3D_1$ data over the 1 GeV range. A smaller boundary radius, consistent with bag constants of recent hadron models, will improve the residue fit.

INTRODUCTION

Quantum chromodyanics (q.c.d.) is not readily soluble, but has two simple limits. At high momentum transfer the quarks and gluons are asymptotically free, with perturbative corrections due to gluon exchange. At low momentum transfer color singlet clusters of quarks and gluons (hadrons) are confined and interact via the exchange of other hadrons. These features are partly incorporated into the MIT bag model[1] in which free quarks and gluons are confined in a hole in the vacuum in equilibrium with the pressure due to the volume energy of the vacuum. The equilibrium radius depends on the number, type and state of the confined quarks and gluons. In the chiral and cloudy bag models[2] interaction with an external pion field at the bag surface mediates long range interactions. In this work[3] we regard quark clustering to be well approximated by hadrons at separations less than that of full confinement but large enough for high order qcd to modify the free behavior of quarks and gluons. This enables the exchange of all hadrons to mediate interactions between bags.

A measure of the radius outside of which hadron clustering is an adequate approximation is given by the correspondence of the NN hadron exchange potential[4] to the data for $E_L \lesssim 350$ MeV, which is good for $r > 0.8$ fm. This agreement can be extended to $E_L = 800$ MeV by inclusion of transition potentials to Δ and $N^*(1440)$ isobar channels with theoretical one pion exchange terms.[5]

One can infer the range within which asymptotic freedom is a good approximation from the SU_3 lattice gauge theory calculation of the ratio of successive Wilson Loop expectation values.[6] These closely approach the weak and strong coupling limits with a narrow but finite transition region. Choosing the lattice size to produce confinement at the MIT bag radius (R = 1.3 fm for a q^6 bag) indicates that asymptotic freedom is accurate for R < 1.0 fm corresponding to $r \leq 1.14$ fm. (see Figure 1) Therefore a spherical boundary at

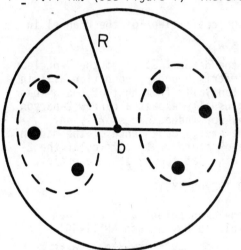

Figure 1. The q^6 bag

0.8 fm $\leq r_0 \leq 1.14$ fm, corresponding to a q^6 bag radius 0.7 fm < R < 1.0 fm, can satisfy the conditions of R-matrix theory.[7] Unlike the equilibrium bag radius, R_{eq}, the radius R is only a formal boundary within which a complete set of internal wave functions are generated, and at which internal and external wave functions are to be matched (see Figure 2).

This generalizes the method of Jaffe and Low[8] in which the two body wave function is constrained to vanish at $r_0 = b$, where $b = 1.14 R_{eq}$ for the 2-baryon (q^6). The radius r_0 may be chosen to optimize the accuracy of the internal and external Hamiltonians. We also extend the approach of Ref.8 by including the external hadronic interaction and extending the fit to the data away from the pole region by using the boundary condition[8,9]

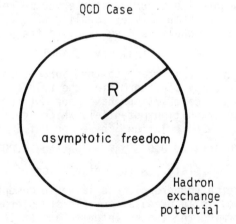

QCD Case

Figure 2. Two regions connected by R-matrix formalism

$$f(w) \equiv r_0 \, P(w) \equiv r_0 \frac{d\psi}{dr_0} \frac{1}{\psi(r_0)} \approx f^0 + \sum_{i=1}^{i_{max}} \frac{(Res)_i}{w-w_i} \cdots \qquad (1)$$

with i_{max} chosen so that $w_{i_{max}+1}$ is substantially beyond the fitted energy range. Eq.1 for $f(w)$ becomes exact as $i_{max} \to \infty$. For multichannel systems the quantities of Eq.1 are matrices. The residues are given by[8]

$$(Res)_i^{\alpha\beta} = - \frac{\partial s}{\partial r_0} \frac{r_0}{2w_i} \xi_\alpha \, \xi_\beta \cdots \qquad (2)$$

where ξ_α is the fractional parentage coefficient of the channel in the q^6 state. When $r_0 < b$ then $s = w_i^2$.

It is important to note that the determination of the 2-nucleon S-matrix from the above f (or P) -matrix boundary condition will in general shift the energy pole substantially from that of the q^6 bag state.[8] Consequently one does not usually expect a narrow 2-baryon resonance near the q^6 state energy, nor indeed necessarily any structure at all. Nevertheless the energy dependence of the hadronic partial waves is determined by the q^6 states. In particular the 2-hadron wave functions will vanish at r_0 at the q^6 state energies.

APPLICATION TO THE NN SYSTEM

The field theoretical Feshbach-Lomon potential[10] has been supplemented by transition potentials to NΔ, $\Delta\Delta$ and NN*(1440) channels to fit the NN data for $E_L \le 800$ MeV.[5] The $^1D_2$ and $^3F_3$ resonant-like structures observed in recent data were found to be generated naturally by coupling to the NΔ channel. The coupling to isobar channels is also vital to the proper energy dependence of several other partial waves, including the $^3S_1 - ^3D_1$. The sensitivity of this state to the isobar coupling provides the first opportunity of determining the role of isobars in the deuteron. Previous calculations have made ad hoc assumptions about the short range transition potentials which are now fixed by the scattering data. A calculation[11] based on Ref.5 produces a deuteron with 4.0% NN($^3D_1$)state, 0.67% $\Delta\Delta(^3S_1)$ state, 0.80% $\Delta\Delta(^7D_1)$ state and 0.84% NN*($^3S_1$) state. A full discussion of the properties of this deuteron (quadrupole and magnetic moments, and asymptotic D/S state ratio) is in preparation. This deuteron model may be used to determine the isobar intermediate state contribution to eD and other reactions.

This fit to the data, like the earlier version with no isobar channels[10] that was fitted to E_L < 400 MeV data, was obtained with a pole-free (energy independent) f-matrix at r_0 = 0.71 fm. This is consistent with the fact that for such a small r_0 there are no q^6 states for w < 3.26 GeV (E_L = 800 MeV corresponds to w = 2.24 GeV). It is remarkable that this fit produces poles of the f-matrix at r_0 = 1.45 fm (Corresponding to the equilibrium q^6 bag radius) at w = 2.080 GeV and 2.145 GeV for the $^3S_1$ - $^3D_1$ and $^1S_0$ states respectively.[3,12] The corresponding standard MIT bag model states are close by at 2.16 GeV and 2.24 GeV. As r_0 decreases the f-pole and bag state energies will get closer, because the equilibrium bag state energies are minima with respect to r_0, while the f-poles have a negative r_0 derivative. It can therefore be expected that the f-pole and bag state energies will be equal for some r_0 between 0.71 fm and 1.45 fm. In Fig.3 the r_0 dependence is plotted and shows that the crossover occurs at r_0 = 1.20 fm for the $^3S_1$ - $^3D_1$ state and at 1.25 fm for the $^1S_0$ state. These values are only a little outside the radius at which asymptotic freedom can be expected to be a very good approximation, and well within the range at which the hadron exchange potential is successful. Unfortunately, the residues predicted by the internal quark states (Eq.2) are about one-fourth the values needed to fit the data away from the pole, even after adjustment of the constant term in the f-matrix.

The remedy for the inconsistency of the residues is a larger vacuum volume energy producing a smaller equilibrium bag radius. The crossover r_0 will then also be smaller, and the resulting residues predicted by the interior quark states will be larger. But for smaller r_0 the residues required to fit the data are decreased because the isobar coupling effect, which like the f-pole produces attraction increasing

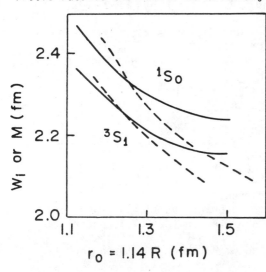

Figure 3. Boundary radius dependence of $^3S_1$ and $^1S_0$ f-poles ---, bag states ——

with energy, becomes stronger. Not only the chiral and cloudy bag models[2] but also a recent revision of the MIT bag model[13] fit the hadron spectrum with larger vacuum energies and substantially smaller equilibrium radii than the standard MIT bag model.[1] We are now examining the consequences for the crossover values of r_0 and the data fits with pole residues predicted by the internal quark states.

We have concentrated on the $^1S_0$ and $^3S_1$ - $^3D_1$ partial waves because these are the only ones which have, for $r_0 < 1.5$ fm, f-poles below $E_L = 1$ GeV (where enough data is available to determine the f-matrix). However, several other partial waves ($^3P_0$, $^1S_2$, $^3F_3$ among them) have f-poles approaching $r_0 = 1.5$ fm at $E_L = 1$ GeV[3]. As the amplitude analyses become available from data taken at Saturne II (up to 3 GeV) further tests of the R-matrix approach will be possible. The present analysis will determine the bag model parameters to be used. Therefore comparison with many partial waves (including the S-waves) for $E_L > 800$ MeV will provide stringent tests. Some of the f-poles predicted may imply narrow structures for experimental observation.

1. A. Chodos et al., Phys. Rev. D9, 3471 (1974); D10, 2599 (1974).
2. F. Myhrer, G.E. Brown and Z. Xu, Nucl. Phys. A362, 317 (1981); A.W. Thomas, S. Théberge and G.A. Miller, Phys. Rev. D24, 216 (1981).
3. E.L. Lomon, in Comptes Rendus des Neuvièmes Journées d'Etudes de la Division Physique Théorique, Aussois, 1980, edited by H. Sazdjian and M. Veneroni (Institut de Physique Nucléaire, Orsay, 1981); Proc. Workshop on Nuclear and Particle Physics up to 31 GeV, Los Alamos, NM, 1981. LA-8775-C
4. M. Lacombe et al., Phys. Rev. C21, 861 (1980).
5. E.L. Lomon, Phys. Rev. D26, 576 (1982).
6. M. Creutz, Phys. Rev. Lett. 45, 313 (1980).
7. E.P. Wigner and L. Eisenbud, Phys. Rev. 72, 29 (1947).
8. R.L. Jaffe and F.E. Low, Phys. Rev. D19, 2105 (1979).
9. H. Feshbach and E.L. Lomon, Ann. Phys. (NY) 29, 19 (1964).
10. E.L. Lomon and H. Feshbach, Ann. Phys. (NY) 48, 94 (1968).
11. E.L. Lomon, private communication.
12. Using perturbative long range interactions similar results wave obtained by R.L. Jaffe and M.P. Shatz, C.I.T. preprint CALT-68-775.
13. C.E. Carlson, T.H. Hansson and C. Peterson, MIT preprint CTP 1020.

DENSITY-DEPENDENT CHARGE ASYMMETRY AND
THE COULOMB ENERGY ANOMALY

J. V. Noble
Department of Physics, University of Virginia,
Charlottesville, Va. 22901

ABSTRACT

The "discrepancy" between measured and calculated isodoublet mass splittings results from an overly optimistic view of the accuracy of present theory. In fact, charge symmetry breaking interactions are both sufficiently large and sufficiently uncertain that the "best" theoretical estimates of the isodoublet splittings bracket the measured values within a fairly broad uncertainty.

The energy splittings between members of nuclear isodoublets have been considered anomalous because "standard" nuclear theory has been unable to account for them to an accuracy of better than 10%.[1-3] If we focus on an isodoublet like Sc(41)-Ca(41) for definiteness, to a first approximation, the mass difference between the two is given by

$$\Delta M_{41} = \Delta E_{coul} + \Delta E_{exch} + \Delta E_{s.o.} + \Delta E_{pol} + \Delta M_{pn} \tag{1}$$

where ΔE_{coul} is the Coulomb energy of the valence proton in the field of the Ca(40) "core", ΔE_{exch} is the correction arising from antisymmetrization of the valence - with the core proton, $\Delta E_{s.o.}$ is the electromagnetic spin-orbit interaction (which takes into account the opposite signs of the neutron and proton magnetic moments), and ΔE_{pol} is the effect of core polarization by the valence particle (in a self-consistent Hartree-Fock calculation using all 41 particles, this would be 0, up to 2p1h admixtures). Finally, ΔM_{pn} is the proton-neutron mass difference. As we see from Table I below, there is a discrepancy of some 0.7 to 1.1 MeV between theory and experiment, or about 10-15%.

Table I Contributions to A = 41 isodoublet splitting from
"standard" nuclear theory (MeV)

$\Delta E(coul)$	$\Delta E(exch)$	$\Delta E(s.o.)$	$\Delta E(pol)$	ΔM_{pn}	$\Delta M_{Th'y}$	ΔM_{41}
6.9	-0.3	-0.1	-0.3 to +0.1	-1.3	4.9 to 5.3	6.0

(Although certain authors feel the discrepancy may be as little as 0.5 MeV, I believe this results from the use of density-dependent interactions[4] in Hartree-Fock calculations, and this degree of density-dependence has been shown to be in disagreement with nuclear spectra.[5] Hence I conclude that such small discrepancies are wishful thinking.)

In this paper I consider three additional sources of charge asymmetry which will affect the isodoublet splitting. To his great credit, Negele has maintained for a long time that charge asymmetry in the nuclear forces will prove to be the source of the discrepancy.[6] However, until recently it appeared that all the available theoretical sources of hadronic charge-symmetry breaking (CSB) were far too small to explain the anomaly. For example, all calculations in which a meson-nucleon vertex or exchange diagram is modified by the inclusion of an additional photon are too small by at least an order of magnitude. (An example of this is my own, unpublished, calculation of the effect of the nuclear Coulomb field on the charged-pion propagators, which leads to a mass-splitting of at most 20 keV under the most optimistic assumptions.) Langacker and Sparrow[7] have recently proposed that the same mass splitting between "up" and "down" quarks which is invoked to explain the large rate of eta decay into 3 pions, as well as the p-n mass difference, will lead both to $\pi°$-η and ρ-ω mixing, which in turn generates a CSB interaction at the level of one boson exchange. This calculation gives several hundred keV of additional splitting, in the right direction. On the other hand, Coon and Scadron have criticized it for being overly optimistic, especially in employing a value of the ηNN coupling constant which is at the extreme upper end of the published values, and which is generally believed to be much too large.[8]

Several years ago, Riska and Chu examined the CSB interactions which could be generated in the 2 nucleon system by either meson + photon exchange, or by two pion exchange (2PE).[9] They concluded that the diagrams with explicit photons are very small, in agreement with all other such results; but when they considered the 2PE diagrams shown below in Fig. 1, taking account of the p-n mass difference they found a substantial CSB interaction leading to mass splittings of

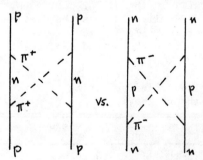

several hundred keV in the wrong direction! Recently I was led to recalculate the resulting NN potential as a function of relative separation and/or spin orientation, but merely evaluated the diagrams at low energy and zero momentum transfer. This gives only the volume integral of the CSB potential, but that is all we need to know for evaluating the isodoublet splitting. My

Fig. 1. Two pion exchange diagrams illustrating the origin of a CSB interaction in the p-n mass difference.

results look identical to the formulae in their paper, and I agree with their statements about the various pieces of the calculation in qualitative terms of sign and magnitude. However, our final resules disagree — mine gives several hundred keV in the right direction. Since their results are not tabulated, and since their computer programs have long since vanished into the shredder, I have no way to trace down this discrepancy. I can only suggest that I have confidence in my own calculation because it is algebraically much simpler than theirs and hence easier both to check and to evaluate.

The final, and to my mind, major, source of hadronic CSB comes from the renormalization of the nucleon mass in nuclear matter. As we shall see in Section E of this Workshop, in the relativistic quantum field-theoretic approach to nuclear structure, the mass of the nucleon is renormalized by hundreds of MeV in nuclear matter. This must at least cast some doubt on the validity of the assumption that the p-n mass difference remains unrenormalized when the nucleons are placed in nuclei. If the mass difference were entirely electro-magnetic (or electroweak), we should expect a nontrivial quenching of it, consistent with the apparent increase in the nucleon's charge radius.[10] If the p-n mass difference must be ascribed at least partially to the u-d quark mass difference, then all bets are off until we know how these masses, and their splittings by flavor, are generated. That is, if we suppose the quarks acquire masses by a Higgs mechanism, then the extent to which their masses will be quenched in nuclei will depend on the masses and couplings of the necessary Higgs bosons. In Table II below, I tabulate all the new terms, together with what I consider to be their theoretical uncertainties.

Table II Contribution of "non-standard" terms to isodoublet mass splittings (MeV)

Anomalous CSB	2PE-CSB	Mass Shift	"Standard"	Total
0.1 to 0.3	0.2 to 0.3	-0.7 to -1.3	4.9 to 5.3	5.2 to 6.5

As you can see, the theory now has such a large uncertainty that it includes the experiment. Hence there is no longer a Nolen-Schiffer-Okamoto anomaly, in the sense of a disagreement between experiment and a very accurate theory. In fact, the theory is not at all accurate, and includes substantial contributions from density-dependent CSB interactions which are of such a complex nature that our present theoretical technology does not suffice to evaluate them with the requisite accuracy.

I should note that the renormalization of the p-n mass differ-ence in particular is interesting for several reasons. First, it does not contradict the recent experiment of Austin's group,[11] on the difference between the scattering of neutrons and protons from ^{40}Ca. The CSB NN forces do not have this desirable characteristic, if we take them large enough to explain the anomaly by themselves. Second, the mass renormalization is density-dependent, and therefore is negligible in A = 3. Here the 2PE and anomalous isospin effects are each large enough separately to explain the missing 100 keV of energy splitting, if Langacker and Sparrow are correct, but not if

Coon and Scadron are right. In the latter case, we will need to invoke 2PE CSB to explain the data.

REFERENCES

1. K. Okamoto, Phys. Lett. 11, 150 (1964).
2. J. A. Nolen and J. P. Schiffer, Ann. Rev. Nucl. Sci. 19, 471 (1969).
3. N. Auerbach, J. Hufner, A. K. Kerman and C. M. Shakin, Rev. Mod. Phys. 44, 48 (1972).
4. N. Auerbach, V. Bernard and N. van Giai, Phys. Rev. C21, 744 (1980).
5. L. Zamick, Phys. Lett. B39, 471 (1972).
6. J. W. Negele, Comments on Nuclear and Particle Physics A6, 15 (1974).
7. P. Langacker and D. A. Sparrow, Phys. Rev. C25, 1194 (1982).
8. S. A. Coon and M. Scadron, Phys. Rev. C26, 562 (1982).
9. D. O. Riska and Y. H. Chu, Nucl. Phys. A235, 499 (1974).
10. J. V. Noble, Phys. Rev. Lett. 46, 412 (1981); and "Beyond the Impulse Approximation" in Proc. Conf. New Horizons in Electromagnetic Physics, Univ. of Va., April 1982.
11. R. P. DeVito, S. M. Austin, W. Sterrenburg and U. E. P. Berg, Phys. Rev. Lett. 47, 628 (1981).

MODEL STUDY OF INTERMEDIATE STATE BLOCKING IN FIRST-ORDER OPTICAL POTENTIAL THEORY

Khin Maung Maung and P.C. Tandy

Department of Physics
Kent State University
Kent, Ohio 44242

Restrictions on the intermediate states allowed for nucleon-nucleon scattering operators embedded in many-nucleon systems can arise in several circumstances. The most familiar is Pauli blocking of the occupied ground-state levels in the nucleon-nucleon G-matrix for nuclear matter. The corresponding projection operator is $Q_{NM} = \theta(p_1-k_F)\theta(p_2-k_F)$.

The first-order optical potential from Watson multiple scattering theory involves a nucleon-nucleon scattering operator in which intermediate states of the struck nucleon corresponding to the target ground state are projected out. The corresponding projection operator is $Q = 1-|\phi_o><\phi_o|$ where $|\phi_o>$ is the target ground state. These states are introduced elsewhere in the theory, namely, when the optical potential is used in the wave equation for elastic scattering. These latter modifications of intermediate states are not related to the Pauli principle since they operate also for a system of bosons, yet when applied to the scattering problem the operator Q_{NM} is qualitatively similar to Q.

A simple model of the free NN t-matrix is employed to study and compare the effects of intermediate-state modifications through both these operators. Particular attention is paid to the unitarity properties of the resulting modified NN scattering operator for first-order optical potentials.

0094-243X/83/970087-01 $3.00 Copyright 1983 American Institute of Physics

SESSION B
ELASTIC SCATTERING
AND THE OPTICAL MODEL

PROTON AND NEUTRON ELASTIC SCATTERING BETWEEN 80 AND 1000 MEV.

P. Schwandt
Indiana University, Bloomington, IN 47405

INTRODUCTION

The task set for me in this review talk is threefold. First, I was asked to review the available experimental data for nucleon-nucleus elastic scattering in the intermediate-energy regime (which, in nuclear physics, conventionally ranges from pion production threshold up to ~ 1 GeV; I will actually show and discuss data for energies down to 80 MeV). Secondly, I intend to discuss the description of these data in terms of conventional optical-model phenomenology, i.e., the "standard" optical model based on Woods-Saxon potentials. Thirdly, I shall describe the failures of the standard optical model and improvements of the phenomenology.

After a brief introduction to conventional, non-relativistic optical model phenomenology, an overview of available differential cross-section and analyzing power data in terms of angular extent and distribution in energy is given, principally for protons; the lack of corresponding neutron data and its consequences for the neutron optical potential at intermediate energies are discussed. A sampling of the proton data and standard optical model fits to these data are presented, along with a discussion of characteristic systematic features of the data and their interpretation, followed by a brief review of other observables (spin rotation, depolarization, reaction cross section). Then the results of standard optical model analyses are given, with emphasis on energy systematics of potentials and their volume integrals and RMS radii.

Peculiarities and problems with both central and spin-orbit potentials in this conventional phenomenology encountered in the energy region from about 150 to about 500 MeV are pointed out, with particular emphasis on the question of potential shape ambiguities. Apparent resolution of many of these difficulties in the phenomenological approach through the use of unconventional (non-Woods-Saxon) potential shapes is demonstrated. Purely empirical alternative parametrizations of potential formfactors as well as formfactors guided by (if not based on) microscopic nuclear matter or relativistic theories of the optical potential are discussed. Finally, various implications of these findings (e.g., for the use of these optical potentials in direct-reaction calculations) are pointed out.

CONVENTIONAL OPTICAL MODEL PHENOMENOLOGY

The first topic is a very quick review of what we mean by conventional optical model phenomenology. Standard local optical model potentials involving Woods-Saxon form factor for the complex central potential are written in this familiar form:

$$U_c(r) = V_0 f_R(r) + iW_0 f_I(r) \tag{1}$$

with Woods-Saxon (two-parameter Fermi) functions

$$f_X(r) = \{1+\exp[(r-R_X)/a_X]\}^{-1}. \tag{2}$$

Figure 1a below shows a typical example of such a potential form factor for the case of p+Ca at 160 MeV. The standard spin-orbit potential also is a complex potential of the conventional Thomas form which involves the derivative of a Woods-Saxon function:

$$U_{so}(r) = [V_{so}g_R(r) + iW_{so}g_I(r)] \; \vec{\sigma}\cdot\vec{L} \tag{3}$$

with $\qquad g(r) = \chi_\pi^2(1/r)(d/dr)f(r).$

For the same case of p+Ca and a grazing partial wave of L=10, the real and imaginary spin-orbit potentials are displayed in Figure 1b.

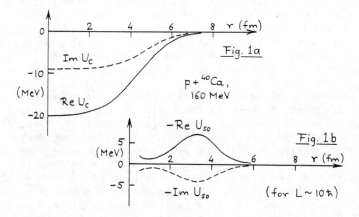

Fig. 1a

p + ^{40}Ca, 160 MeV

Fig. 1b

(for L ~ 10ℏ)

In the conventional optical model calculation these potentials are then inserted into a Schroedinger equation which at these energies incorporates some minimal relativity, i.e., involves at least relativistic kinematics, but in addition frequently involves either (1) a replacement of the reduced mass by a reduced total energy or (2) relativistic corrections which arise in the reduction of the Dirac equation for the upper component of the wave function under certain assumptions about the nature of the Dirac potential. For example, with U(r) taken as the time-like component of a Lorentz 4-vector potential, the reduced wave equation has the form

$$[\nabla^2 + \beta T - \alpha U(r)]\Psi(r) = 0 \tag{4}$$

where $T = T_1+T_2$, $\alpha = 2m_2(E-m_2)/E$, $\beta = m_2(E-m_2+m_1)/E$, $E = T+m_1+m_2$. In either case, one arrives at essentially identical radial partial wave equations which look very much like the non-relativistic equation except for a renormalization of the potential terms.

THE EXPERIMENTAL DATA

Let me now begin the discussion of the available intermediate-energy elastic scattering data by showing, first of all, an up-to-date graphical tabulation of the measured observables presently available for protons as a function of target mass, 12<A<208, and proton kinetic energy, 65<T_p<1000 MeV. In Figure 2, differential cross-section data are presented by ●, polarizations or analyzing powers by ☐, or both cross sections and analyzing powers by ■.

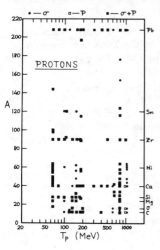

Further details and references for all data presented in this figure can be found in Table 1 appended to this review. Not included here is the profusion of data which exists below 65 MeV. As can be seen from this figure, for protons over this energy region up to 1 GeV there is now a fair amount of data, particularly for the lighter nuclei (up to A = 60 or so) and particularly up to 200 MeV, and then again at 800 MeV where there is a nice set of data for a wide range of nuclei. Also, for Ca, Zr, Sn and Pb in particular there are measurements over a wide range of energies.

Fig. 2

When we look at the same situation for neutrons, the corresponding picture is presented in Figure 3. In drawing this graph I had to change the ground rules: in order to have anything on this graph, I have to show data below 65 MeV, which I did not show for protons. The recent, high-quality neutron differential cross-section

measurements included here come principally from the Ohio State University[1], Livermore[2], TUNL[3], and the Michigan State University[4]. In the absence of these lower-energy data Figure 3 would be essentially empty; the tentative points indicated at 155 MeV are small-angle cross-section and polarization data[5] of relatively low quality measured with the Rochester synchrocyclotron many years ago. This deplorable lack of neutron data at intermediate energies means that I will not have very much to say about the neutron optical potential. However, I want to make some remarks about what one can do, in the absence of neutron data, about the neutron optical potential at medium energies for which there is a clear need in (p,n) reaction calculations, for example. Because there is a fair amount of recent good data up to 40 MeV (although mostly cross sections) as well as recent

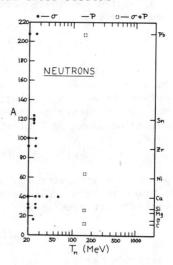

Fig. 3

analyses[1,6,7] of these data in terms of conventional optical potentials, we are reasonably knowledgable about the neutron optical potential up to about 40 MeV. The problem I want to address here, then, is the extrapolation of this knowledge to higher energy.

If one writes the nucleon-nucleus optical potential at nucleon energy E in the Lane model formulation, i.e., in terms of an isospin independent and an isospin dependent term,

$$U_N(E) = U_0(E) + (4/A)U_1(E) \; \vec{\tau} \cdot \vec{T}, \tag{5}$$

then one is dealing with an isoscalar term, an isovector term of opposite sign for protons and neutrons, and a Coulomb correction term for protons which arises from the difference in the kinetic energies for a proton and a neutron inside a nucleus:

$$U_N(E) = U_0(E) \pm \varepsilon U_1(E) + \Delta U_C, \; \pm \text{ for} \begin{cases} \text{protons} \\ \text{neutrons} \end{cases} \tag{6}$$

with $\varepsilon = (N-Z)/A$ and $\Delta U_C \sim -\bar{U}_C \delta U/\delta E$, $= 0$ for neutrons.

For neutrons, then, the potential is simply the difference of these two terms,

$$U_n(E_n) = U_0(E_n) - \varepsilon U_1(E_n), \tag{7}$$

and from the analysis by Rappaport[7] up to 40 MeV one has some quantitative information about the energy dependence of the isovector term which, for the real part of the potential, decreases with increasing neutron energy (in MeV) at the rate

$$\text{Re}U_1(E_n) \sim 23 - 0.2E_n \quad \text{(MeV)} \tag{8}$$

and for the imaginary part is essentially constant, independent of neutron energy:

$$\text{Im}U_1(E_n) \sim 10\text{-}13 \quad \text{(MeV)}. \tag{9}$$

Now, what does one do about the neutron potential for energies much larger than 40 MeV? Due to the lack of data one is forced to deduce the neutron potential U_n from the proton potential U_p which one can certainly do formally, either in the following prescription which includes the Coulomb correction term,

$$U_n(E_n) = U_p(E_n) - 2\varepsilon U_1(E_n) - \Delta U_C, \tag{10}$$

or in a prescription where one does not include the Coulomb correction term explicitly but inserts the proton potential at a Coulomb-corrected energy,

$$U_n(E_n) = U_p(E_p = E_n + \Delta E_C) - 2\varepsilon U_1(E_n) \tag{11}$$

where ΔE_C may be taken to be, for example, the Coulomb displacement energy of the target isobaric analog state. The problem with these prescriptions is that, first of all, there are uncertainties in defining the Coulomb correction term (especially its radial dependence), and one is probably forced to resort to microscopic models since phenomenology has very little to say about this. Second, since one knows from low-energy analyses that the isovector term is energy dependent, one has to somehow obtain this term at the higher energies, for example, from intermediate-energy proton scattering on nuclei with different asymmetry (N-Z) terms. That procedure, unfortunately, is ill-determined because of shape and parameter ambiguities and because of uncertainties in defining the Coulomb correction term. One can thus obtain a first-order approximation to the neutron potential by this method, but it is probably not a very reliable approach at energies of several hundred MeV.

PROTON SCATTERING OBSERVABLES $\sigma(\theta)$ AND $A(\theta)$

I shall now dispense with any further mention of neutron scattering and neutron optical potentials and return to the case of proton scattering for the rest of my discussion. Let me first show what typical data sets look like in terms of the angular extent (or extent in momentum transfer) over the intermediate-energy region. In Figure 4 we have cross-section angular distributions indicated by solid bars, polarization or analyzing power angular distributions by broken bars, as a function of energy for the particular case of Ca;

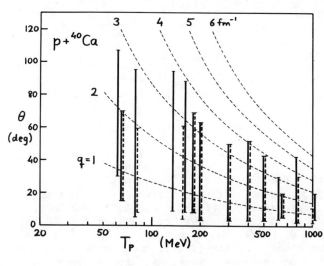

Fig. 4

for other nuclei graphs of the angular range over which the data have been measured would look very similar to this figure. The dashed curves in this graph are contours of constant momentum transfer. The most noteworthy fact here is that over this wide energy region, from <100 to 1000 MeV, the measurements typically extend out to 3 or 4 fm^{-1} in momentum transfer. Beyond this nearly universal "limit of convenience", measurements indeed become more difficult because the cross sections are very small (<100 nb/sr), but this is an unfortunate limitation from the point of view of impact on optical model analyses; in the energy region from 150 MeV to 500 MeV in particular, measurements over a substantially larger range of momentum transfer (out to 5 or 6 fm^{-1}) would greatly enhance our ability to sample the interior structure of the optical potential.

Next, I would like to present a sampling of the existing angular distributions. Figure 5 shows the differential cross section (divided by the Rutherford cross section) for Ca, Zr, Pb at energies between 60 and 180 MeV[8]. The curves that are drawn here (as well as in all subsequent data figures unless noted otherwise) are not just guides to the eye but are in fact the results of optical model fits to cross-section and analyzing power data. We see that over this energy range the standard optical model clearly provides an excellent representation of the data. One characteristic feature of

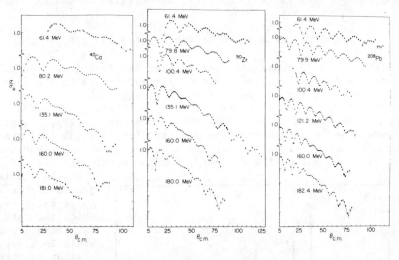

Fig. 5

the data I would like to call attention to at this point is the peculiar change in the nature of the diffraction scattering observed at energies beyond 100-150 MeV where over some limited angular region the regular diffractive oscillations are smoothed out; I will return to this feature later for further discussion.

Typical analyzing power measurements corresponding to these cross-section data are presented in Figure 6 for Ca on the left, for Zr in the middle, and for Pb to the right, at energies from 80 to 180 MeV. Most of these data are from recent measurements[9] at the Indiana University Cyclotron Facility (IUCF), but included are some older measurements from other laboratories, as noted in the figure. Here I would like to point out that in this energy region, and particularly for the lighter nuclei, the analyzing power is predominantly positive, oscillatory at the higher energies but approaching unity (+1) at the lower energies at large angles. For heavier nuclei one sees a more oscillatory pattern with brief excursions into the negative region. Again, I will come back to a discussion and interpretation of these systematics in a moment.

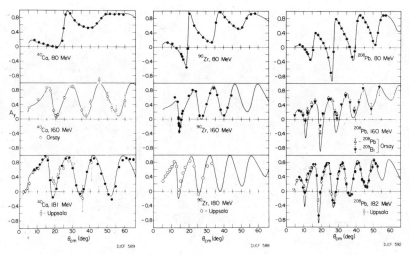

Fig. 6

Figure 7 illustrates similar results for a light nucleus, Si, at energies from 80 to 180 MeV[10]. Shown here now is the absolute cross section in mb/sr so that one can see that, typically, the data span about 8 orders of magnitude in cross section over a 4 fm^{-1} range in momentum transfer. The analyzing powers, again, are predominantly positive.

The data I have presented so far are representative of all available measurements up to 200 MeV. Let me now show some of the available results at energies above 200 MeV. At TRIUMF, measurements have been made[11] with the MRS at 200, 300, 400 and 500 MeV for Ca and Pb.

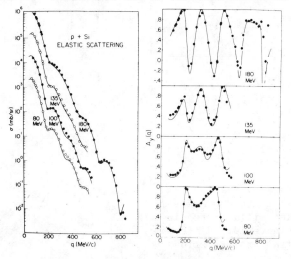

Fig. 7

Figure 8 presents the results for Ca at 500 MeV, both differential cross sections and analyzing powers. More recently, data for the same nucleus have also been obtained at the same energy with the HRS at LAMPF. As can be seen from Figure 9, the LAMPF data[12] are of much better quality but only extends out to about 30 degrees, while the TRIUMF data extend out to about 45 degrees, i.e., to significantly larger momentum transfers. The curves in both Figures 8 and 9 are relativistic Dirac-Hartree (DH) model fits to the data[18].

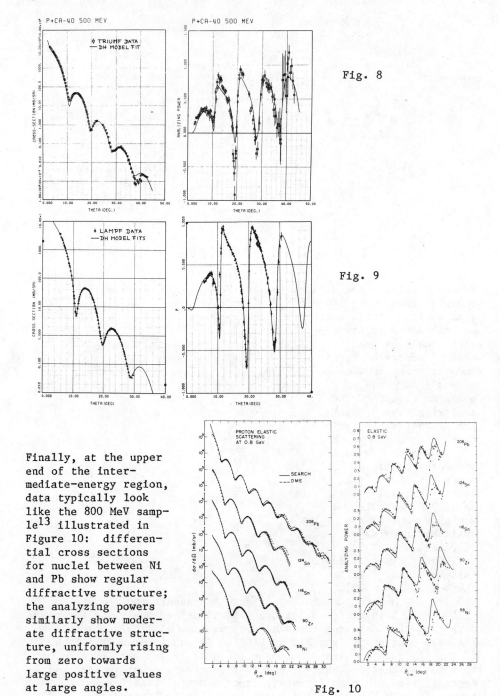

P+CA-40 500 MEV

◊ TRIUMF DATA
— DH MODEL FIT

Fig. 8

P+CA-40 500 MEV

♦ LAMPF DATA
— DH MODEL FITS

Fig. 9

Finally, at the upper end of the inter-mediate-energy region, data typically look like the 800 MeV sample[13] illustrated in Figure 10: differential cross sections for nuclei between Ni and Pb show regular diffractive structure; the analyzing powers similarly show moderate diffractive structure, uniformly rising from zero towards large positive values at large angles.

PROTON ELASTIC
SCATTERING
AT 0.8 GeV

— SEARCH
--- DME

ELASTIC
0.8 GeV

Fig. 10

CHARACTERISTIC FEATURES OF THE DATA AND "SPIN-CHANNEL DOMINANCE"

I have collected most of these data for Pb over the whole intermediate-energy region on two graphs. In Figure 11 are plotted the
angular distributions of the differential cross section between 120
MeV and 1 GeV as a function of momentum transfer to illustrate again
the point I made earlier about the characteristic behavior one
observes for the differential cross section at energies around 200
MeV where the regular diffractive oscillations are washed out.
Otherwise, both low-energy and high-energy data appear extremely
similar. However, if one looks closely, one finds at 800 MeV one
less oscillation over this range of momentum transfers (between zero
and 4 fm^{-1}) than at low energies. That is a real effect which
reflects the fact that the incident proton effectively sees a
somewhat smaller nucleus at 800 MeV than it does at 120 MeV (this

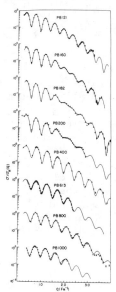

result is readily appa-
rent also in the optical
potentials, as I shall
show later). Figure 12
is a sketch of the cor-
responding analyzing
powers for Pb over this
same energy region. I
did not bother to plot
the data points in this
figure but instead re-
represented them by the
smooth curves. Again,
to amplify an earlier
observation, I should
like to point out that
at low energies the
analyzing powers at the
larger angles (larger
momentum transfers)
rapidly approach unity;

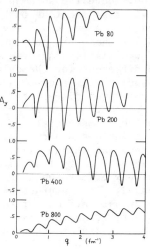

Fig. 12

at medium energies they oscillate strongly; at
somewhat higher energies they become largely positive again until, at 800 MeV, the oscillations
are much reduced in amplitude and ride on an essentially linear rise
with increasing momentum transfer.

Fig. 11

This systematic behavior can be understood more readily if one
looks not at the differential cross section and the analyzing power
but at equivalent observables, namely the partial differential cross
sections for scattering of spin-up protons (↑) and of spin-down
protons (↓). I have (again for the case of Pb) carried out the
transformation to these equivalent observables for energies of 80,
200 and 400 MeV to illustrate the important points about the energy
systematics, which are as follows (see Figure 13): at energies below
100 Mev we find the spin-up cross section dominating beyond some
angle such that indeed the unpolarized differential cross section

is essentially the spin-up cross section
while the analyzing power clearly is large
and positive at the larger angles. This
is typical of what one finds in a semi-
classical model for scattering from a predominantly real potential, hence what we
observe is essentially the nuclear rainbow
effect: since at these lower energies we
are in fact dealing with optical potentials
(both central and spin-orbit components)
where the real part dominates over the imaginary part, we get a significant contribution at the larger angles from orbiting
(rainbow scattering). In the energy region
around 200 MeV we find a different situation: here the spin-up and spin-down partial cross sections both oscillate strongly
and have about equal magnitudes, but are
out of phase or, more precisely, the period
of oscillation in terms of scattering angle
is different for the spin-up and spin-down

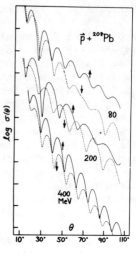

Fig. 13

cases. This leads to a wash-out in the angular structure of the
unpolarized differential cross section (which is essentially the sum
of these two quantities) and a strongly oscillating analyzing power
(which is essentially the difference of these two quantities) with
oscillations about zero. In terms of the optical potential, the
reason for this behavior is that at these energies the real and
imaginary central potentials are roughly comparable which means the
rainbow effect (which gives rise to spin-up dominance) has disappeared. On the other hand, the spin-orbit potential for the
important surface partial waves is comparable to the real central
potential, which means that a spin-up proton sees a potential of
larger radial range than a spin-down proton. That size differential
gives rise to the shift in the diffraction patterns, i.e., the
different periodicities for the two spin-channel cross sections. If
one goes to even higher energies, one finds that again spin-up
scattering dominates over spin-down scattering, leading again to
largely positive analyzing powers. In this case the two partial
cross sections are in phase, and hence one finds regular diffractive
structure in the unpolarized differential cross section. But now
the reason for the spin-up dominance is very different from that at
low energies: because the imaginary central potential completely
dominates the scattering at energies above 400 MeV, the differentia-
tion between spin-up and spin-down scattering is largely determined
by differential absorption of spin-up and spin-down protons from an
essentially black nucleus.

OTHER DIFFERENTIAL AND INTEGRAL OBSERVABLES

So far I have mentioned only differential cross sections and
analyzing powers (or polarizations) as observables in elastic
scattering. Of course, we know that even if we scatter from a

spin-zero nucleus we have really three independent observables
because for a nuclear interaction potential consisting of a central
and a spin-orbit part the scattering amplitude has two complex terms

$$M(\theta) = F(\theta) + G(\theta) \; \vec{\sigma}\cdot\hat{n}, \tag{12}$$

hence there are three real numbers involved (in addition to an
over-all phase) and therefor three observables can be defined:
besides the differential cross section

$$\sigma(\theta) = |F|^2 + |G|^2 \tag{13}$$

there are the polarization (analyzing power) P and a spin rotation
function Q which are defined by the complex relation

$$P(\theta) + iQ(\theta) = 2FG^*/\sigma(\theta) \tag{14}$$

While Q can always be measured in principle, very few measurements
of spin rotation (which involve double-scattering of a polarized
beam with in-plane spin quantization axis) exist because of
experimental difficulties, particularly at lower energies. At higher
energies (where more efficient second-scattering polarimeters can be
constructed) spin-rotation measurements can be performed in practice
relatively more straightforwardly. The spin rotation function Q is
related to the spin rotation angle β illustrated in the diagram
below by the relation

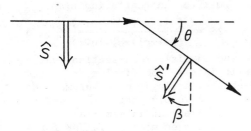

$$Q(\theta) = [1-P^2(\theta)]^{1/2} \sin\beta \tag{15}$$

Of course, one does not measure
β directly; rather, one measures the familiar Wolfenstein
parameters R and A, and expres-
ses Q in terms of these as

$$Q(\theta) = R(\theta)\sin\theta + A(\theta)\cos\theta \tag{16}$$

Such measurements have recently been made
at LAMPF at 500 MeV[14] and, I believe, now
also at 800 MeV. The 500 MeV data for Ca
are illustrated in Figure 14. The utility
of this kind of measurement is illustrated
in Figure 15 where the differential cross
section σ, polarization P and spin rotation function Q are presented, along with
several calculations. The solid line
(which essentially passes through the data
points) is not a standard optical model
fit but a Dirac-Hartree model fit to the
data which I do not want to discuss (for
details of the Dirac-Hartree phenomenology

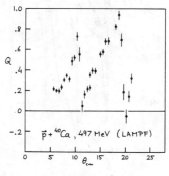

Fig. 14

see the contribution to this workshop by B.C. Clark). The point
which I do want to illustrate here is represented by the dashed
curves which are the results of a standard optical model fit to the
cross section and analyzing power data. Such a standard model fit
gives a reasonable account of σ and P at these energies, but the

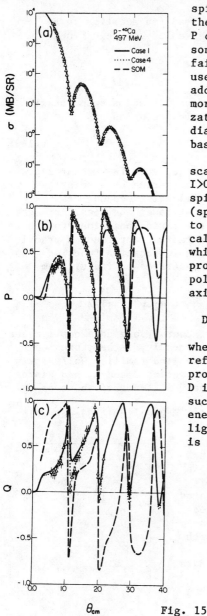

spin rotation function Q predicted by
the standard model which fits the σ and
P data is way off the mark in comparison to the Q data. I see this as a
fairly convincing indication of the
usefulness of measurements of this
additional observable Q for determining
more realistic optical model parametrizations of the scattering at intermediate energies than is possible on the
basis of σ and P data alone.

If one is dealing with proton
scattering from a nucleus that has spin
I>0, then the resulting possible spin-
spin interactions of the type $\vec{\sigma}\cdot\vec{I}$
(spherical) or $3(\vec{\sigma}\cdot\hat{r})-\vec{\sigma}\cdot\vec{I}$ (tensor) lead
to an additional observable, the so-
called depolarization parameter D(θ),
which is related to the spin-flip
probability S(θ) in the scattering of a
polarized beam with spin quantization
axis normal to the scattering plane:

$$D = 1-2S = \frac{(\sigma\uparrow\uparrow-\sigma\uparrow\downarrow+\sigma\downarrow\downarrow-\sigma\downarrow\uparrow)}{(\sigma\uparrow\uparrow+\sigma\uparrow\downarrow+\sigma\downarrow\downarrow+\sigma\downarrow\uparrow)} \quad (17)$$

where the partial-cross-section indices
refer to initial and final state spin
projections. Again, the measurement of
D involves a double scattering, and one
such measurement at intermediate
energies has been made at TRIUMF for a
light nucleus, ^{9}Be, at 225 MeV[15] which
is illustrated in Figure 16 along with

Fig. 16

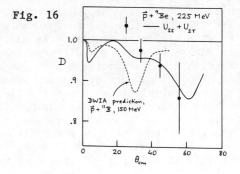

Fig. 15

a calculation that involves both spherical and tensor spin-spin
interaction potentials whose magnitudes are of order of 1 MeV, i.e.,
fairly weak in relation to the spin-orbit interaction. I do not
wish to elaborate on this particular result, except to remark that
measurements of this quantity are fairly difficult to make with the
precision required to obtain conclusive information on the spin-spin
interaction terms in the optical potential.

One integral observable I should like to mention briefly is the
proton total reaction cross section, σ_R, whose energy dependence is
illustrated in Figure 17 over the intermediate-energy range. The
various points shown at a given energy are for different nuclei
because what is presented here is the measured reaction cross
section normalized to a geometrical cross section, πR^2, where R is
defined in terms of the nuclear mass number A as $R=1.25A^{1/3}$ (fm).
The measurements are seen to cluster within a fairly narrow band
which is essentially reproduced by the dashed curves which outline
the region of results obtained from optical model calculations. We

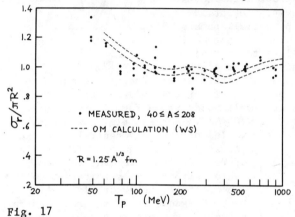

thus see that the
optical model can re-
produce these data
reasonably well over
most of the energy
range; the kink in the
optical model results
around 400-500 MeV is
most likely an arti-
fact of the model
parametrization which
(from other evidence)
is known to fail at
those energies.

Fig. 17

RESULTS OF STANDARD OPTICAL MODEL ANALYSES

Let us now take a
quick look at the
potentials that one
finds in a conven-
tional, standard
optical model analy-
sis over this energy
region. Remember,
we are talking here
about simple Woods-
Saxon parametriza-
tions of the optical
potentials. The real
central potential,
shown in the left
panel of Figure 18,

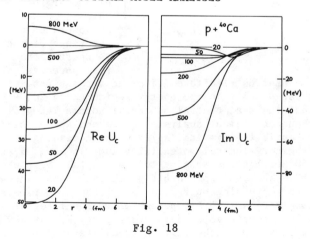

Fig. 18

becomes repulsive above about 600 MeV and, since in this parametrization the potential has a monotonic radial dependence, it is
obviously either attractive everywhere or repulsive everywhere. In
the right panel of Figure 18 is shown the imaginary part of the central potential which is seen to increase monotonically with energy.

The phenomenological spin-orbit potentials one finds in this energy region are
illustrated in Figure 19. The real spin-
orbit term is attractive while the imaginary part is repulsive, in the sense of
having opposite signs (of course, whether
they are individually attractive or repulsive for any particular J-state depends on
whether J=L+1/2 or J=L-1/2). Generally,
the real spin-orbit potential decreases
with increasing energy, while the imaginary spin-orbit potential grows with increasing energy. Note, however, that at 500
MeV the real spin-orbit potential appears
larger than at 200 MeV. That is a peculi-
arity which I do not believe to be real
but rather to be one manifestation of the

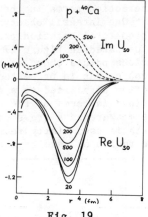

Fig. 19

failure of the standard optical model in this energy region. There
are other indications of problems with the standard optical model
parametrization in terms of simple Woods-Saxon form factors above

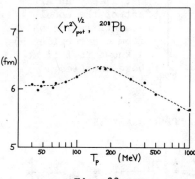

Fig. 20

200 MeV. For example, the RMS
radius of the real central potential,
illustrated in Figure 20 for the case
of Pb over the intermediate-energy
region, typically exhibits a peculiar
non-monotonic behavior, indicating
that the geometry of the real central
potential appears to be changing
quite substantially with energy. At
high energies we find an RMS radius
which is considerably smaller than
at lower energies, indicating that
the range of the repulsive potential
is shorter than that of the attractive potential at low energies.

Let us also take a look at various potential volume integrals.
First of all, in the energy region up to 200 MeV, Figure 21 shows
that the real central potential volume integral J_R obtained in a
phenomenological analysis[9] (dots and shaded bands) decreases
essentially logarithmically with increasing energy, while the
imaginary central potential volume integral J_I changes little up to
150 MeV but then begins to rise rapidly as the energy approaches 200
MeV. The real spin-orbit potential volume integral K_R also falls
with increasing energy, at about the same rate as J_R, while the
imaginary spin-orbit potential volume integral K_I increases to
values compar-able to the real part, K_R, at 200 MeV. Looking at the
situation over the full intermediate-energy region in Figure 22

(note the logarithmic energy scale; results of phenomenological single-energy analyses[16] are indicated by dots which are connected by the heavy solid curves) we note the following characteristic behavior: the imaginary central potential volume integral J_I is seen to rise very rapidly above 200 MeV while the corresponding real quantity J_R falls and crosses zero near 600 MeV (left panel). Clearly, the imaginary potential dominates above 300 MeV or so. Here we also see again (right panel) the very peculiar behavior exhibited by the spin-orbit potential, both

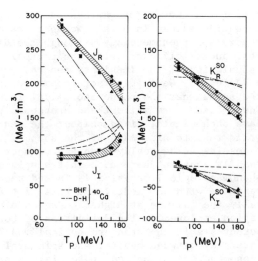

Fig. 21

the real and imaginary parts, in the vicinity of 200 MeV in terms of its volume integrals: with increasing energy, the real part falls sharply and seems to have a minimum near 200 MeV before resuming its decrease beyond 400 MeV. Similarly, the imaginary part peaks at 200 MeV, decreases rapidly again and even changes sign near 400 MeV.

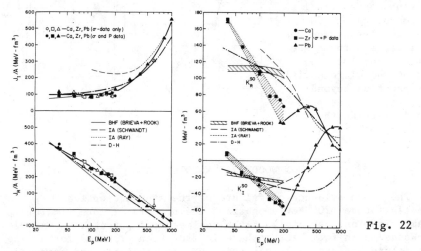

Fig. 22

This behavior is clearly not very realistic in the sense that no model calculation, be it a microscopic BHF model[17] or some other semi-phenomenological model such as the relativistic Dirac-Hartree (DH) model[18], or an impulse-approximation (IA) calculation, can even qualitatively reproduce these empirical results of the simple Woods-Saxon parametrization of the optical potential. These various theoretical model predictions are also illustrated in Figure 22 by

the broken curves for various energy regions. For the central
potential volume integrals we do not see drastic differences between
theory and phenomenology although the differences that do exist are
real and significant (with the exception of the very unrealistic IA
calculation of the imaginary central potential between 100 and 500
MeV) because we are in fact comparing apples and oranges here, in
the sense that the phenomenological results are restricted to
Woods-Saxon form while the microscopic potentials certainly do not
have such simple, fixed shapes; the central potential volume integrals tend to hide much of the shape difference in the potentials
themselves. The really important, major differences which are
readily apparent here concern the spin-orbit potential where the
phenomenological behavior is indeed seen to be drastically at
variance with any model, and this, I think, indicates most dramatically the failure at higher energies of the Woods-Saxon parametrization employed in the definition of both central and spin-orbit
potentials in the standard optical model (because of the strong
interplay between central and spin-orbit components in the
phenomenological optical potential, a problem with one component is
substantially reflected in the other component). Figure 23
illustrates that failure
of the standard model
parametrization directly
by presenting an example
of the quality of the
best fit one can obtain
with the conventional
phenomenology at 400
MeV: we already see the
problems, and these are
serious problems, in
fitting the differential
cross-section beyond
~20°, and similar prob-
lems in fitting the
analyzing powers.

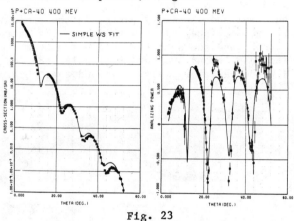

Fig. 23

NON-WOODS-SAXON POTENTIALS

The above evidence of failure of the standard optical
potential above ~200 MeV emphasizes the need for a more flexible
parametrization, certainly of the real central potential. Various
non-Woods-Saxon form factors have in fact been proposed for a long
time; there has been a long history of non-monotonic and energy-
dependent potential shapes, derived from microscopic models or
other considerations, for energies near 200 MeV or beyond.
Starting in 1966 Elton[16], in analyzing 180 MeV proton scattering
from [56]Fe, found empirically the potential shape illustrated by
curve 1 in Figure 24 in order to fit the data. Later Humphreys[20],
in terms of a relativistic model involving vector and scalar interactions, tried to explain the Elton result and obtained the type

of potential shown by curve 2 in Figure 24. A more modern version of this is a relativistic DH calculation in the local density approximation made very recently by Jaminon[21] (for ^{40}Ca, curve 3) which already qualitatively resembles the Elton result. Then there are some BHF calculations, an early one by Mahaux[22] and later one by von Geramb, Brieva and Rook[23] (curves 4 and 5, respectively). Hence, there clearly is is supporting evidence from various microscopic or semi-microscopic models for major deviations from the Woods-Saxon parametrization. In fact, if one looks at the real potential calculated in infinite nuclear matter (Figure 25) by non-relativistic BHF methods (due to Mahaux[24] and Brieva and Rook[25]) or variational methods (due to Pandharipande[26]) or a very recent calculation by Shakin[27]

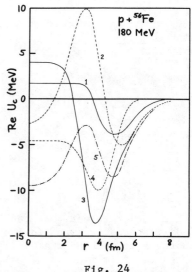

Fig. 24

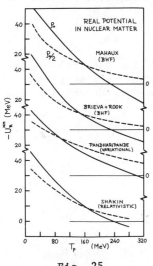

Fig. 25

in the relativistic DHF framework, one sees in all cases that at lower energies the potential in nuclear matter at full density ρ_0 is more attractive than at half that density, for example. At higher energies (above 200 MeV), on the other hand, the inverse is true so that the potential at high nuclear density becomes repulsive long before the potential at a lower density becomes repulsive. If one employs these nuclear matter results in a very simple-minded local density approximation and calculates the radial potential distribution for a finite nucleus, one comes up with the curves in Figure 26 which, of course, are not realistic in detail but illustrate correctly the dominant gross feature of the nuclear potential, namely that above 200 MeV the interior of the nucleus becomes repulsive while the tail region remains attractive (up to 600 or 700 MeV). Another illustration of this is from the relativistic Dirac-equation model which, when transformed into a Schroedinger-equivalent form, yields energy-dependent, effective real central potentials[28] which (for ^{40}Ca) have the shapes presented in Figure 27. Again, at higher energies one still finds a small attractive tail outside of a strongly repulsive interior. This type of potential can indeed give excellent fits to the data, certainly greatly superior to any fit with standard Woods-Saxon potentials, as examplified in Figure 15 (solid curves) for ^{40}Ca at 500 MeV[29].

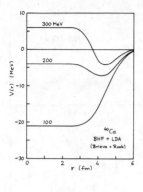

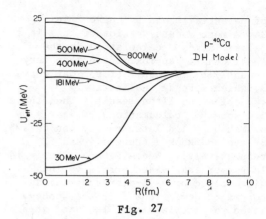

Fig. 26

Fig. 27

In order to see what can be done in terms of relatively simple modifications or extensions of the standard phenomenological approach to mock up these microscopically-based, non-Woods-Saxon potential form factors we may, for example, add to the conventional attractive Woods-Saxon form factor a repulsive Woods-Saxon term of different geometry, or a Woods-Saxon-squared term, depending on ones taste:

$$\mathrm{Re}U_c(r) = V_1 f_1(r) - V_2 [f_2(r)]^n, \quad n = 1 \text{ or } 2 \qquad (18)$$

We then have six parameters to play with to define the real central potential, and so we can come up with a variety of shapes (see Figure 28) which not only resemble the microscopic potential shapes very nicely but have the advantage of simple mathematical representation. As illustrated in Figure 29, at energies below 200 MeV or so the available data do not differentiate very well between a Woods-Saxon form (dashed curve) or a non-Woods-Saxon shape (solid curve). In this energy region we are thus dealing with a serious optical potential shape

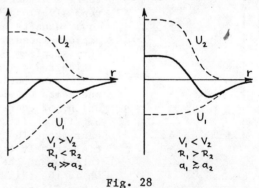

Fig. 28

ambiguity which we can hope to resolve only with additional high-momentum-transfer data. If, however, we go to energies of several hundred MeV we can get clearly superior fits to the data with the above modification of the Woods-Saxon parametrization. This is illustrated in Figure 30 for Ca at 400 MeV where the simple Woods-Saxon form factor failed to give a satisfactory representation of the data (cf. Figure 23); with an additional repulsive Woods-Saxon-squared term, for example, one obtains the excellent fit shown here.

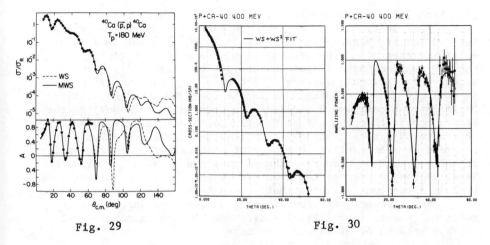

Fig. 29   Fig. 30

The potentials corresponding to the fits shown in Figures 29 and 30 are presented in Figure 31. At 180 MeV, the simple Woods-Saxon form and the modified Woods-Saxon form for the real potential are not drastically different in shape; a significant difference is found, however, for the central strengths of the real and imaginary potentials: a 40% reduction in V(0) is accompanied by an increase in W(0) by more than a factor of two. At 400 MeV the standard Woods-Saxon potential, which does not give a very good fit to the data, is very shallow but still attractive, while the Woods-Saxon plus Woods-Saxon-squared form factor, which fits the data very well, has the characteristic repulsive core

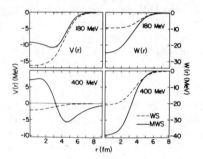

Fig. 31

followed by a significant attractive tail. Again, correlated with the reduced interior attraction of the non-Woods-Saxon real part one always finds a much stronger central absorption.

While these drastically different alternative potentials appear to produce only relatively small differences in the elastic scattering observables (i.e., give rise to largely phase-equivalent asymptotic wavefunctions), the respective wavefunctions in the nuclear interior are indeed very different in magnitude and phase. Obviously, the use of Woods-Saxon versus non-Woods-Saxon distorting potentials can seriously influence the results of DWBA or DWIA calculations for reactions such as (p,n) or (p,2p) which are not surface localized. Another consequence of the more strongly-absorbing non-Woods-Saxon potentials, as pointed out in a recent paper by Meyer et al.[30], is that the proton mean free path in the nuclear interior deduced from the imaginary central potential is considerably shorter than conventionally assumed on the basis of the standard optical model.

Finally, these non-monotonic potentials (which change sign in
the nuclear surface) deduced from the data at energies of several
hundred MeV are clearly not obtainable in any conventional IA or
multiple-scattering expansion model of the optical potential such as
the KMT approach with a free NN interaction t-matrix[31]. Even
second-order KMT calculations, which have been reasonably
successful[32] in describing data in the 800-1000 MeV region, have
failed dramatically when applied[33] to data below ~500 MeV for the
simple reason that the strong density-dependence of the isoscalar
effective NN interaction in the nuclear medium (which gives rise to
the repulsive/attractive character of the realistic central optical
potential) must be properly taken into account, e.g., by using a
density-dependent effective t-matrix in the KMT model. Recent work
in this direction (cf. contributions to this workshop by J. Kelly
and L. Ray) has met with considerable success and indeed yields
potential shapes in qualitative agreement with phenomenological
non-Woods-Saxon or microscopic results.

CONCLUSION

In summary, the fairly extensive proton-nucleus elastic
scattering data of high quality (both differential cross sections
and analyzing powers) which have become available throughout the
intermediate-energy region in recent years have helped us enormously
in gaining new and significant insights into the local proton
optical potential and its energy dependence over a very wide energy
range. The more realistic non-Woods-Saxon central potential form
factors deduced from these data in phenomenological analyses are
generally found to be in at least qualitative agreement with recent
results of various sophisticated theoretical models of the optical
potential, in either a non-relativistic BHF framework or a
relativistic Dirac-equation context. This convergence of phenomenology and theory is exciting and gratifying, not only because it
signifies a better and deeper understanding of the propagation and
scattering of nucleons in the nuclear medium, but also because it
promises the practical benefit of more realistic treatment of
distored waves in the description of reaction processes.

REFERENCES

1. J. Rapaport, V. Kulkarni and R.W. Finlay, Nucl. Phys. A330, 15 (1979).
2. L.F. Hansen et al., Bull. Amer. Phys. Soc. 27, 721 (1982).
3. R.S. Pedroni et al., Bull. Amer. Phys. Soc. 27, 722 (1982).
4. R.P. DeVito et al., Phys. Rev. Lett. 47, 628 (1981).
5. R.S. Harding, Phys. Rev. 111, 1164 (1958).
6. C. Wong, S.M. Grimes and R.W. Finlay, to be published.
7. J. Rapaport, to be published.
8. 61 MeV: C.B. Fulmer et al., Phys. Rev. 181, 1565 (1969); 80-182 MeV: A. Nadasen et al., Phys. Rev. C23, 1023 (1981).
9. P. Schwandt et al., Phys. Rev. C26, 55 (1982).
10. C. Olmer (IUCF), private communication.

11. D.A. Hutcheon et al., Polarization Phenomena in Nulear Physics 1980, AIP Conf. Proc. No. 69, 454.
12. G.W.Hoffmann et al., Phys. Rev. Lett. 47, 1436 (1981).
13. L. Ray, W.R. Coker and G.W. Hoffmann, Phys. Rev. C18, 2641 (1978).
14. A. Rahbar et al., Phys. Rev. Lett. 47, 1811 (1981).
15. G. Roy et al., to be published.
16. P. Schwandt, IUCF Ann. Rep. 1980, p.7.
17. F.A. Brieva and J.R. Rook, Nucl. Phys. A307, 493 (1978), and references therein.
18. L.G. Arnold et al., Phys. Rev. C23, 1949 (1981).
19. L.R.B. Elton, Nucl. Phys. 89, 69 (1966).
20. R. Humphreys, Nucl. Phys. A182, 580 (1972).
21. M. Jaminon, Phys. Rev. C26, 1551 (1982).
22. J.P. Jeukenne, A. Lejeune and C. Mahaux, Journ. Physique 35, suppl. 11, C5 (1974).
23. H.V. von Geramb, F.A. Brieva and J.R. Rook, Lecture Notes in Physics 89 (Springer,1979), p.104.
24. C. Mahaux, Common Problems in Low- and Medium-Energy Nuclear Physics, edited by B. Castel et al. (Plenum Press, 1979), p.265.
25. F.A. Brieva and J.R. Rook, Nucl. Phys. A291, 299 (1977).
26. B. Friedman and V.R. Pandharipande, Phys. Lett. 100B, 205 (1981)
27. L.S. Celenza, W.S. Pong and C.M. Shakin, to be published.
28. L.G. Arnold et al., Phys. Rev. C25, 936 (1982).
29. B.C. Clark, R.L. Mercer and P. Schwandt, to be published.
30. H.O. Meyer and P. Schwandt, Phys. Lett. 107B, 353 (1981).
31. A.K. Kerman, H. McManus and R.M. Thaler, Ann. Phys. 8, 551 (1959).
32. G.W. Hoffmann et al., Phys. Rev. C24, 541 (1981), and references therein.
33. D.A. Hutcheon et al., Phys. Rev. Lett. 47, 315 (1981); G.W. Hoffmann et al., ibid., p.1436.

ADDENDUM

Table 1. Data reference table for proton elastic scattering from nuclei with $12 \leqslant A \leqslant 209$ at bombarding energies of $65 \leqslant E_p \leqslant 1044$ MeV.

Target	Energy (MeV)	Angular Range σ	A	Source[*]	Reference[**]
^{12}C	65	15–115°	15–115°	RCNP	NIM 169, 589 (1980)
	100	5–90°	----	McGill	CJP 44, 2961 (1966)
	122	7–143°	7–143°	IUCF	PR C27 (in print)
	152	----	8–65°	Orsay	NP 80, 625 (1966)
	156	4–65°	----	Orsay	NP A221, 403 (1974)
	160	8–143°	8–143°	IUCF	PR C27 (in print)
	185	7–85°	2–65°	Uppsala	NP A319, 377 (1979)
	200	7–157°	7–115°	IUCF	PR C27 (in print)
	400	3–29°	3–29°	LAMPF	unpublished
	613	4–31°	----	Saclay	Ref. (a)
	800	4–67°	----	LAMPF	PR C23, 2599 (1981)
	800	2–41°	2–30°	LAMPF	PR C18, 1756 (1978)
	1000	----	3–19°	LNPI	PL 90B, 364 (1980)
	1040	4–30°	----	Saclay	PL 45B, 119 (1973)
^{13}C	135	9–82°	----	IUCF	IUCF Ann.Rep. 1980,21
	200	7–115°	7–115°	IUCF	PR C23, 616 (1981)
	800	10–40°	----	LAMPF	PR C18, 1436 (1978)
^{14}N	144	6–62°	----	IUCF	PR C21, 25 (1980)
	160	8–48°	8–48°	IUCF	IUCF Ann.Rep. 1980,24
	800	6–32°	----	LAMPF	PR C25, 2550 (1982)
^{16}O	65	18–82°	18–72°	RCNP	PR C26, 944 (1982)
	100	6–84°	----	McGill	CJP 48, 765 (1970)
	135	8–91°	8–91°	IUCF	PRL 45, 2012 (1980)
	200	6–143°	6–143°	IUCF	IUCF Ann.Rep. 1981,1
	613	4–34°	----	Saclay	Ref. (a)
	800	6–26°	6–26°	LAMPF	PRL 43, 421 (1979)
	1000	----	5–20°	LNPI	PL 90B, 364 (1980)
	1040	3–39°	----	Saclay	Ref. (a)
^{24}Mg	65	14–87°	14–72°	RCNP	PR C26, 944 (1982)
	100	10–60°	----	McGill	NP A193, 438 (1972)
	135	8–57°	8–57°	IUCF	PR C26, 55 (1982)
	155	----	4–58°	Orsay	NP A112, 417 (1968)
	800	5–31°	----	LAMPF	PR C25, 422 (1982)
^{26}Mg	65	20–80°	15–80°	RCNP	RCNP Ann.Rep. 1979,20
	800	5–31°	----	LAMPF	PR C25, 422 (1982)
^{27}Al	65	20–80°	15–80°	RCNP	RCNP Ann.Rep. 1979,20
	156	4–44°	----	Orsay	NP A221, 403 (1974)
	185	8–72°	4–40°	Uppsala	PS 4, 235 (1971)

Target	Energy (MeV)	Angular Range σ	A	Source[*]	Reference[**]
^{28}Si	65	14-90°	14-74°	RCNP	PR C26, 944 (1982)
	80	10-88°	10-88°	IUCF	unpublished
	99	10-82°	10-82°	IUCF	unpublished
	100	10-60°	----	McGill	NP A193, 438 (1972)
	135	7-89°	8-62°	IUCF	PR C26, 55 (1982)
	155	----	4-58°	Orsay	NP A112, 417 (1968)
	180	9-92°	9-92°	IUCF	IUCF Ann.Rep. 1981,1
^{30}Si	650	5-20°	5-20°	LAMPF	BAPS 27, 730 (1982)
	800				
^{32}S	155	----	4-58°	Orsay	NP A112, 417 (1968)
^{34}S	650	5-20°	5-20°	LAMPF	BAPS 27, 730 (1982)
	800				
^{40}Ca	65	16-80°	16-66°	RCNP	PR C26, 944 (1982)
	80	5-95°	8-59°	IUCF	PR C26, 55 (1982)
	135	9-95°	----	IUCF	PR C23, 1023 (1981)
	155	----	4-61°	Orsay	NP A112, 417 (1968)
	156	5-53°	----	Orsay	NP A221, 403 (1974)
	160	8-88°	----	IUCF	PR C23, 1023 (1981)
	182	8-69°	8-69°	IUCF	PR C26, 55 (1982)
	185	5-62°	2-41°	Uppsala	PS 4, 235 (1971)
	200	9-63°	9-63°	IUCF	unpublished
	200	3-55°	3-55°	TRIUMF	Ref. (b)
	300	3-50°	3-50°	TRIUMF	Ref. (b)
	400	3-52°	3-52°	TRIUMF	Ref. (b)
	500	3-43°	3-43°	TRIUMF	Ref. (b)
	500	6-30°	6-30°	LAMPF	PRL 47, 1437 (1981)
	613	3-30°	----	Saclay	Ref. (a)
	650	4-26°	----	LAMPF	unpublished
	800	2-42°	2-31°	LAMPF	PR C25, 2563 (1982)
	1000	----	2-14°	LNPI	PL 90B, 364 (1980)
	1044	4-20°	----	Saclay	NP A274, 443 (1976)
^{42}Ca	650	5-20°	5-20°	LAMPF	BAPS 27, 730 (1980)
	800	3-23°	4-20°	LAMPF	PL 81B, 151 (1979)
	1044	4-18°	----	Saclay	NP A274, 443 (1976)
^{44}Ca	65	16-68°	16-70°	RCNP	PR C26, 944 (1982)
	800	3-24°	3-24°	LAMPF	PL 81B, 151 (1979)
	1044	4-18°	----	Saclay	NP A274, 443 (1976)
^{48}Ca	65	24-68°	24-70°	RCNP	PR C26, 944 (1982)
	500	6-28°	6-28°	LAMPF	PRL 47, 1437 (1981)
	800	3-23°	3-21°	LAMPF	PL 81B, 151 (1979)
	1044	4-20°	----	Saclay	NP A274, 443 (1976)
^{48}Ti	65	16-68°	16-70°	RCNP	PR C26, 944 (1982)
	156	5-42°	----	Orsay	NP A221, 403 (1974)
	1044	4-20°	----	Saclay	NP A274, 443 (1976)

Target	Energy (MeV)	Angular Range σ	A	Source[*]	Reference[**]
^{54}Fe	65	16–68°	16–70°	RCNP	PR C26, 944 (1982)
	800	3–22°	3–20°	LAMPF	PL 79B, 376 (1978)
^{56}Fe	65	14–68°	14–70°	RCNP	PR C26, 944 (1982)
	156	8–42°	----	Orsay	NP A221, 403 (1974)
	185	10–66°	4–32°	Uppsala	PS 4, 235 (1971)
	800	3–32°	3–24°	LAMPF	PL 79B, 376 (1978)
^{58}Ni	65	14–68°	14–70°	RCNP	PR C26, 944 (1982)
	100	8–129°	----	Maryland	NP A301, 349 (1978)
	160	5–45°	----	Harvard	PR 140, B1237 (1965)
	178	6–60°	6–38°	Uppsala	NP A322, 285 (1979)
	800	2–32°	2–21°	LAMPF	PL 79B, 376 (1978)
	1044	4–24°	----	Saclay	PL 67B, 402 (1977)
^{60}Ni	65	14–68°	14–70°	RCNP	PR C26, 944 (1982)
	1044	3–18°	----	Saclay	PL 67B, 402 (1977)
^{62}Ni	65	13–66°	13–68°	RCNP	PR C26, 944 (1982)
	156	6–41°	----	Orsay	NP A221, 403 (1974)
^{64}Ni	65	13–68°	13–70°	RCNP	PR C26, 944 (1982)
	800	3–25°	3–25°	LAMPF	PL 79B, 376 (1978)
	1044	4–16°	----	Saclay	PL 67B, 402 (1977)
^{89}Y	65	13–58°	13–60°	RCNP	PR C26, 944 (1982)
	156	8–55°	----	Orsay	NP A221, 403 (1974)
	185	10–40°	5–36°	Uppsala	NP A305, 333 (1978)
^{90}Zr	65	13–58°	13–60°	RCNP	PR C26, 944 (1982)
	80	6–91°	8–58°	IUCF	PR C26, 55 (1982)
	99	----	8–59°	IUCF	PR C26, 55 (1982)
	100	15–77°	----	Maryland	NP A301, 349 (1978)
	135	6–126°	11–59°	IUCF	PR C26, 55 (1982)
	156	8–41°	----	Orsay	NP A221, 403 (1974)
	160	5–82°	12–44°	IUCF	PR C26, 55 (1982)
	180	6–86°	----	IUCF	PR C23, 1023 (1981)
	185	10–40°	5–33°	Uppsala	PS 4, 235 (1971)
	500	6–30°	6–30°	LAMPF	PRL 47, 1437 (1981)
	800	2–30°	2–21°	LAMPF	PR C18, 1756 (1978)
	1000	4–17°	3–11°	LNPI	PL 90B, 364 (1980)
^{92}Zr	104	12–51°	12–51°	IUCF	PR C26, 55 (1982)
	800	3–18°	3–18°	LAMPF	unpublished
^{116}Sn	800	2–22°	2–22°	LAMPF	PL 76B, 383 (1978)
^{120}Sn	100	23–55°	----	Maryland	NP A301, 349 (1978)
	104	8–51°	8–51°	IUCF	PR C26, 55 (1982)

Target	Energy (MeV)	Angular Range σ	A	Source*	Reference**
^{120}Sn	156	5–41°	----	Orsay	NP A221, 403 (1974)
	160	5–45°	----	Harvard	PR 140, B1237 (1965)
^{124}Sn	800	2–22°	2–22°	LAMPF	PL 76B, 383 (1978)
^{144}Sm ^{176}Yb	800	5–20°	----	LAMPF	PR C22, 1168 (1980)
^{208}Pb	65	14–72°	14–70°	RCNP	PR C26, 944 (1982)
	80	6–93°	8–58°	IUCF	PR C26, 55 (1982)
	98	10–76°	10–76°	IUCF	PR C26, 55 (1982)
	100	19–79°	----	Maryland	NP A301, 349 (1978)
	121	5–83°	----	IUCF	PR C23, 1023 (1981)
	155	----	5–40°	Orsay	JP 30, 13 (1969)
	156	6–40°	----	Orsay	NP A221, 403 (1974)
	160	6–77°	----	IUCF	PR C23, 1023 (1981)
	160	5–45°	----	Harvard	PR 140, B1237 (1965)
	182	6–80°	8–53°	IUCF	PR C26, 55 (1982)
	185	4–38°	4–38°	Uppsala	PR C10, 307 (1974)
	200	3–59°	3–59°	TRIUMF	Ref. (b)
	300	3–49°	3–49°	TRIUMF	Ref. (b)
	400	3–51°	3–51°	TRIUMF	Ref. (b)
	500	3–42°	3–42°	TRIUMF	Ref. (b)
	500	5–32°	5–32°	LAMPF	PRL 47, 1437 (1981)
	613	4–22°	----	Saclay	Ref. (a)
	650	4–22°	----	LAMPF	unpublished
	800	3–42°	2–32°	LAMPF	PR C24, 541 (1981)
	1000	2–15°	2–11°	LNPI	PL 90B, 364 (1980)
^{209}Bi	65	13–72°	13–72°	RCNP	PR C26, 944 (1982)
	153	----	5–56°	Orsay	NP 80, 625 (1966)
	156	7–55°	----	Orsay	NP A221, 403 (1974)

* Laboratory Abbrev.: IUCF – Indiana Univ. Cyclotron Facility, Bloomington, IN
LAMPF – Los Alamos Meson Physics Facility, Los Alamos, NM
LNPI – Leningrad Nuclear Physics Institut, Gatchina, USSR
RCNP – Research Center f. Nuclear Physics, Osaka, Japan
TRIUMF – Tri-University Meson Facility, Vancouver, B.C.

** Journal Abbrev.: BAPS– Bulletin of the American Physical Society
CJP – Canadian Journal of Physics
JP – Journal de Physique

** Journal Abbrev. (cont'd):

    NIM – Nuclear Instruments and Methods
    NP – Nuclear Physics
    PL – Physics Letters
    PR – Physical Review
    PRL – Physical Review Letters
    PS – Physica Scripta

References: (a) CEN Saclay Report DPh-N/ME/78-1
           (b) AIP Conf. Proc. No. 69 (Polarization Phenomena in
               Nuclear Physics 1980), p.454

PHYSICS LIMITATIONS OF THE OPTICAL MODEL

N. Austern
University of Pittsburgh, Pittsburgh, PA 15260

ABSTRACT

Aspects of optical potential theory are discussed.

The title of this talk was suggested by the organizers. Although I am not strongly involved in OP theory, I accepted the topic as a chance to learn what has been going on and to be a little critical about it. I will try to approach the OP from its applications. One result is to recognize that the idea of OP is inherently a little vague--even for nucleons.

My earliest appreciation of the OP was in connection with deuteron stripping theory, in regard to which I want to mention the premature death of its pioneer, Stuart Butler.[1] In their first DWBA calculations of stripping Tobocman and Kalos[2] didn't know which distortions to use, so they experimented with many possibilities. It was only when the OP was recognized as the correct starting point that the true value of DWBA became clear. Of course, the OP had been in use earlier, for the analysis of energy-averaged neutron total cross sections, my point is that it took a while to make the connection with reaction theory.

The early history illustrates two applications of the OP: (a) It describes elastic scattering, (b) It provides wavefunctions for reaction calculations. I will argue below that these two applications tend to be a little contradictory. There is also a more theoretical application of the OP: (c) In microscopic calculations it allows a helpful intermediate step between the analysis of target nucleus excitations, caused by two-body interactions with the projectile, and the determination of the projectile wavefunction. Individual excitations of the nucleus are often weak enough to be treated perturbatively, so that the OP can be derived as a simple sum over these excitations; however this sum can then strongly distort the wavefunction of the projectile.

Introductory discussions often describe the OP as an extrapolation of the shell model single-particle potential (SMP) into the continuum. This is false. In bound-state calculations the SMP generates zero-order configurations. Residual interactions then mix and split these configurations, to produce the actual nuclear states. On the other hand, in applications of the OP to elastic scattering it is typical to demand a perfect description of the scattering of a physical nucleon by an actual nucleus. Obviously this must incorporate all the residual interactions of shell model theory--at the very least they are responsible for the imaginary part of the OP! A problem arises when we come to reaction theory, the residual interactions now appear both in transition matrix elements for the reaction and in the excitations (which may be collective) that determine projectile wavefunctions in the

0094-243X/83/970115-06 $3.00 Copyright 1983 American Institute of Physics

116

entrance and exit channels. A unified perturbative description of
a reaction, for example, must allow these two kinds of effects to
occur in any order, thereby blurring the OP description of the
channel wavefunctions. Thus the concern for an excellent description of elastic scattering conflicts with the desire to unify
different nuclear processes.

Microscopic theories that emphasize elastic scattering define
the OP as the single-particle (s·p·) interaction operator that
identically gives the projection of the entire projectile-nucleus
wavefunction on the exact wavefunction of the target nucleus. This
means it govern the single-particle wavefunction $u(\vec{r})$ in the
projected antisymmetrized product

$$P\Psi(0,\ 1,\ 2,\dots A) = A[u(\vec{r}_o)\phi_o(1,\dots A)]\ .$$

Various formal derivations of a s·p· potential for $u(\vec{r})$ have
been devised, from a time-dependent s·p· Green's function, to a
projected stationary Schrodinger operator, to multiple scattering
theory. In principle these derivations yield the "generalized
optical potential", which must then be energy averaged. Fortunately
the Indiana workshop concerns high enough energies so that
absorption into open channels automatically smooths the scattering
and eliminates the need for averaging.

One convenient formal expression for the OP is the multiple
excitation series of Feshbach and collaborators[3],

$$OP' = U_{00} + \sum_{\alpha \neq 0} U_{0\alpha} \frac{1}{E^+ - \varepsilon_\alpha - K - U_{\alpha\alpha}} U_{\alpha 0} + \dots$$

where E and ε_α are the projectile and intermediate nuclear
energies, respectively, and where

$$U_{\alpha\beta} = (A-1)<\phi_\alpha|\tau|\phi_\beta>.$$

Here $|\phi_\alpha>$ is the state function of an excited state α, and τ
is given by the usual KMT expression

$$\tau = v + v \frac{A}{E^+ - K - H_N} \tau\ .$$

It is familiar that this formulation does not quite give the OP
that I have been discussing, but rather the KMT auxiliary
potential OP'. The scattered amplitudes for these two potentials
are related by $T = (A/A-1)T'$.

OP analyses of elastic scattering experiments frequently
concern one nucleus at one energy. Sometimes the purpose of an
analysis is to fit a phenomenological potential to the data, and
thereby to determine the elastic S matrix and extrapolate it off

the energy shell to generate the wavefunction $u(\vec{r})$. Such analyses
should be regarded as preparation for reaction calculations.
Alternatively, careful studies of elastic scattering in a single
system, especially at large q, can be precision tests of microscopic
predictions.

More generally, the OP is used empirically in elastic scattering
to describe relationships among different target nuclei and different
energies. The dependence on mass number A and isospin T tends to be
smooth, as if the projectile were only affected by some typical
nuclear properties in the <u>interior</u> region and <u>surface</u> region.
Indeed, the dependence on A usually reduces to a dependence on the
nuclear geometry. Clearly this simple relation to geometry would be
spoiled if the OP for particular nuclei were to receive large
contributions from special excitations of those systems. Fortunately
the strong high-energy excitations: giant resonances, quasi-elastic
knockout, have properties that do vary smoothly with A, so one can
expect an orderly description of elastic scattering at high energy.

On the other hand the real part of the optical potential is
known to change its sign at medium energies[4-6]. In the transition
energy region (Schwandt's phrase) from about 200-600 MeV, calculations of the real potential not only show that it is weak, they also
show interesting irregularities in the radial dependence, as if
different regions of the nucleus were more or less sensitive to the
reversal of the average two-nucleon interaction. The irregularities
in the transition region seem to be related to interference between
independent competing physical processes. Several kinds of interference have been suggested[7]: (1) Competition between vector and
scalar contributions in a Dirac formulation, (2) Competition between
direct and exchange terms in a BHF formulation, (3) Effects that
arise in a hypernetted chain calculation. It would be surprising
if OP irregularities associated with competing physical effects
would behave in an orderly geometrical way as A and E vary. Under
these conditions, in the transition energy region the OP probably
is not an effective interpolation device. Of course, the irregularities of the real OP might be worth special study, as indicators of
the contributing physics. Effects of this kind are probably best
studied near 200 MeV, where the imaginary potential has not yet
become so strong as to dominate the scattering.

Another indication that the OP need not have a simple relation
to nuclear geometry is the occasional evidence[8,9] for an ℓ-dependent
part in the OP. The quantum number ℓ does not have a local relation
to the nuclear density, it refers to the overall dimensions and
shape of the system.

Our formal expression for the OP shows that it is nonlocal:
because τ and the states $|\phi_\alpha>$ are antisymmetrized, because of
propagation in intermediate states, because the elementary τ
operator is itself nonlocal. I have little to say about this,
except that theory is our only source of information about non-
locality, and that it cannot be safe at high energy to make

approximate calculations with dubious "equivalent local potentials."
It is interesting that an extensive linear combination of nonlocal
contributions from many intermediate excitations may be approximately
local[9,10], because of cancellations among long range parts of
Green's functions. In 30 MeV calculations by Coulter and Satchler[9]
this kind of suppression of nonlocality seems to occur for intermediate inelastic channels, but not for intermediate pickup channels,
perhaps merely because the latter are less numerous. It is
interesting to speculate that nonlocalities might have some special
sensitivity to A and E that complicates the relation between the
nonlocal OP and nuclear geometry.

In the approximate replacement of a nonlocal potential $U(r,r')$
by a local equivalent $U_{local}(r)$, we all know that averaging over the
nonlocality causes U_{local} to depend strongly on E. Even so, may I
remind you that the effects of $U_{local}(r)$ are only partially
equivalent to those of $U(r,r')$? From a WKB point of view U_{local} is
chosen to have the same momentum function

$$p(r) = \sqrt{2m[E - U_{local}(r)]}$$

as the wavefunction generated by the nonlocal $U(r,r')$. But the
normalization of a WKB wavefunction is governed by the velocity
function $[v(r)]^{-1/2}$. Horiuchi[11] has pointed out that with a nonlocal
potential we have

$$\vec{v} = (i/\hbar)[H, \vec{r}] \neq \vec{p}/m.$$

This is the origin of the "Perey effect" for the wavefunction.
Recent Dirac equation calculations[7,12] find that the Darwin term,
proportional to $\nabla u(\vec{r})$, causes a reduction of $u(\vec{r})$ in the nuclear
interior. This can be regarded as another example of the Perey
effect.

In our formal expression for the OP, the operator τ tends to
express collisions of the projectile with single nucleons in the
target nucleus. At high energy the amplitude for such collisions
is a very rapidly decreasing function of q, the momentum transfer.
Clearly any effects that would allow some elementary collisions
between the projectile and objects that have larger masses could
enhance the scattering at high q. Illustrations of such effects[13,14]
are found in two recent articles by von Geramb and collaborators,
which treat excitations of a "collective" nature, which thereby allow
the projectile to exchange momentum with the entire nucleus in a
single step. As a result there are significant contributions in the
region of (small) elastic cross sections at high q. Both articles
apply simple coupled channel calculations, using a standard
phenomenological OP that fits the data at low q, supplemented by

some phenomenological coupling to the additional, special excitation
that is under study. For proton energies in the range 25-50 MeV the
analysis[13] emphasizes excitations of giant multipole resonances. For
proton energies in the range 122-200 MeV the analysis emphasizes[14]
the intermediate formation of a Δ particle. Because the Δ has
140 MeV less kinetic energy than the incident proton, a large
momentum transfer is entailed. Of course, both processes could
perfectly well be included as modifications of the single-channel
OP, at a cost of some additional nonlocality. The coupled channel
calculations might be preferred on the argument that they display
the physics in more detail and use only familiar expressions for
potentials. Thus we see that unusual nuclear excitations either
can be included into the single channel OP, using suitable Green's
functions, or they can be reserved for special treatment by coupled
equations. Of course, what is more to the point is whether the
underlying physics of such excitations is available.

Other processes that bring in collision partners with large
mass and cause scattering at high q can be imagined. For example,
Cannata, Dedonder and Lenz[15] analyze contributions to the OP based
on microscopic three-body amplitudes, in which the projectile
interacts simultaneously with two target nucleons. Scattering by
dibaryon admixtures can also be imagined.

ACKNOWLEDGEMENT

Research support was provided by the National Science
Foundation.

REFERENCES

1. May 15, 1982.
2. W. Tobocman and M. Kalos, Phys. Rev. 97, 132 (1955).
3. H. Feshbach, A. Gal and J. Hüfner, Ann. Phys. (N.Y.) 66, 20
    (1971).
4. F. A. Brieva, in Microscopic Optical Potentials, ed.
    H. V. von Geramb (Springer, 1979), p.84;
    H. V. von Geramb, F. A. Brieva and J. R. Rook, ibid,
    p. 104. J. Kelly, et al., Phys. Rev. Letts. 45, 2012 (1980).
5. L. Ray, G. S. Blanpied, W. R. Coker, Phys. Rev. C20, 1236 (1979).
6. A. Nadasen, P. Schwandt, et al., Phys. Rev. C23, 1023 (1981).
7. For a summary see, L. G. Arnold, et al., Phys. Rev. C25, 936
    (1982).
8. R. S. Mackintosh and A. M. Kobos, in Microscopic Optical
    Potentials, ibid, p. 188, and references given therein.
9. P. W. Coulter and G. R. Satchler, Nucl. Phys. A293, 269 (1977).
10. H. Dermawan, F. Osterfeld, V. A. Madsen, Phys. Rev. C25, 180
    (1982).
11. H. Horiuchi, Prog. Theor. Phys. 63, 725 (1980); 64, 184 (1980).
12. For additional references see session E of this Workshop.
13. M. Pignanelli, H. V. von Geramb, R. DeLeo, Phys. Rev. C24, 369
    (1981).

120

14. H. V. von Geramb, M. Coz, H. O. Meyer, J. R. Hall,
     W. W. Jacobs and P. Schwandt, preprint 1982.
15. F. Cannata, J. P. Dedonder, F. Lenz, preprint 1982.

FIRST ORDER INTERPRETATION OF OPTICAL POTENTIALS

L. Ray
Department of Physics, The University of Texas at Austin
Austin, Texas   78712

ABSTRACT

The theoretical understanding of intermediate energy proton-nucleus observables in terms of nucleon-nucleon phenomenology and non-relativistic multiple scattering theory is discussed. Specifically, the ability of local, second order, spin-dependent Kerman, McManus and Thaler optical potential models to describe proton-nucleus elastic scattering data from 300-800 MeV is reviewed. Calculations which rely on the impulse approximation are fairly successful at 800 MeV but fail significantly for energies less than 500 MeV. Evidence is presented which indicates that medium effects are quite important in modifying the true proton-nucleon effective interaction from that given by the impulse approximation throughout the intermediate energy range, even at 800 MeV. It is concluded that the trends of the intermediate energy optical potentials can be accounted for qualitatively in terms of non-relativistic multiple scattering theory and medium modified effective interactions. The Glauber model and Dirac phenomenological approaches are briefly discussed.

I. INTRODUCTION

Understanding the elastic scattering of protons from nuclei at intermediate energies has received much attention during the last decade.[1-5] The motivation for this effort is two-fold. First, one would like to quantitatively account for the proton-nucleus (p-A) scattering observables in terms of nucleon-nucleon (N-N) phenomenology. Second, after having obtained a successful working model of the proton-nucleus interaction, one would then proceed to infer new nuclear structure information from analyses of the data.

The achievement of these goals has been hampered in the past by a combination of problems including; poor quality proton-nucleus data, inaccurately known proton densities for the target nuclei, and lack of knowledge of the on-shell nucleon-nucleon scattering amplitudes. These and other difficulties have been overcome in recent years through extensive experimental and theoretical efforts.

Precision elastic scattering data are now available for over 30 nuclei at 800 MeV and for 4 targets at 500 MeV from the high resolution spectrometer (HRS) laboratory of the Clinton P. Anderson Meson Physics Facility (LAMPF). In some cases a complete set of elastic scattering observables (i.e., $d\sigma/d\Omega$, analyzing power $A_y(\theta)$, and spin rotation Q)[6] are available. High momentum transfer electron scattering data are now available for many of these same nuclei permitting sufficiently accurate knowledge of the proton density distributions. The nucleon-nucleon (N-N) free scattering amplitudes determined from phase shift analyses have stabalized in the last few years, thanks to vigorous experimental and analytic efforts. In addition, a number of important theoretical corrections to the proton-nucleus interaction model have been recently included in numerical calculations.

With the availability of precise data and accurate input information it has been possible in just the last two years to subject the proton-nucleus interaction models (optical potentials) to absolute accuracy tests. These results, which will be summarized in section III, have for the first time permitted quantitative statements to be made concerning the accuracy of the impulse approximation (IA)[7] (reaction mechanism) and of extracted matter density differences (new nuclear structure information). Through such tests the role of the nuclear medium in modifying the proton-nucleon effective interaction from that predicted by the IA as a function of bombarding energy and momentum transfer has been quantitatively studied and has been found to be surprisingly significant throughout the entire intermediate energy range.[8] Evidence, to be presented here, indicates that the trends of the empirical proton-nucleus interaction (optical potential) in the intermediate energy range will be, at least qualitatively, understood when medium modified N-N effective interactions are computed. Accordingly, the proton-nucleus elastic scattering problem should be viewed with renewed interest, particularly by theorists.

In order to discuss our present understanding of proton-nucleus elastic scattering a model is needed. Three will be considered here to varying extents. A few aspects of the Glauber model[9] will be discussed in section II; particularly how errors arising from the eikonal approximation might affect extraction of the underlying effective interactions and nuclear structure. The status of non-relativistic optical potential approaches, such as that of Kerman, McManus, and Thaler (KMT)[7], is briefly reviewed in section III. The present level of understanding of the phenomenological optical potential trends is also discussed in section III. Relativistic approaches[10,11], which have primarily been phenomenological, are considered in section IV. Finally some conclusions are offered in section V.

II. CONSEQUENCES OF THE EIKONAL APPROXIMATION

The Glauber model as applied to projectile-nucleus scattering provides a direct, integral relationship between the free projectile-target nucleon scattering amplitudes, the nuclear wave function, and the projectile-nucleus scattering amplitudes.[9] Inherent in this approach is the eikonal approximation which treats the relative proton-nucleus wave function diffractively, by assuming extreme forward angle peaking of the differential cross section. In this section we shall entirely ignore questions relating to the exact nature of the proton-nucleus interaction and concentrate instead on consequences of the eikonal approximation. Realistic p-A optical potentials will be used however.

The Schrödinger equation for the scattering of a projectile proton from a local optical potential, $U^{opt}(r)$, is given by

$$(\nabla^2 + k^2)\psi(\vec{r}) = (2\varepsilon/(\hbar c)^2) \, U^{opt}(r)\psi(\vec{r}), \qquad (1)$$

where k is the relativistic p-A center-of-momentum system wave number and ε is the reduced energy.[5] The optical potential is assumed to be spin-dependent according to,

$$U^{opt}(r) = U_{cen}(r) + U_{so}(r)\vec{\sigma}\cdot\vec{\ell} \qquad (2)$$

where U_{cen} and U_{so} are the central and spin-orbit parts of the complex optical potential. Equation (1) can be solved by the usual partial wave methods, yielding the elastic differential cross section, polarization, and spin rotation function.

In the eikonal approximation $\psi(\vec{r})$ is expressed as,[9]

$$\psi(\vec{r}) = e^{ikz}\rho(\vec{r}) \qquad (3)$$

for plane waves incident along the z-axis. The eikonal approximation consists of neglecting the second derivatives of $\rho(\vec{r})$, assuming the back scattering at $\theta = 180°$ is negligible, and presuming the momentum transfer to be perpendicular to the incident beam direction. The result for the proton-nucleus scattering amplitude,

$$F(q) = f(q) + g(q)\vec{\sigma}\cdot\hat{n}, \qquad (4)$$

is given in the eikonal approximation by,[9]

$$f(q) = ik\int_0^\infty bdb\, J_0(qb)[1 - e^{i\chi_N(b)}\cos(kb\chi_s(b))], \qquad (5)$$

$$g(q) = ik\int_0^\infty bdb\, J_1(qb)\, e^{i\chi_N(b)}\sin(kb\chi_s(b)), \qquad (6)$$

$$\chi_N(b) = -\frac{\varepsilon}{(\hbar c)^2 k}\int_{-\infty}^{\infty} dz'\, U_{cen}(\vec{b} + \hat{k}z'), \qquad (7)$$

and

$$\chi_s(b) = -\frac{\varepsilon}{(\hbar c)^2 k}\int_{-\infty}^{\infty} dz'\, U_{so}(\vec{b} + \hat{k}z'). \qquad (8)$$

The spin-independent and dependent p-A amplitudes are represented by $f(q)$ and $g(q)$, respectively, n is the unit vector normal to the scattering plane, q is the momentum transfer, and b is the impact parameter. Coulomb interactions are omitted here so as to not obscure the effects of the eikonal treatment of the nuclear interaction.

Equations (1) and (4-8) have each been applied to the case of 800 MeV p + ^{208}Pb. The optical potential in Eqs. (2), (7) and (8) has been computed using the impulse approximation, realistic densities, and second order correlation corrections according to Ref. 5. This optical potential, together with the Coulomb interaction, has

been shown to provide a realistic description of the 800 MeV p +
^{208}Pb data of Ref. 12.

In Figs. 1-3 comparisons of the optical model and eikonal predictions for the differential cross section, polarization and spin
rotation for p + ^{208}Pb at 800 MeV are shown (with no Coulomb). Overall, the eikonal approximation is found to be reasonably accurate to
surprisingly large scattering angles. As a result, matter densities,
inelastic transition strengths, and other bulk properties of nuclei,
which are inferred from theoretical analyses of p-A scattering data,
would not be seriously affected by the eikonal approximation. Furthermore, the differences shown in Figs. 1-3 are smaller than the
medium effects to be discussed below. Within this context the present discussion will focus on the small differences displayed in
Figs. 1-3.

As has been noted before[13] the eikonal approximation predicts
deeper minima in the differential cross section than that of the optical model. In a similar fashion, more structure is predicted for
A_y and Q by the eikonal calculation than is actually present.[13] A
new feature of the eikonal approximation to be emphasized here is the
gradually increasing discrepancy which occurs in the positions of the
diffractive maxima and minima in $d\sigma/d\Omega$ as the scattering angle increases. The diffractive pattern predicted with the eikonal approximation in Fig. 1 is "stretched" outward in angle relative to the optical model (correct) result. Chi-square fits to the large angle
$(2^o - 42^o$ c.m.) 800 MeV p + ^{208}Pb angular distribution based on
local optical potentials tend to emphasize the forward angle data
and produce curves which are "out-of-phase" with the back angle data
(i.e., the computed minima occur at slightly smaller angles than
that seen in the data).[12] It is therefore possible, and in fact has
been suggested by earlier Glauber model calculations,[14] that eikonal
models might be able to describe these large angle data with greater
facility than similar optical model calculations. Unless non-eikonal
corrections[4,15] are carefully included, important high momentum
transfer $(q \gtrsim 3.5$ fm$^{-1})$ physics, such as non-locality and off-shell
effects, could be overlooked in Glauber model analyses.

Other kinds of subtle but interesting, p-A physics could also
be obscured in eikonal model analyses. The depths of the diffractive minima in $d\sigma/d\Omega$ (Fig. 1) are affected by a broad variety of
corrections; primarily spin-orbit terms,[16] medium corrections to the
real spin-independent N-N effective interaction,[17] and intermediate
Δ-isobars.[4] The depths of the minima in the analyzing power (see
Fig. 2) are very sensitive to medium corrections to the real spin-
orbit potential,[8] while the heights of the maxima at large angles
are strongly affected by correlations.[12] The enhanced structure
predicted by the eikonal model for Q relative to the optical model
result is quite similar in shape and magnitude to the effects produced by medium modifications to the real spin-independent amplitudes.[17]

Eikonal calculations with the Coulomb interaction were also
conducted.[18] Larger differences were found between the eikonal and
optical model calculations with Coulomb terms included than are shown
in Figs. 1-3. Differences large enough to possibly affect matter

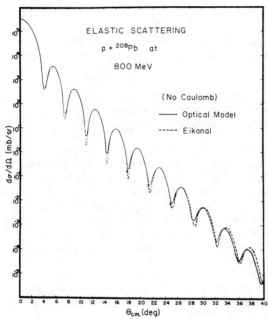

Fig. 1. 800 MeV
p + ^{208}Pb elas-
tic differential
cross section.

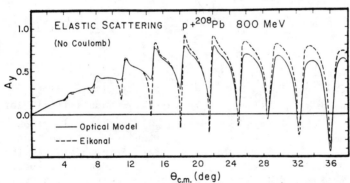

Fig. 2. 800 MeV p + ^{208}Pb elastic analyzing power.

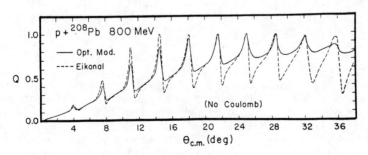

Fig. 3. 800 MeV p + ^{208}Pb elastic spin rotation.

density determinations are noted. The special problem of treating the long range Coulomb interaction in an eikonal approach is a delicate one and if handled inadequately might even obscure the extraction of the underlying effective interactions or bulk nuclear structure properties.

III. SURVEY OF RECENT MICROSCOPIC OPTICAL POTENTIAL STUDIES OF 300-800 MeV PROTON-NUCLEUS ELASTIC SCATTERING

As stated in the introduction the primary concern here is to review the status of "first principles" understanding of the proton-nucleus interaction. Most theoretical efforts have concentrated on the 800 MeV HRS-LAMPF data and these results will be discussed first. The theoretical model used in this section for relating N-N and p-A phenomenology is based on the Kerman, McManus and Thaler optical potential formalism of Ref. 7. The numerical calculations employ local spin-dependent, second order optical potentials as described in Ref. 5.

The nucleon-nucleon free scattering amplitudes are perhaps the most important input quantity needed in proton-nucleus optical potential calculations.[5,7] In general the nucleon-nucleon amplitudes can be represented by five terms given by[19]

$$f_{NN} = a + c(\sigma_{1n} + \sigma_{2n}) + m\sigma_{1n}\sigma_{2n} + g[\sigma_{1P}\sigma_{2P} + \sigma_{1K}\sigma_{2K}]$$
$$+ h[\sigma_{1P}\sigma_{2P} - \sigma_{1K}\sigma_{2K}], \qquad (9)$$

where $\sigma_{1\ell} \equiv \vec{\sigma}_1 \cdot \hat{\ell}$, $\hat{\ell} = \hat{n}$, $\hat{P}$ or $\hat{K}$, $\hat{n} = (\vec{k}_i \times \vec{k}_f)/|\vec{k}_i \times \vec{k}_f|$, $\hat{P} = (\vec{k}_i + \vec{k}_f)/|\vec{k}_i + \vec{k}_f|$, $\hat{K} = (\vec{k}_f - \vec{k}_i)/|\vec{k}_f - \vec{k}_i|$, and similarly for $\sigma_{2\ell}$. For many years very little information was available for these quantities at energies near 1 GeV,[20,21] particularly with respect to the real spin-independent term, the single spin-flip parts, and the range of the imaginary part of the spin-independent amplitude. The reason for this was due to the fact that the only nucleon-nucleon data available at energies near 1 GeV were the total cross sections, differential cross sections, and polarizations.[20,21] Because the lowest order p-A analysis requires only a and c (Ref. 5) attempts were made to obtain just these two amplitudes from the available N-N data by assuming m=g=h=0. Even with this severe truncation of the amplitude set, large discrete ambiguities were found to occur in the determination of a and c which resulted in unacceptably large differences in the predicted proton-nucleus observables.[5]

Fortunately, a great deal of N-N data[22] have been acquired during the last few years which permit unambiguous and reasonably accurate phase shift analyses to be performed in this energy range, resulting in reliable N-N amplitudes for use in proton-nucleus calculations. Several sets of phase shifts are available, including those of Hoshizaki[23], Bugg[24], Bystricky[25] and Arndt[22]. In the calculations reported here the Arndt phase shifts have been used exclusively.[22]   Comparison of the different N-N amplitudes has been presented in a recent review article by Wallace.[4]

With respect to the proton-nucleus optical potential model, a number of important refinements have also been incorporated into the analyses in recent years. Spin-dependent correlation corrections[5] (second order optical potential terms) have been shown to be important in understanding the large angle analyzing power data for $\vec{p}$ + ^{208}Pb at 800 MeV (Ref. 12) as shown in Fig. 4.

The electromagnetic spin-orbit (EMSO) potential which arises from the coupling of the incident proton's magnetic moment with the Coulomb field of the nucleus, has been included by several authors (Refs. 12, 26 and 27) and is found to make significant contributions to the forward angle A_y for sufficiently energetic projectiles. Writing the N-N amplitudes which are used in lowest order p-A calculations with the Coulomb dependent terms made explicit yields,[12]

$$f_{pp}(q) = a_{pp}^{C}(q) + a_{pp}^{CN}(q) + [c_{pp}^{C}(q) + c_{pp}^{CN}(q)]\vec{\sigma}_p \cdot \hat{n}$$

and
$$f_{pn}(q) = a_{pn}(q) + c_{pn}^{CN}(q)\vec{\sigma}_p \cdot \hat{n},$$

(10)

where $a_{pn}(q) \overset{\sim}{=} a_{pn}^{CN}(q)$. The a_{pp}^{C} and c_{pp}^{C} are purely electromagnetic amplitudes where the EMSO term, $c_{pp}^{C}(q)$, is proportional to $|q|^{-1}$ at small momentum transfer. The remaining terms represent Coulomb-distorted nuclear amplitudes and give rise to the usual nuclear optical potential quantities. Figures 5 and 6 display the effects of including the EMSO term for 800 MeV p + ^{40}Ca and ^{208}Pb, respectively. Both sets of calculations rely on the impulse approximation, use Arndt amplitudes, include spin-dependent correlations and assume realistic densities. These solid curves may be regarded as absolute KMT-IA predictions of the analyzing powers and are observed to be quite successful, except around $\theta_{cm} = 10^{\circ}$ (q = 1 fm^{-1}). Similar calculations at 500 MeV indicate a reduced EMSO effect.

Other Coulomb correction terms to the optical potential have been investigated and found to be surprisingly significant. Proper inclusion of Coulomb effects in the KMT optical potential formalism was shown in Ref. 28 to have very large effects in 800 MeV p + ^{208}Pb elastic scattering predictions at high momentum transfer of the order of 4 fm^{-1}.

The absolute KMT-IA prediction of the overall slope of the diffractive envelope and the angular positions of the diffractive minima for the 800 MeV p + ^{40}Ca differential cross section using the proton density from electron scattering analysis[29] and assuming a Hartree-Fock neutron density[30] is in slight disagreement with the data.[31] Adjusting the neutron density to optimize the fit to the cross section data (see Fig. 7) results in a neutron distribution which has a root-mean-square (rms) radius $\sim$ 0.1 fm smaller than is expected theoretically.[30]   Thus, the overall size of the 800 MeV imaginary central optical potential (for heavy nuclei) is predicted by the above KMT-IA model to be about 1-2% too big in radial extent (rms radius).

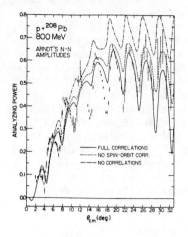

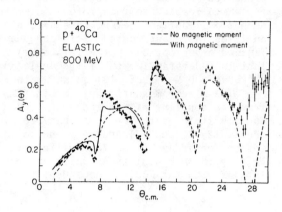

Fig. 4. KMT-IA calculation with and without spin-dependent correlation terms.

Fig. 5. Electromagnetic spin-orbit (EMSO) effects for 800 MeV p + ^{40}Ca analyzing power.

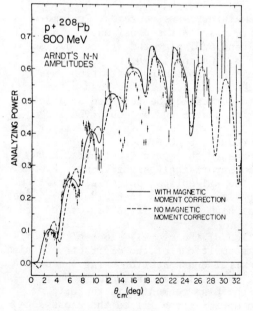

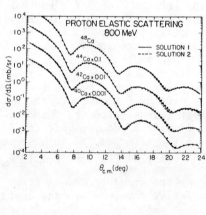

Fig. 6. Same as Fig. 5, except for 800 MeV p + ^{208}Pb.

Fig. 7. Typical KMT-IA fit to the 800 MeV p + Ca data of Ref. 31.

   In summary, the local spin-dependent, second order KMT optical
potential model of Ref. 5 with the IA and the above EMSO and Coulomb
corrections provides an accurate prediction of the shape and structure of the differential cross sections (except as noted in the pre-
ceding paragraph and at large angles, see Fig. 8) and of the analyzing powers (except near q = 1 fm^{-1} as shown in Figs. 5 and 6). Preliminary 800 MeV p + ^{40}Ca spin rotation data have been obtained at
the HRS-LAMPF laboratory.[32] The KMT-IA parameter free prediction
using the above model with EMSO terms is also in good agreement with
the data. Overall, we have found that the local KMT-IA model is
fairly successful at 800 MeV, with the important exceptions being as
noted above. The discussion of these minor discrepancies will be re-
sumed below.

   The success of the absolute, parameter free predictions at 800
MeV provides encouragement for nuclear structure studies based on
the abundant, high quality HRS-LAMPF data at 800 MeV. Within the
context of elastic scattering and in light of the above discussion,
it can be seen that absolute neutron distributions, neutron rms
radii, matter radii, etc. cannot be obtained by analyses using current models. However, relative differences in matter distributions
(i.e., neutron isotopic density differences) should be considered
a reliable and accurate product of the local second order KMT-IA
model analysis.

   To verify this a test analysis was conducted for 800 MeV p +
^{48}Ca and ^{54}Fe (Ref. 33) for which the N = 28 neutron distributions
are expected to be negligibly different.[30] The extracted isotonic
proton density difference (ΔZ = 6 protons) can be compared directly
with the accurate electromagnetic measurements.[34] The difference in
the rms radii of the ^{54}Fe and ^{48}Ca proton densities is found to be
0.26 ± 0.07 fm compared to 0.20 ± 0.01 fm obtained from combined
analysis of electron scattering and muonic atom data.[34] The comparison of the corresponding differences in the radial proton density
distributions is shown in Fig. 9. This good agreement provides impetus for viewing the neutron isotopic density differences extracted
from KMT analyses of the 800 MeV p-A data as reliable, new nuclear
structure information. Examples can be found in Refs. 5, 31 and 35.

   Turning attention to lower energies it is found that the local
KMT-IA model fails to reproduce the structure of the 500 MeV p +
^{40}Ca differential cross section (minima are too deep), analyzing
power (most dramatically near 10-15°, q $\sim$ 1 fm^{-1}),[8] and spin rotation[36] (too small at all angles) as can be seen in Figs. 10-12, respectively. The imaginary central optical potential rms radius is
again predicted to be too large, corresponding to the deduced ^{40}Ca
neutron rms radius being about 0.2 fm too small.[8]

   Many aspects of the model were explored in an attempt to resolve
these large discrepancies. Variation in the neutron density models,
N-N amplitude uncertainties, correlation effects, EMSO terms, proper
relativistic transformation of the N-N amplitudes to the p-A Breit
frame[4], Fermi motion averaging[37], non-locality due to the total momentum dependence of the interaction[37], spin-unsaturated sub-shell
effects[38], and Coulomb corrections[28] were explored and found to be
incapable of explaining the data. It has been concluded[8] that

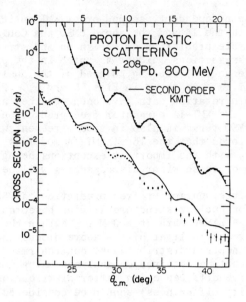

Fig. 8. KMT-IA fit to the 800 MeV p + ^{208}Pb data of Ref. 12.

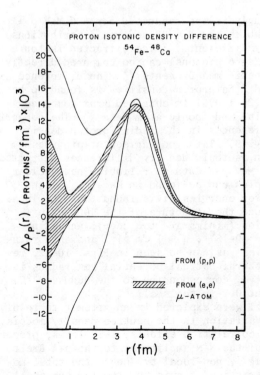

Fig. 9. Proton isotonic density difference envelopes from KMT-IA analysis and from electromagnetic results.

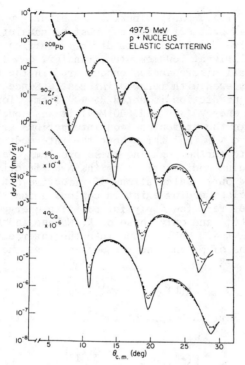

Fig. 10. KMT-IA fits (solid curves) to the 500 MeV data of Ref. 8. Empirical fits are indicated by the dashed curves.

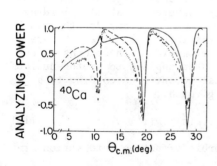

Fig. 11. KMT-IA predictions (solid curve) and empirical fit (dashed curve) to the 500 MeV p + ^{40}Ca analyzing power data of Ref. 8.

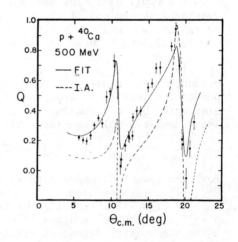

Fig. 12. KMT-IA predictions (dashed curve) and empirical fit (solid curve) to the 500 MeV p + ^{40}Ca spin rotation data of Ref. 36.

medium effects (i.e., Pauli blocking and binding potential effects in intermediate N-N scattering states) must be causing the true proton-nucleon effective interactions to differ significantly from the IA value, particularly at low momentum transfer, $q \sim 1$ fm^{-1}. In light of this the small discrepancies encountered in the analysis of 800 MeV data were viewed with renewed interest. Since a similar breakdown in the IA prediction is noted at $q \cong 1$ fm^{-1} for each target nuclei at each energy (500 and 800 MeV) it was concluded that medium modifications to the proton-nucleon interactions are likely to be significant at $q \sim 1$ fm^{-1} throughout the intermediate energy range.

In order to further examine these problems, detailed computation of the medium modifications to the N-N effective interactions must be carried out. Calculations of effective N-N interactions in an infinite nuclear medium exist in copious amounts in the literature.[39] However very few apply for positive energy projectiles above 100 MeV[40,41] and only those of K. Nakano and H. V. von Geramb[42] (for E < 400 MeV) extend into the intermediate energy range. The calculations of Nakano and von Geramb solve the Bethe-Goldstone equation,

$$t(E) = v + vG^{(+)}(E)t(E), \qquad (11)$$

where v is the Paris N-N interaction[43], $G^{(+)}(E)$ is given by

$$G^{(+)}(E) = \frac{Q(\vec{k}_1', \vec{k}_2')}{E - \varepsilon(k_1') - \varepsilon(k_2') + i\delta}, \qquad (12)$$

Q is the Pauli exclusion operator[40], E is the N-N initial energy parameter (assumed positive), $\varepsilon(k)$ contains the kinetic plus potential energy in intermediate states, and $+i\delta$ selects outgoing wave boundary conditions. These $t(E)$ are only computed on-shell, yielding local effective spin-dependent interactions. Because of the Pauli operator and the potential energy terms in $G^{(+)}(E)$, the $t(E)$ are dependent on the density of the (uniform) medium. These density dependent, effective interactions have been applied to proton-nucleus scattering by constructing the optical potential according to a local density, folding prescription given by,

$$U(\vec{r}) = \int d^3 r' t(|\vec{r} - \vec{r}'|, \rho(\vec{r})) \, \rho(\vec{r}') . \qquad (13)$$

Both the spin-independent and spin-orbit effective interactions in Eq. (11) have been used.

In Figs. 13-16 results for the elastic cross sections and A_y for p + ^{40}Ca at 300 MeV and p + ^{208}Pb at 400 MeV are compared to data[37] using the density dependent effective interactions of von Geramb (solid curves) and the IA (dashed curves). The local KMT spin-dependent second order optical potential model (Ref. 5) is used in each case. Hartree-Fock neutron densities are also assumed.[30] The diffractive maxima and minima in the computed differential cross sections are

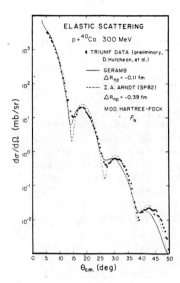

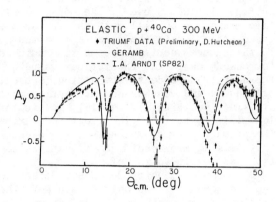

Fig. 13. KMT-IA and von Geramb density dependent fits to the 300 MeV p + ^{40}Ca preliminary data of Ref. 37.

Fig. 14. Same as Fig. 13, except analyzing power.

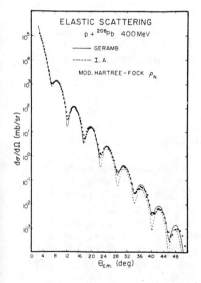

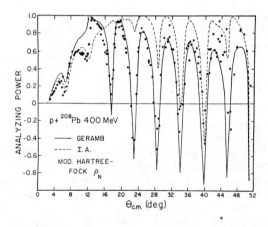

Fig. 16. Same as Fig. 15, except analyzing power.

Fig. 15. Same as Fig. 13, except 400 MeV p + ^{208}Pb.

forced to line up with the data by adjusting the surface radius and diffuseness of the Hartree-Fock neutron density. Other than causing the diffractive envelope to line up with the data this adjustment has no further influence on the description of these data. As seen in Figs. 13 and 15 use of the Geramb amplitudes causes the diffractive minima to be filled in and in the case of p + ^{208}Pb at 400 MeV the overall fit to the data is much improved. As indicated in Fig. 13 the deduced neutron-proton rms radius difference is significantly increased toward the theoretical value of -0.05 fm (Ref. 30) when the density dependent interaction is used. The overall improvement offered by the von Geramb interaction in the predicted A_y is quite striking (see Figs. 14 and 16), particularly in the forward angle ($q \sim 1$ fm^{-1}) region. The clear indication of these results is that the primary physics which is not present in KMT-IA calculations is that arising from medium modifications to the N-N effective interaction, as concluded in Ref. 8.

Since the density dependent effective interactions of Nakano and von Geramb extend only to 400 MeV, the 500 MeV p + ^{40}Ca cross section, A_y and spin rotation data were fit phenomenologically by adjusting the N-N effective interaction away from the IA value yielding a spin-dependent isoscalar proton-nucleon interaction appropriate for 500 MeV.[17] The local second order KMT optical model of Ref. 5 was used in the analysis. The ^{40}Ca proton density was obtained from electromagnetic measurements[29] while the neutron density was held fixed according to the prescription,

$$\rho_n(r) = \rho_p(r)\Big|_{(e,e)} + \rho_n(r)\Big|_{HF} - \rho_p(r)\Big|_{HF}, \tag{14}$$

where HF denotes Hartree-Fock densities.[30] The phenomenological effective interactions are assumed to be functions of momentum transfer only and are not taken to have any explicit density dependence since the improvement provided by the von Geramb amplitudes is due primarily to changes in the q-dependence. The fits are shown in Figs. 10-12. The description of the forward angle A_y data is still somewhat deficient, however the overall structure of the data is reproduced sufficiently well in order that the trends of the empirical amplitudes are indicative of the physics missing in the KMT-IA calculations.

The significance of these 500 MeV N-N effective interactions is made evident when compared to the medium modified interactions of von Geramb at 300 and 400 MeV. In Fig. 17 the 500 MeV isoscalar effective interactions obtained above are compared to the corresponding Arndt phase shift amplitudes. The analogous comparison between the Geramb isoscalar amplitudes for nuclear densities corresponding to a Fermi momentum of 1.4 fm^{-1} and the free N-N amplitudes are shown for 300 and 400 MeV in Figs. 18 and 19, respectively. The comparison between Figs. 17-19 is striking. The suppression of Im(a) at low q, which is required in the 500 MeV analysis, is predicted by the medium effects calculation of Geramb. Similarly the overall constant reduction in Re(a) and the low momentum transfer enhancements of both the Re(c) and Im(c) are seen in each figure.

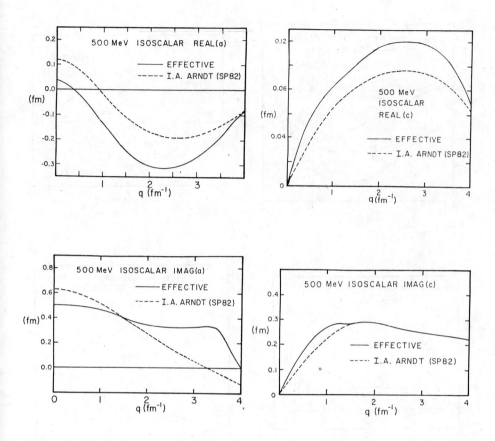

Fig. 17. 500 MeV N-N interactions; empirical (solid curves), IA
values (dashed curves).

136

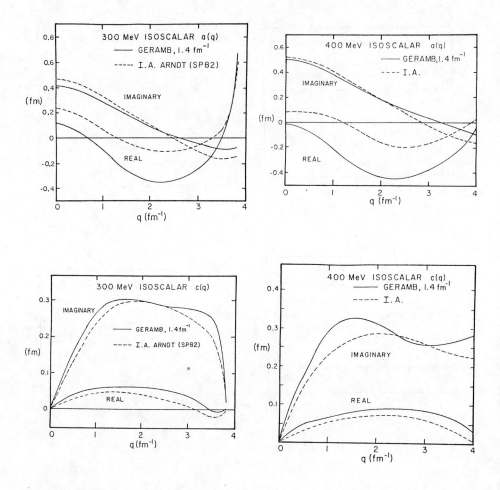

Fig. 18. 300 MeV N-N interaction; von Geramb 1.4 fm^{-1} density (solid curves), IA values (dashed curves).

Fig. 19. Same as Fig. 18, except 400 MeV.

It therefore seems very reasonable to expect that medium modified amplitudes in the local KMT model will provide at least qualitative descriptions of the proton induced elastic differential cross sections, analyzing powers, and spin rotation data for heavy nuclei throughout the intermediate energy range.

It needs to be emphasized, however, that much theoretical work remains. An effective N-N interaction calculation similar to that of von Geramb is currently being carried out for E > 400 MeV[44] where the N-N coupled-channels potential model of Lomon[45] replaces the Paris interaction. Eventually, however, finite nuclear medium effects, off-shell corrections, as well as other non-locality effects must be dealt with. The results obtained so far are very encouraging and suggest that parameter free descriptions of proton-nucleus elastic scattering are well within reach.

Finally, the geometry of the real, spin-independent optical potential in the intermediate energy range is discussed. Figures 20 and 21 display the complex, spin dependent optical potentials for p + ^{40}Ca at 300 MeV and p + ^{208}Pb at 400 MeV, respectively. The solid curves are obtained using the full density dependent Geramb amplitudes as in Eq. (13) while the dashed curves are derived from the IA amplitudes. Figure 22 displays the complex optical potential for p + ^{40}Ca at 500 MeV obtained phenomenologically (solid curves) and predicted by the IA (dashed curves). The optical potentials in Figs. 20-22 correspond to the elastic observables shown in Figs. 10-16. Much discussion has been given recently to the somewhat bizarre shape of the real, spin-independent optical potential in the 300-500 MeV energy range.[46-49] In Figs. 21 and 22 it is observed that the IA predicts that this part of the optical potential has a non Woods-Saxon shape. Medium effects merely tend to enhance this so-called "wine-bottle" structure. The radial forms follow directly from the momentum transfer dependence of the real spin-independent, isoscalar amplitudes shown in Figs. 17-19, and are not significantly affected by the geometry of the matter density or by second order optical potential terms. At low energies $Re[a(q)]$ is predominantly positive giving rise to an attractive t_{eff} and a $Re[U^{opt}(r)]$ which follows the matter density in shape. As the energy increases the node in $Re[a(q)]$ moves in toward q = 0. At high energies ($\sim$ 1 GeV) $Re[a(q)]$ is primarily negative, t_{eff} is overall repulsive , and again the geometry of $Re[U^{opt}(r)]$ generally follows that of the matter density (except for a very small attractive tail). At intervening energies (300-500 MeV) the t_{eff} has both a long range attractive part and a short range repulsive component which when folded with the (Woods-Saxon) matter density gives rise to the wine-bottle shapes displayed in Figs. 20-22. The medium modifications bring about this phenomenon at lower energies than is predicted by the IA. Note the similarity between the Re(a) of Geramb at 300 MeV and the IA value for Re(a) at 500 MeV. The comparison of the corresponding $Re[U^{opt}(r)]$ at 300 and 500 MeV in Figs. 20 and 22 is even more enlightening.

138

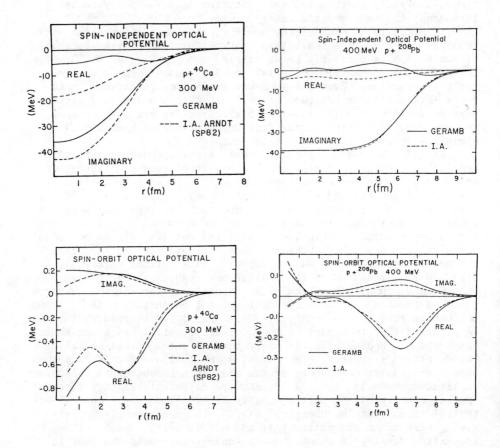

Fig. 20. 300 MeV p + ^{40}Ca optical potentials.

Fig. 21. 400 MeV p + ^{208}Pb optical potentials.

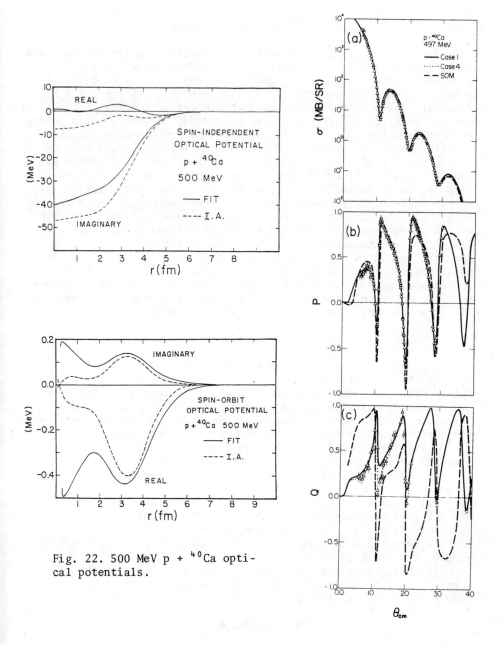

Fig. 22. 500 MeV p + ⁴⁰Ca optical potentials.

Fig. 23. Dirac phenomenological
fits to the 500 MeV p + ⁴⁰Ca data
of Refs. 8 and 36 (solid and dotted curves) and Schrödinger equation optical model fit (dashed
curves). From Ref. 10.

IV. RELATIVISTIC APPROACHES

Since relativistic approaches to understanding the intermediate energy proton-nucleus interaction will be the subject of a later session in this workshop, the present discussion will be extremely brief. In a recent preprint by B. C. Clark, R. L. Mercer, and P. Schwandt[10] excellent fits to the complete set of 500 MeV p + ^{40}Ca elastic scattering observables are presented. Figure 23 is a reproduction from this work.[10] The calculations are based on the Dirac equation where, in addition to the Coulomb interaction, a Lorentz scalar and a time-like component of a Lorentz vector potential are used. The significant feature of these calculations is that given a good fit to $d\sigma/d\Omega$ and A_y (or spin rotation), the spin rotation (or A_y) is <u>correctly predicted by the Dirac phenomenology</u>. Such is certainly not the case in the above discussed non-relativistic KMT approach[16] nor in standard phenomenological optical model analyses based on the Schrödinger equation.[10] The Dirac model fit to the A_y data in Fig. 23 is also seen to be significantly better at forward angles than that displayed in Fig. 11. The facility with which these proton-nucleus elastic data have been fitted using the Dirac phenomenology recommends its continued use and points out very dramatically the need for a relativistic analogue of KMT.

V. CONCLUSIONS

The conclusions of this presentation are:

(1) Although the eikonal approximation has been shown to be fairly accurate at intermediate energies, the somewhat subtle effects due to medium corrections and correlation phenomena could be obscured in Glauber model calculations which do not include non-eikonal corrections when careful fits to the data are required.

(2) Empirical p-A optical potentials are qualitatively understood in terms of multiple scattering theory and medium modified effective interactions at the lower end of the intermediate energy range, E < 400 MeV.

(3) General understanding of the optical potential trends above 400 MeV will probably be accessible once the infinite medium modified N-N effective interactions are computed.

(4) Precise interpretation of p-A data and optical potentials remains a distant goal.

(5) Calibrated effective interactions together with the KMT optical potential are capable of yielding accurate nuclear structure information, such as isotopic neutron density differences.

ACKNOWLEDGEMENTS:

The author would like to thank the following: R. Fergerson and Dr. G. W. Hoffmann for providing the preliminary p + ^{40}Ca spin rotation data at 800 MeV prior to publication, Dr. M. Barlett for providing the 500 MeV effective amplitudes, Prof. D. A. Hutcheon for kindly allowing the preliminary 300 MeV p + ^{40}Ca TRIUMF data to be shown prior to publication, and Prof. B. C. Clark for discussions regarding Dirac optical potential phenomenology. This work was supported in part by the U. S. Department of Energy and the Robert A. Welch Foundation.

REFERENCES

1. A. Chaumeaux, V. Layly, and R. Schaeffer, Ann. Phys. (N.Y.) 116, 247 (1978).
2. G. D. Alkhazov, S. L. Belostotsky, and A. A. Vorobyov, Phys. Rep. 42C, 89 (1978).
3. H. Feshbach, A. Gal, and J. Hüfner, Ann. Phys. (N.Y.) 66, 20 (1971).
4. S. J. Wallace, in Advances in Nuclear Physics, Vol. 12, edited by J.W. Negele and E. Vogt (Plenum, New York, 1981), page 135.
5. L. Ray, Phys. Rev. C19, 1855 (1979).
6. R. J. Glauber and P. Osland, Phys. Lett. 80B, 401 (1979).
7. A. K. Kerman, H. McManus, and R. M. Thaler, Ann. Phys. (N.Y.) 8, 551 (1959).
8. G. W. Hoffmann, et al., Phys. Rev. Lett. 47, 1436 (1981).
9. R. J. Glauber, in Lectures in Theoretical Physics, edited by W. E. Brittin and L. G. Dunham (Interscience, New York, 1959), page 315.
10. B. C. Clark, R. L. Mercer, and P. Schwandt, Ohio State Univ. preprint OSU-TR-192 (1982), unpublished.
11. L. S. Celenza, W. S. Pong, and C. M. Shakin, Brooklyn College preprint BCINT 82/082/116 (1982), unpublished.
12. G. W. Hoffmann, et al., Phys. Rev. C24, 541 (1981).
13. G. Fäldt and A. Ingemarsson, Uppsala preprints GWI-PH2/82 and GWI-PH3/82 (1982), unpublished.
14. G. K. Varma and L. Zamick, Nucl. Phys. A306, 343 (1978).
15. E. Bleszynski, et al., Phys. Rev. C25, 2563 (1982).
16. L. Ray, in Proc. of the Fifth International Symposium on Polarization Phenomena in Nuclear Physics, eds. G. G. Ohlsen, R. E. Brown, N. Jarmie, W. W. McNaughton, and G. M. Hale (American Institute of Physics Press, New York, 1981), page 295.
17. M. L. Barlett, L. Ray, and G. W. Hoffmann, Bull. Am. Phys. Soc. 27, 729 (1982).
18. I. Ahmad, Nucl. Phys. A247, 418 (1975).
19. M. J. Moravcsik, The Two-Nucleon Interaction (Clarendon, Oxford, 1963), pages 11-18.
20. J. Bystricky, F. Lehar, and Z. Janout, Saclay Report No. CEA-N-1547(E) (1972), unpublished.
21. O. Benary, L. R. Price, and G. Alexander, Lawrence Berkeley Laboratory Report No. UCRL-20000NN (1970), unpublished.
22. R. A. Arndt and D. Roper, VPI & SU Scattering Analysis Interactive Dialin program and data base.
23. N. Hoshizaki, Prog. Theor. Phys. (Japan) 60, 1796 (1978).
24. D. V. Bugg, et al., Phys. Rev. C21, 1004 (1980).
25. J. Bystricky, C. Lechanoine, and F. Lehar, Saclay preprint D Ph P E 79-01 (1979), unpublished.

26. P. Osland and R. J. Glauber, Nucl. Phys. A326, 255 (1979).

27. G. Fäldt and A. Ingemarsson, Uppsala preprint GWI-PH 5/81 (1981), unpublished.

28. L. Ray, G. W. Hoffmann, and R. M. Thaler, Phys. Rev. C22, 1454 (1980).

29. I. Sick, et al., Phys. Lett. 88B, 245 (1979).

30. J. W. Negele and D. Vautherin, Phys. Rev. C5, 1472 (1972); the DME code of Negele provided the numerical results given here.

31. L. Ray, et al., Phys. Rev. C23, 828 (1981).

32. R. Fergerson and G. W. Hoffmann, private communication.

33. L. Ray and G. W. Hoffmann, submitted to Phys. Rev. C.

34. H. D. Wohlfahrt, et al., Phys. Rev. C22, 264 (1980); Phys. Rev. C23, 533 (1981); and H. J. Emrich, private communication.

35. G. Pauletta, et al., Phys. Lett. 106B, 470 (1981).

36. A. Rahbar, et al., Phys. Rev. Lett. 47, 1811 (1981).

37. D. A. Hutcheon, et al., Phys. Rev. Lett. 47, 315 (1981); and D. A. Hutcheon, private communication.

38. W. G. Love, et al., Phys. Rev. C24, 2188 (1981).

39. H. A. Bethe, Ann. Rev. Nucl. Sci. 21, 93 (1971), and references therein.

40. J. P. Jeukenne, A. Lejeune, and C. Mahaux, Phys. Rev. C10, 1391 (1974).

41. F. A. Brieva and J. R. Rook, Nucl. Phys. A291, 299 and 317 (1977).

42. K. Nakano and H. V. von Geramb, private communication.

43. M. Lacombe, et al., Phys. Rev. C21, 861 (1980).

44. L. Ray, unpublished.

45. E. L. Lomon, Phys. Rev. D26, 576 (1982).

46. B. Friedman and V. R. Pandharipande, Phys. Lett. 100B, 205 (1981).

47. H. O. Meyer, et al., Phys. Rev. C24, 1782 (1981).

48. L. G. Arnold, et al., Phys. Rev. C25, 936 (1982).

49. H. O. Meyer, P. Schwandt, W. W. Jacobs, and J. R. Hall, **Phys. Rev. C, Feb. 1983.**

ANOMALY IN THE OPTICAL POTENTIAL FOR DEFORMED NUCLEI

H. Sakaguchi, F. Ohtani, M. Nakamura, T. Noro*, H. Sakamoto,
H. Ogawa, T. Ichihara, M. Yosoi and S. Kobayashi

Department of Physics, Kyoto University, Kyoto 606, Japan

ABSTRACT

Elastic scattering of 65 MeV polarized protons from rare earth nuclei has been measured. The volume integral of the real central part of the optical potential (J_R) and the volume integral of the spin orbit part of the potential (J_{ls}) show anomalous behaviors for these deformed nuclei.

Recent progress in the application of nuclear matter theory to nuclear reaction, and high accuracy measurements in the medium energy have brought about renewed interests in the optical potential. The central problem among them is how the projectile interacts with nucleon in the nucleus. Several global behaviors observed in the elastic proton scattering were described in our previous reports.[1],[2] We have measured the elastic scattering of 65 MeV polarized protons from Sm-isotopes, ^{164}Dy and ^{172}Yb. The experiment was performed using the polarized proton beam of RCNP Osaka Cyclotron and the high resolution Spectrograph RAIDEN with momentum dispersion of 20000. The beam polarization was monitored by a sampling polarimeter. In Fig. 1 we show energy spectra of the elastic and inelastic scattering for the spin-up mode and the spin-down mode respectively. The level energy of the 1st excited state of ^{154}Sm is about 82 kev. The separation of the elastic and inelastic peak was performed using a peak-fit program which automatically reproduces the energy spectrum using a reference spectrum of a single sepa-

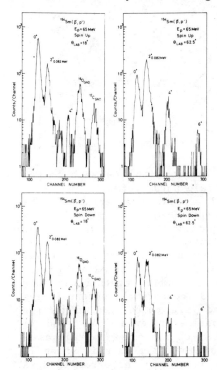

Fig.1 Position spectra of the focal plane counter

* Present Address: Research Center for Nuclear Physics Osaka University, Osaka 567, Japan

144

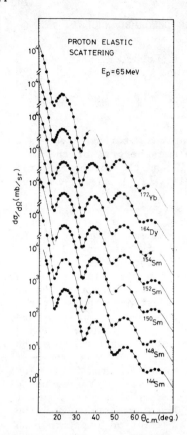

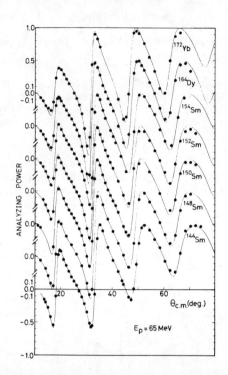

Fig.2 Measured differential cross sections and analyzing powers of the elastic scattering

rated peak obtained from the nearby target with the same target thickness and the same spectrograph field parameters.

We show in Fig. 2 the obtained elastic scattering differential cross sections and the analyzing powers. Solid lines in the figure are conventional optical potential calculations. From the best fit parameter we have calculated the volume integrals of the real central part of the optical potential and the spin-orbit part of the optical potential, J_R/A values for deformed nuclei are plotted as a function of the target mass number together with our previous data.[1] Error bars in Fig. 3 indicate the uncertainties in the optical potential fitting and were calculated using the following procedure. First, all parameter except V_R and r_C (Coulomb radius) were searched for to obtain the $\chi^2_{min}(V_R)$ as a function of V_R. A $\chi^2_{min}(V_R)$ curve for the ^{46}Ti case is shown in Fig. 4. The curve resembles a parabola. Then the error in J_R/A was calculated from the $(J_R/A)_{pot.1}$ and $(J_R/A)_{pot.2}$ values, which were obtained with the parameter sets at $\chi^2_{min}(V_R) = 1.25 \chi^2_0$ (best fit). Although observed J_R/A values scatter considerably, as the mass number A increases, J_R/A decreases rapidly to the minimum in the Fe-Ni region and then increases gradually towards the Pb-Bi region.

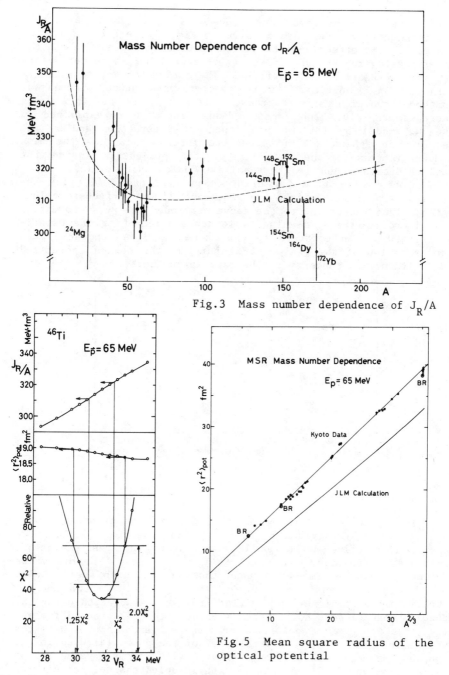

Fig.3 Mass number dependence of J_R/A

Fig.4 An example of the procedure
for deducing the error bar is shown

Fig.5 Mean square radius of the
optical potential

This global trend is remarkably reproduced by the JLM model   calculation shown by the dashed curve in Fig. 3.   According to the JLM model, the effective interaction is density dependent.   In the lower density region, the effective interaction is stronger due to the smaller Pauli blocking effect.   The surface-to-volume ratio is large in light nuclei.   As the target mass number increases, surface-to-volume ratio decreases as $A^{-1/3}$ and the J_R/A value decreases rapidly.   The second gradual increase is explained in the JLM model by the isospin dependent interaction and by the velocity dependence of the effective interaction.   (As the mass number increases, the velocity of the projectile inside the nucleus decreases due to the repulsive Coulomb potential.)

On the other hand for deformed nuclei, the J_R/A value decreases rapidly.   Deformation increases the surface-to-volume ratio.   Thus, according to the above mentioned theory, deformation must increases the J_R/A-value, which contradicts with the experimental data.   In contrast to the anomalous J_R/A tendency, mean square radius of the real central part of the optical potential keeps its normal tendency in the rare earth nuclei.

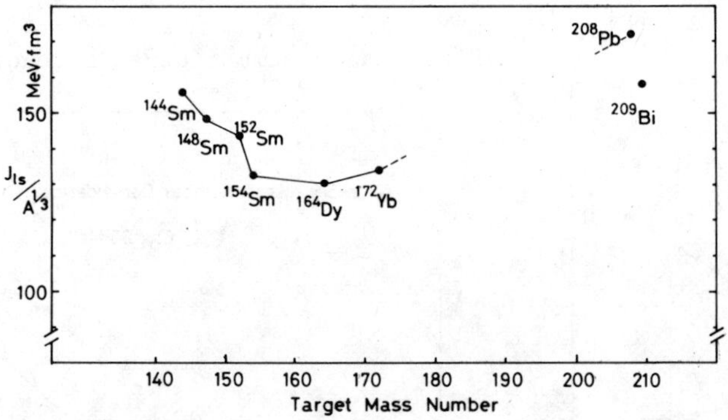

Fig.6 Deformation effect in the spin orbit part of the optical potential

In Fig. 6 we show the volume integral of the spin-orbit part of the potential divided by $A^{1/3}$ for these deformed nuclei.   We notice this value decreases from ^{144}Sm as the mass number A increases.   For nuclei such as ^{154}Sm, ^{164}Dy and ^{172}Yb this value becomes to 130 MeV fm lower than that for the open shell nuclei in the case of shell effect.   The J_{ls}/A value increases for nuclei in the Pb-Bi region.[1] Our preliminary calculation using the coupled channel code ECIS, which fits both the elastic and inelastic scattering with polarization observables, shows that the volume integral of real central part of the potential used in the coupled channel calculation recovers its normal value.   The $J_{ls}/A^{1/3}$-value of the potential in the coupled channel calculation also recovers its normal value.

It is very difficult to include the collective excitation dynamics in the microscopic optical potential. But by using the coupled channel calculation for collective excitation, the anomalous deformation effect in the optical potential seems to disappears. Thus various properties such as J_R/s, $J_{l\hat{s}}/A^{1/3}$, $<r^2>_{pot}$ of the remaining potential could be directly compared with the microscopic optical potential without collective excitation.

REFERENCES

1) H. Sakaguchi et al., Phys. Lett. 89B, 152(1979); Phys. Lett. 99B, 92 (1981); Phys. Rev. C26, 944 (1982).
2) T. Noro et al., Nucl. Phys. A336, 189 (1981).
3) P. Jeukenne, A. Lejeune and C. Mahaux, Phys. Rev. C10, 1391 (1974); Phys. Rep. 25C, 83(1976); Phys. Rev. C15, 10(1977); Phys. Rev. C16, 80(1977).

148

MICROSCOPIC ANALYSIS OF PROTON ELASTIC SCATTERING
IN THE RANGE 80-200 MeV*

F. S. Dietrich
Lawrence Livermore National Laboratory
University of California, Livermore, CA 94550

F. Petrovich
Florida State University
Tallahassee, FL 32306

We have undertaken a systematic comparison of differential cross-sectic
and analyzing-power data on ^{12}C, ^{28}Si, ^{40}Ca, ^{90}Zr, and ^{208}Pb at 8(
200 MeV with calculations based on the single-step folding-model approach
the optical potential. In these calculations, proton densities have bee
inferred from electron scattering results, with neutron densities either th
same as for protons (^{12}C, ^{28}Si, ^{40}Ca) or with a small neutron sk
consistent with 800-MeV proton scattering results (^{90}Zr, ^{208}Pb). T
effective two-body interactions that have been used are the Love-Franey
matrix, a density-dependent interaction based on the Paris potential (calc
lated by von Geramb), and finally the Brieva-Rook density-dependent centra
interaction used with the spin-orbit part of the Love-Franey interaction.

Calculations with the Love-Franey interaction (i.e., the impulse approx
mation) have been made for all of the nuclei, whereas up to the present tir
the two density-dependent interactions have been used only for ^{208}Pb. Cor
parison with data is made both by exhibiting the calculated cross sections a
analyzing powers, and also by least-squares fitting normalizing parameters f
the four components of the optical potential (central and spin-orbit, real a
imaginary).

The calculations without fitting show that the Love-Franey interaction
qualitatively reasonable at forward angles (first three maxima) but ris
above the data at back angles ($\gtrsim 50°$) above 150 MeV for ^{90}Zr and ^{208}P
Including medium corrections by using the Brieva-Rook central interacti
yields significantly worse results in ^{208}Pb for both cross sections a
analyzing powers, whereas the Paris-potential-based interaction is comparab
to the Love-Franey for differential cross sections, but somewhat better f
analyzing powers. For all three effective interactions, the weakening of t
real part of the optical potential toward the higher energies is represent
by a near cancellation of direct and exchange contributions; the manner
which this occurs is different for the three interactions and will be shown.

* Work performed under the auspices of the U.S. Department of Energy by t
Lawrence Livermore National Laboratory under contract no. W-7405-ENG-48.

MOMENTUM-SPACE OPTICAL POTENTIAL SND ELASTIC SCATTERING CALCULATIONS

D.H. Wolfe[a], M.V. Hynes[b], A. Picklesimer[bcd],

P.C. Tandy[ab] and R.M. Thaler[bc]

[a] Department of Physics, Kent State University, Kent, OH 44242
[b] Los Alamos National Laboratory, Los Alamos, New Mexico, 87545
[c] Department of Physics, Case Western Reserve University, Cleveland, OH 44106
[d] Department of Physics and Astronomy, University of Maryland, College Park, MD 20742

Initial results are presented for proton-nucleus elastic scattering observables calculated with a newly developed microscopic momentum-space code. This is the first phase of a program to treat elastic and inelastic scattering consistently via an integral equation approach. A number of microscopic features which are often approximated or ignored are quite amenable to exact treatment within this approach, e.g. non-local effects in elastic scattering, and inelastic effects which are non-linear in the NN t-matrix and target densities but nevertheless confined to one participating nucleon[1].

The elastic scattering results presented here employ an optical potential from first order multiple scattering theory including Coulomb and spin-orbit effects. The optical potential matrix element $<\vec{k}\phi_0|t|\phi_0\vec{k}>$ is here calculated with both the Love and Franey[2] and the Picklesimer and Walker[3] models for the NN t-matrix and comparisons are made. The effects of recoil and three types of factorization prescriptions, which allow estimates of the role of off-shell and non-local effects, are studied. Calculations including full folding of t with the target wave functions and various non-standard spin-dependent effects are in progress. The results presented are those obtained to date. It is hoped that this work will eventually map out a more complete understanding of the capabilities of first order multiple scattering theory so that the physical implications of intriguing discrepencies between present theory and data may be more confidently judged.

1. A. Picklesimer, P.C. Tandy, and R.M. Thaler, Phys. Rev. C25, 1215 (1982); C25, 1233 (1982).

2. W.G. Love and M.A. Franey, Phys. Rev. C24, 1073 (1981).

3. A. Picklesimer and G.E. Walker, Phys. Rev. C17, 237 (1978).

SEPARABLE REPRESENTATION OF THE NON-LOCALITY
OF OPTICAL MODEL POTENTIALS

G. Rawitscher, University of Connecticut, Storrs, CT 06268

The non-locality which is due to the coupling of the elastic channel to nuclear excited states is usually written in terms of Green's functions and inelastic transition potentials. The Green's functions can in turn be expanded in a sum of separable terms if positive energy Weinberg states[1] are used as discrete basis sets in all channels, and the resulting expansion for the non-local optical potential becomes separable. If in addition a projection onto the elastic channel Weinberg states is performed, then the number of terms in the separable expansion is reduced to a practical number, as will be discussed. Applications to nucleon-nucleus scattering at intermediate energies are envisaged.

[1]G. Rawitscher, Phys. Rev. C25, 2196 (1982).
0094-243X/83/970150-01 $3.00 Copyright 1983 American Institute of Physics

A THEORY FOR PROTON-^{3}He SCATTERING

M.J. Paez[*] and R.H. Landau[†]
Department of Physics, Oregon state University, Corvallis and
University of Surrey, Guildford, Surrey

The elastic scattering of proton from ^{3}He and ^{4}He is calculated with a microscopic momentum space optical potential. The theory is unique in its inclusion of the full structure of spin 1/2 x 1/2 scattering. It is equivalent to a co-ordinate space potential with central, spin-orbit and several tensor terms each described as non-local, and velocity dependent. Also included are nucleon recoil and binding, antisymmetrized NN amplitudes, a Lorentz invariant angle tranformation (no small angle approximation), and realistic form factors for the matter and spin distributions of nucleons within the nucleus.

The NN amplitudes contain S-K partial waves and are determined (on-shell) from the Saclay phase shifts[1]. The off-shell extrapolation derives from the Graz potentials[2]. We are thus able to calculate single and multiple scattering at all angles throughout the energy range 10-600 MeV.

Calculations were performed with a version of the momentum space code LPOTT[3] modified to incorporate the NN amplitudes and a variety of spin observables. Our latest results will be presented along with a comparison of previous calculations of $\pi^{\pm}$, K$^+$ and p reactions.

1. J. Bystricky et al., Saclay Report D Ph PE 79-01.
2. K.R. Schwarz et al. (to be published).
3. R.H. Landau, (Computer Physics Communications, to be published).

[*]On leave from and supported in part by University de Antioquia.
[†]Supported in part by the N.S. National Science Foundation and the U.K. Science Research Council.

SESSION C
THE NN INTERACTION
IN REACTIONS

PROTON SCATTERING TO COLLECTIVE STATES: WHAT WE LEARN ABOUT THE EFFECTIVE INTERACTION IN THE NUCLEAR MEDIUM

James J. Kelly
Department of Physics and Laboratory for Nuclear Science
Massachusetts Institute of Technology, Cambridge, MA   02139

ABSTRACT

Electron scattering measurements of the normal parity iso-scalar "collective" excitations of N = Z nuclei completely specify the nuclear structure information required for the analysis of complementary proton scattering experiments. The analysis of the proton scattering data is then interpreted as a study of the medium modifications of the two-nucleon effective interaction. It has been found that the isoscalar spin-independent central component of the effective interaction is strongly dependent upon density and is well described by the local density approximation based upon nuclear matter effective interactions.

The results are sensitive to the difference between effective interactions based upon the Hamada-Johnston and Paris potentials, and favor an intermediate interaction.

The qualitative features of the medium corrections can be adequately represented by a few physically motivated parameters, which can be chosen so as to reproduce either the Hamada-Johnston or Paris results.

I. INTRODUCTION

The description of nucleon induced reactions may be divided into three dominant components: a) the reaction mechanism; b) the structure of the target nucleus; and c) the two-nucleon effective interaction in the nuclear environment. In this talk, we shall study situations where the reaction mechanism and the nuclear structure are under control in order to ascertain the sensitivity of the data to the medium modifications of the two-nucleon inter-action.

We assume that strongly excited transitions are dominated by direct, one-step excitation and may be described in the distorted wave approximation. Discussion of circumstantial evidence for the accuracy of this assumption may be found in Ref. 1.

We also assume that the effective interaction may be adequate-ly described by a local operator that depends upon the separation, spin, and isospin of the two interacting nucleons and upon the local density of the surrounding medium, but not upon the degrees of freedom unique to the finite nucleus. The applications of this local density approximation (LDA) are studied here and in Refs. 1-3. It is also convenient to make a local momentum approximation for the exchange contribution.[4] With these restrictions, the required nuclear structure information consists of point nucleon transition densities.

In order to minimize the nuclear structure uncertainties, the present analysis is restricted to a class of transitions for which the relevant target densities can be accurately measured over a large range of momentum transfer. When the transverse form factor is negligible, normal parity isoscalar transitions of an N = Z nucleus can be described by a single transition density ρ_J^{fi}. These properties are characteristic of the "collective" states that are the main theme of this talk. The proton component of this density is measured by electron scattering. Charge symmetry for an N = Z nucleus ensures, to a high degree of accuracy, that the neutron and proton densities are equal for isoscalar transitions. As electron scattering is mediated by the well-understood electromagnetic interaction, electroexcitation data can be readily interpreted as a measurement of this transition density, with minimal uncertainty.

In Section II, we discuss the concept of radial localization and how it guided the choice of examples for the present study of medium effects. In Section III, impulse approximation (IA) calculations are compared to the data. This comparison identifies the characteristic signatures of density dependence. In Section IV, we discuss the general properties of effective interactions. In Section V, LDA calculations using effective interactions derived for nuclear matter are compared with the data. Our conclusions are presented in Section VI.

II. RADIAL LOCALIZATION

Given the success of mean field theories, it is not unreasonable to suppose that the many-body modifications of the short-ranged two-nucleon interaction may be adequately described by the local properties of the nuclear medium in the vicinity of the interacting pair. In the low density regions of the nuclear surface, we expect that the medium modifications are relatively small, while they may be much larger in the high density nuclear interior.

In the energy regime E_p = 100-300 MeV, the nucleus is relatively transparent to nucleons, permitting significant penetration of the probe into the high density interior. In this situation, elastic scattering is sensitive to an average over the entire nuclear volume. A more discriminating study of medium corrections exploits the radial localization of certain inelastic transitions. An inelastic transition whose amplitude is concentrated in the nuclear surface is sensitive to the low density properties of the effective interaction, while an amplitude concentrated in the interior is sensitive to the high density properties of the interaction.

Inelastic monopole and isoscalar electric dipole transitions tend to be good interior states. Inelastic transitions involving large orbital angular momentum transfers tend to be good exterior (surface) states.

For the present analysis, we shall concentrate on the first 1⁻ and 3⁻ states of ^{16}O. The extensive electron scattering data for these states is shown in Figure 1. The transition densities are compared with the ground state density in Figure 2. The 1⁻ transition amplitude is concentrated in the region of highest nuclear density, while the 3⁻ state is concentrated on the surface.

III. IMPULSE APPROXIMATION

The simplest assumption one can make for the effective interaction is that the medium corrections are unimportant. In this impulse approximation (IA), the effective interaction is represented by the Love-Franey (LF) parameterization of the free two-nucleon t-matrix.[10] The normal parity isoscalar transitions are driven by the isoscalar spin-orbit and spin-independent central components illustrated in Figure 3. At low-q, the spin-orbit component is negligible. The central component is attractive at low-q. As the momentum transfer increases, the central interaction is reduced in absolute magnitude, and then changes sign and reaches a repulsive maximum. The spin-orbit component increases with q and fills in the minimum in the central interaction.

Before comparing IA calculations with data, we note that, in a fully microscopic description, the distorting potential should also be calculated microscopically. The term "consistent" distorted waves shall be used to mean that the distorted waves were generated from the microscopic elastic scattering potential calculated from the same effective interaction that is assumed to have induced the inelastic transition. However, in cases where the microscopic optical potential fails to reproduce the elastic scattering data and the reaction cross section, the traditional approach is to replace the microscopic optical potential by a phenomenological potential fitted to the elastic scattering data. It is instructive to discuss the sensitivity of the inelastic data to the choice of distorted waves.

The IA predictions for the excitation of the lowest 1⁻ and 3⁻ states of ^{16}O by 135 MeV protons are compared with data[3] in Figure 4. The solid curves employ consistent microscopic distorted waves. The long-dashed curves use distorted waves generated by a standard Woods-Saxon phenomenological optical potential fitted to the elastic scattering data. The short-dashed curves also use phenomenological distorted waves but omit the inelastic spin-orbit contribution, retaining just the central. We discuss the calculations using phenomenological distorted waves first.

For the 1⁻ state, the IA overestimates the forward angle cross section by as much as a factor of four. At larger momentum transfers, the IA falls well below the cross section data. At forward angles, the IA analyzing power predictions resemble the data, but for momentum transfers beyond the zero in the central interaction, the analyzing power data show a deep negative excursion that the IA fails to reproduce. This feature clearly correlates with the underestimation of the high-q cross section data. These defects

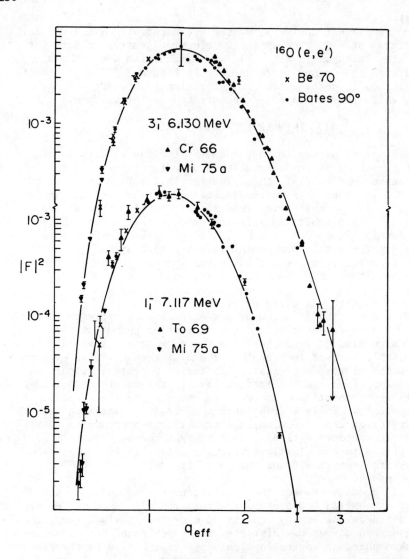

Figure 1

Longitudinal form factors for the lowest 1^- and 3^- states of ^{16}O. Solid points - Ref. 5; inverted triangles - Ref. 6; crosses - Ref. 7; 3^- triangles - Ref. 8; 1^- triangles - Ref. 9

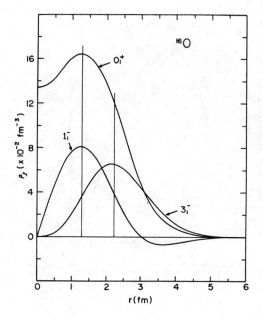

Figure 2

Point nucleon transition densities for ground state and lowest 1^- and 3^- states of ^{16}O.

Figure 3

Moduli of the isoscalar spin-orbit and spin-independent central components of the free two-nucleon t-matrix.

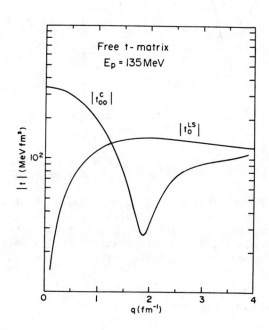

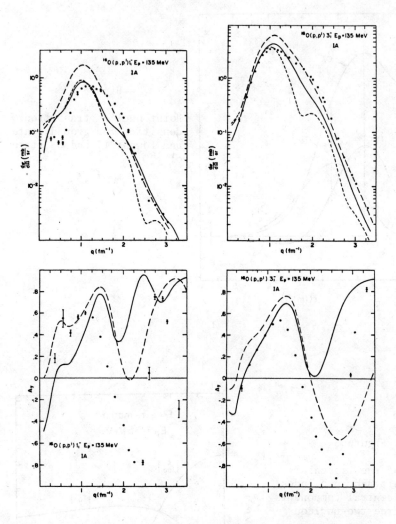

Figure 4

IA calculations for the excitation of the lowest 1⁻ and 3⁻
states of ^{16}O by 135 MeV protons. Long-dashed and short-
dashed curves use phenomenological distorted waves. The
inelastic spin-orbit contribution is omitted from the short-
dashed curves. Solid curves use consistent distorted waves.

are less severe, though still present, in the IA description of the surface 3⁻ state.

The low-q overestimation is readily interpreted in terms of the neglect, by the IA, of the Pauli inhibition of forward scattering. This problem is most severe for the 1⁻ state, whose transition density peaks in the high density nuclear interior where the medium effects should be largest. The IA description of both the cross section and analyzing power data is considerably better for the 3⁻ state, which peaks in the low density nuclear surface.

The use of consistent distorted waves appears to partially alleviate the low-q deficiencies of the free interaction. This reduction is achieved by the absorption produced by the IA imaginary optical potential, whose interior depth is two to three times those of the phenomenological or LDA expectations. The exterior state is less sensitive than the interior state to the difference between the optical potentials because it samples less of the optical potential. However, the IA is known to overpredict the reaction cross section. Moreover, while the IA optical potential appears to improve the description of the low-q cross section, it markedly degrades the description of the analyzing power. The consistent IA analyzing power prediction is yet more positive, in complete contradiction with the data.

While the IA provides a qualitative description of the data, it suffers from the serious limitation of having completely neglected the Pauli blocking of intermediate states mandated by the presence of a finite nuclear density.

IV. EFFECTIVE INTERACTIONS

We assume that the projectile-nucleus interaction is composed of two-body interactions with the A constituent nucleons. The projectile-nucleus scattering amplitude can then be expressed in terms of a multiple scattering series that involves a two-nucleon effective interaction. In principle, the two-nucleon effective interaction appropriate to scattering from an A-nucleon system is complex, nonlocal, energy dependent, and depends upon all of the degrees of freedom of the (A + 1) nucleons. In the local density approximation, we assume that the effective interaction between an energetic nucleon and a nucleon bound in a finite system is essentially the same as that appropriate to infinite nuclear matter with the local density. That is, we assume that in the intermediate energy regime, the properties specific to the finite system are effectively decoupled from the properties characteristic of the underlying medium.

Recently, there has been considerable theoretical effort devoted to the calculation of effective interactions that describe the scattering of a continuum nucleon by a nucleon bound in infinite nuclei.[11,12] The nuclear matter theory and its application to finite nuclei have already been critically reviewed by Mahaux and von Geramb earlier in these proceedings. It is therefore sufficient here to illustrate schematically the salient features. The applications presented in this talk employ effective inter-

actions based upon the Hamada-Johnston[13] and Paris[14] potentials
which have been provided by von Geramb.[15,16]

The first step in the construction of the effective interaction is the solution of a Bethe-Goldstone equation modified so
as to describe the correlated pair wavefunction ψ of a continuum
nucleon and a bound nucleon. This equation has the form

$$\psi = \phi + QG\nu\psi \qquad (1)$$

where ϕ is the uncorrelated wavefunction, ν is the free two-nucleon
potential, Q is the Pauli blocking operator, and $G = (E^+ - K - U)^{-1}$ is
the propagator for kinetic energy K in the presence of the optical
potential U. The density dependence arises through the Pauli
blocking operator and the self-consistent optical potential.

The effective interaction is then constructed to satisfy

$$t\phi = \nu\psi. \qquad (2)$$

The effective interaction is then averaged over the Fermi distribution

$$t = <\frac{\phi^* \nu\psi}{\phi^* \phi}>_{\text{Fermi}}. \qquad (3)$$

The result is a local, density- and energy-dependent effective
interaction with central, spin-orbit, and tensor components.[17]
As the dominant medium modifications occur in the isoscalar spin-
independent central component, we confine the following discussion
to this component.

This procedure illustrates that observation of medium corrections to the effective interaction may be interpreted as observation of the correlated pair wavefunction in nuclear matter involving an energetic nucleon. The qualitative nature of the medium
modifications that we expect, on quite general grounds, is schematically illustrated by the following description. Relative to
the uncorrelated wavefunction for noninteracting nucleons, the
correlated pair wavefunction in a low density environment is
reduced in amplitude, possesses a wound in the repulsive core
region, and is shifted inwards by the overall attraction of the
interaction. The correlated pair wavefunction in a high density
environment differs in several respects. First, Pauli blocking
inhibits scattering and absorption, enhancing the amplitude relative to the unblocked wavefunction. Second, the wound is enhanced
as the Pauli blocking keeps the nucleons apart. Third, the attrac-
tiveness of the interaction is reduced, shifting the wavefunction
to a phase intermediate between the noninteracting case and the
unblocked interacting case.

These effects can be clarified by examining separately the
real and imaginary parts of isoscalar spin-independent central
interaction, shown in Figure 5. As the density increases, the
absorptive part of the interaction is increasingly damped by a
factor that is almost independent of momentum transfer. The

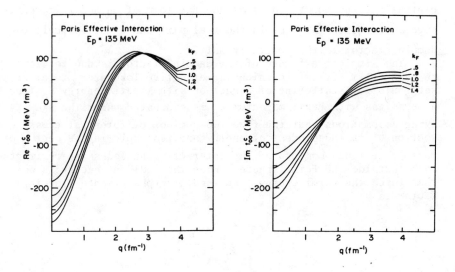

Figure 5

Real and imaginary parts of the isoscalar spin-independent
central component of the 135 MeV effective interaction
based upon the Paris potential.[16]

density dependence of the real part is best described by the addition of a short-ranged repulsive core interaction whose amplitude
increases with density. If this additional repulsion were of zero-
range, the plotted curves would all be parallel. The range of the
repulsive interaction reduces its effectiveness at larger momentum
transfer, so that the curves tend to draw closer together.

These properties of density dependence are of a quite general
nature and can be readily parameterized by the form

$$\text{Re } t_{00}^{c}(q,k_{F}) = \text{Re } t_{00}^{c}(q,o) + a_{R}\left(\frac{k_{F}}{k_{F_{0}}}\right)^{\beta_{R}} f_{R}(\alpha q)$$

$$\text{Im } t_{00}^{c}(q,k_{F}) = \left[1-a_{I}\left(\frac{k_{F}}{k_{F_{0}}}\right)^{\beta_{I}}\right] \text{Im } t_{00}^{c}(q,o) \tag{4}$$

where the wound form factor $f_{R}(\alpha q)$ is a smoothly decreasing function of q and is normalized to unity

$$f_{R}(0) = 1. \tag{5}$$

The medium modifications of the free isoscalar spin-independent central interaction $t_{00}^c(q,o)$ assume the form of an additive repulsive interaction ($a_R > 0$) in the real part and a multiplicative damping factor in the imaginary part.

The simplest estimate of the damping factor is due to Clementel and Villi (CV),[18] in which the effective laboratory total cross section $\bar{\sigma}$ for scattering of a nucleon with kinetic energy E_0 from a Fermi gas with Fermi momentum k_F is evaluated under the simplifying assumptions that the free two-nucleon differential cross section σ_0 is independent of energy and isotropic in the center of mass. The final momenta of the interacting nucleons are restricted to be outside the Fermi sphere. When the incident energy is more than twice the Fermi energy, the resulting phase space blocking factor is

$$\frac{\bar{\sigma}}{\sigma_0} = 1 - \frac{7}{5}\frac{\varepsilon_{F_0}}{E_0}\left(\frac{k_F}{k_{F_0}}\right)^2 , \qquad (6)$$

where $\varepsilon_{F_0} \simeq 37$ MeV and $k_{F_0} \simeq 1.33$ fm^{-1} represent the Fermi energy and momentum at saturation. Using the optical theorem, we obtain

$$\beta_I = 2$$
$$a_I = \frac{7}{5}\frac{\varepsilon_{F_0}}{E_0} . \qquad (7)$$

More sophisticated nuclear matter calculations (e.g., Refs. 11,12, 15,16) produce damping factors whose form closely resembles the CV estimate.

For the strength and form factor of the density dependent repulsive real interaction, there is at present little theoretical guidance of a general nature. The real part of the 135 MeV Paris interaction, shown in Figure 5, can be quite accurately parameterized by

$$f_R(\alpha q) = 1 - \alpha q$$
$$\alpha = .31$$
$$a_R = 83 \text{ MeV fm}^3 \qquad (8)$$
$$\beta_R = 3 .$$

The amplitude of the repulsive interaction appears to be proportional to density. For relatively small momentum transfers, the Paris wound form factor can be equally well represented by the form

$$f_R(\alpha q) = e^{-\alpha^2 q^2} . \qquad (9)$$

In Table I, we compare several estimates of the density dependence of the 135 isoscalar spin-independent central interaction. The exponents of k_F are taken as $\beta_R = 3$ and $\beta_I = 2$. When appropriate, the wound form factor is parameterized by Eq. (9). The damping factors produced by these several nuclear matter calculations are all quite similar to the simple CV estimate. Note that the density dependence of the imaginary part of the effective interaction based upon the Hamada-Johnston (HJ) potential is stronger than that arising from the Paris potential. Similarly, the amplitude of the repulsive interaction is considerably stronger for the HJ than for the Paris potential. This effect is amplified at large momentum transfers by the shorter range of the HJ wound.

The expected manifestations of medium corrections in scattering data are illustrated in Figure 6, which compares the modulus of the 140 MeV isoscalar spin-independent central component of the free (LF) interaction with the density dependent effective interactions based upon the HJ and Paris potentials. As the density increases, the low-q attraction is markedly suppressed and the high-q repulsion enhanced relative to the free interaction. Therefore, one can expect low-q suppression and high-q enhancement of the cross section, in accord with the required modifications of the IA predictions discussed in the previous section. These effects can be expected to be more pronounced for the HJ than for the Paris interaction.

For low density and small momentum transfers, both the HJ and Paris effective interactions are similar to the LF interaction. The low density limit of the Paris interaction is similar to the LF interaction over a much larger range of momentum transfer than is the HJ interaction. However, this is not necessarily a strong argument in favor of the Paris potential because the LF interaction is not necessarily reliable at these larger momentum transfers. Furthermore, the momentum transfer available to the nucleon-nucleon system is limited to 2.5 fm^{-1}. The difference between the HJ and Paris interactions, particularly at larger momentum transfer, reflects the difference between the short-ranged repulsive cores of the input potentials. The hard core of the HJ potential produces a more pronounced repulsive maximum than does the Paris potential.

V. LOCAL DENSITY APPROXIMATION

In the folding model, the scattering potential is obtained by convolution of the transition density with the effective interaction. The local density approximation (LDA) stipulates that this effective interaction t_{eff} is the same as that appropriate to nuclear matter $t_{NM}(\rho_G)$ with the local ground state density ρ_G. As nuclear matter is a uniform medium, there is some ambiguity with respect to the appropriate value of the local density. Is it to be evaluated at the site of the projectile $\vec{r}_1$, the struck particle $\vec{r}_2$, or an average position $(\vec{r}_1 - \vec{r}_2)/2$? Due to the short range of the interaction, the practical consequences of this ambiguity are minor. Therefore, we choose, for numerical convenience,

Table I

Effective Interaction Parameters

Source	Re $t_{00}^c(q,o)$ (MeV fm^3)	a_R (MeV fm^3)	α	Im $t_{00}^c(q,o)$ (MeV fm^3)	a_I
LF[10]	-266.	NA	NA	-213.	NA
CV[18]	NA	NA	NA	NA	.38
JLM[11]	-225.	112.	NA	-166.	.33
HJ[15]	-257.	131.	.10	-187.	.54
Paris[16]	-279.	83.	.30	-239.	.42

NA: parameter not appropriate

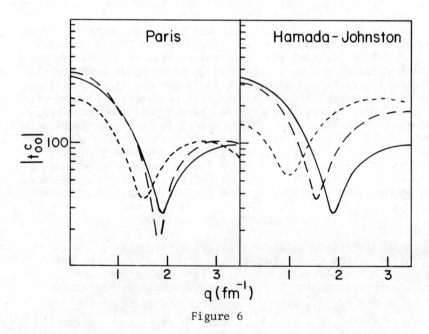

Figure 6

Comparison of $|t_{00}^c(q)|$ at 140 MeV for the LF interaction (solid curve on both sides) with the effective interactions based upon the Paris (left) and HJ (right) potentials. Long (short) dashed curves show low (high) density limits.

$$t_{eff} = t_{NM}(|\vec{r}_1 - \vec{r}_2|, \rho_G(\vec{r}_1)) \ .$$

The elastic central optical potentials for the scattering of 135 MeV protons by ^{16}O through the LF, HJ, and Paris interactions are compared in Figure 7. These potentials all become similar in the low density surface region. The potentials based upon the free two-nucleon interaction are very much deeper in the interior than those based upon density dependent effective interactions. Relative to the IA, the LDA imaginary potentials are smoothly damped in the interior. The Paris imaginary potential is more absorptive than the HJ. The LDA real potentials display more radial structure than the IA or Woods-Saxon potentials.

The structure in the microscopic real central optical potential is produced by competition between the attractive and repulsive components of the interaction. For a uniform density, the net potential is attractive. However, the repulsion grows as the density increases. Therefore, the balance between attraction and repulsion changes in the region of variable nuclear density. The details of the balance depend upon the difference between the gradient of the density and the gradient of the repulsive component. In the energy regime E_p = 100-400 MeV, this competition produces a characteristic depression of the real central optical potential inside the nuclear surface.[19] This characteristic radial feature also arises in several other approaches, including Dirac-Hartree calculations,[20] nonstandard phenomenology,[19] and Dirac phenomenology.[20]

The surface depression produced by the HJ interaction is deeper than that of the Paris interaction for two reasons. First, the low density limit of the HJ interaction is stronger at high-q. Second, the density dependence of the HJ effective interaction is stronger.

The predictions using the LF, HJ, and Paris interactions for the elastic scattering of 135 MeV protons by ^{16}O are compared with the data in Figure 8. The IA substantially overpredicts the forward cross section and possesses too little structure at larger angles. This lack of structure is related to the insufficient surface stiffness of the IA optical potential, which resembles a gaussian more than a Woods-Saxon shape. The HJ potential provides a better estimate of the forward cross section, but the high-q structure in the real central potential produces far too much high-q scattering. The softer Paris potential provides the best description of the elastic scattering.

Note that none of these potentials is strong enough to introduce an additional oscillation in the elastic wavefunctions within the range of the potential. The change in period is quite small. Therefore, there is no phase averaging effect involved in comparing inelastic scattering calculations using distorted waves produced by these various potentials. The primary difference between these sets of distorted waves is their absorption profiles.

The LDA predictions using the effective interactions based upon the HJ potential are compared in Figure 9 with the 135 MeV

166

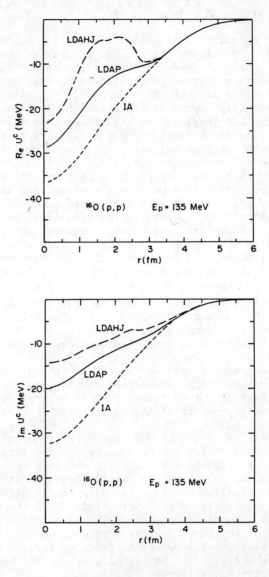

Figure 7

Central optical potentials for the
scattering of 135 MeV protons by ^{16}O.

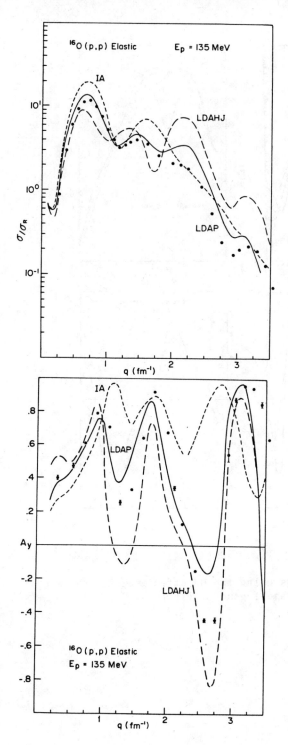

Figure 8

Elastic scattering of
135 MeV protons by ^{16}O.

168

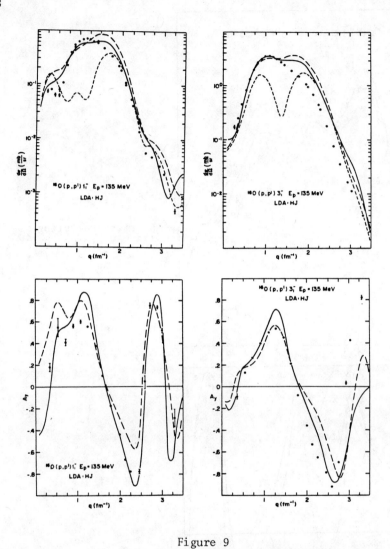

Figure 9

LDA calculations using the HJ effective
interaction. Same legend as Figure 4.

data for the lowest 1^- and 3^- states of ^{16}O. The low-q central contribution to the interior 1^- state is almost completely suppressed by Pauli blocking. The factor of four overestimation using the free interaction is fully explained by density dependence. Similarly, the smaller overestimation of the low-q 3^- cross section is also accounted for. The repulsive maximum of the central contribution to the 1^- cross section is greatly enhanced by the medium corrections. This enhancement is required by both the cross section and analyzing power data. The negative depth of the analyzing power data in the region of the repulsive maximum is explained by the density dependent high-q enhancement. However, the HJ interaction appears to overestimate the magnitude of this effect. In particular, the high-q cross section to the surface 3^- state, which is more sensitive to the low density limit of the effective interaction, is overestimated in the region of this repulsive maximum.

The LDAHJ predictions are far superior, with either choice of distorted waves, to those of the IA and are also less sensitive to this choice than are the IA predictions. The differences between the microscopic and phenomenological optical potentials produce some minor differences between the angular distributions, but little difference in overall magnitude, indicating that the density dependent damping of the absorptive potential is approximately correct. The high-q structure in the real central optical potential reinforces the depth of the negative analyzing powers near 2.5 fm^{-1}, improving the agreement with the data.

One can demonstrate that in this energy regime the magnitudes of inelastic cross sections are primarily sensitive to the imaginary part of the central optical potential, while inelastic analyzing power distributions are sensitive to the real part. Alterations of the strength or radial form of the absorptive potential tend to alter the magnitude of the cross section with little change in its shape and almost no effect upon the analyzing power. However, we note that this sensitivity is reduced in a consistent calculation because a change in absorption is compensated by a change in transition amplitude. Alterations of the radial form of the real central optical potential tend to affect the analyzing power distribution with little effect upon the cross section. The beneficial influence of the surface depression in the real central optical potential upon the inelastic analyzing powers provides good evidence in favor of this feature.

The LDA predictions based upon the Paris potential are compared with the data in Figure 10. With consistent distorted waves, the low-q cross sections are again quite well described by the LDA. For the surface state, the high-q cross section prediction is in much better agreement with the data, indicating that the low density limit of the Paris interaction is better than that of the HJ. However, the Paris interaction still overpredicts the low-q and underpredicts the high-q cross section for the interior state, and its analyzing power fails to become sufficiently negative. These defects indicate that the density dependence of the Paris effective interaction is not sufficiently strong.

170

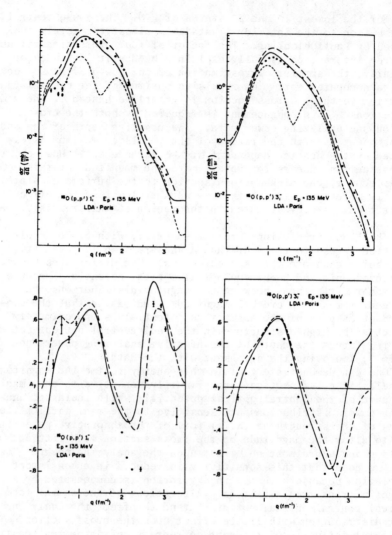

Figure 10

LDA calculations using the Paris effective
interaction. Same legend as Figure 4.

We also observe that the Paris results are more sensitive to
the choice of distorted waves than are the HJ results. The Paris
interaction is more similar to the IA because its density dependence is less than that of the HJ interaction. Part of its low-q
suppression is achieved by a deeper absorptive potential.

Although space does not permit a compilation of the results
for other states of ^{16}O, other nuclei, or other energies, it can
nonetheless be stated that the signatures of density dependence are
characteristic of all normal parity isoscalar transitions excited
by 100-200 MeV protons, as are the properties and quality of the
LDA description of these effects.

VI. CONCLUSION

We have performed a detailed study of a class of inelastic
transitions for which electron scattering measurements completely
specify the nuclear structure required for the analysis of proton
scattering. The analysis of proton scattering data has been interpreted as a study of the two-nucleon effective interaction in the
nuclear medium. An important aspect of this study was the availability of a large range of nuclear transitions with varying radial
localization. Interior transitions provide information about the
interaction at saturation density while surface transitions are
sensitive to the low density limit of the interaction.

We have identified, in the data, several striking signatures
of density dependence in the isoscalar spin-independent central
component of the two-nucleon effective interaction near 150 MeV.
As the density increases, the low momentum transfer attraction of
the effective interaction is suppressed and the high momentum
transfer repulsion enhanced relative to the free interaction.
These medium modifications are manifested by inelastic cross
sections that are correspondingly suppressed at low-q and enhanced
at high-q relative to impulse approximation calculations based
upon the free interaction, and by strong negative analyzing powers
near 2.4 fm^{-1} that the IA is unable to reproduce. These signatures
are characteristic traits of normal parity isoscalar excitations of
nuclei by 100-200 MeV protons.

The characteristic signatures of density dependence are quite
strong and are well described by the local density approximation
based upon effective interactions derived for infinite nuclear
matter. Therefore, in the intermediate energy regime, the medium
modifications are dominated by properties characteristic of the
underlying nuclear medium. The properties specific to the finite
system appear to be of minimal significance.

The success of the local density approximation based upon
effective interactions derived in an independent pair approximation
can be interpreted as the observation of the correlated pair wavefunction describing an energetic nucleon and a nucleon bound in
nuclear matter. The sensitivity of the data to the detailed momentum transfer dependence of the medium modifications can be interpreted as an observation of the radial form of this correlation
function.

The results are sensitive to the choice of free two-nucleon potential. Therefore, within the accuracy of the LDA, the spin-isospin selectivity of nucleon-nucleus reactions can be used to help distinguish between alternative potentials whose descriptions of nucleon-nucleon data are essentially indistinguishable. The Hamada-Johnston potential provides an excellent description of interior transitions, but its low-density limit is clearly too repulsive. The Paris potential provides a superior overall description of both the elastic and inelastic data, but its density dependence is clearly insufficient.

The medium modifications of the effective interaction can be well represented by a very simple parameterization. The imaginary part is subject to a multiplicative damping factor which decreases as the density increases. The modification of the real part can be represented by an additive short-range repulsive interaction whose strength increases with density. This parameterization can reproduce either the HJ or Paris results, or can describe an intermediate case.

Given that the LDA is based upon a truncated nuclear matter theory whose application to finite systems contains several presently untested approximations, one does not expect its estimate of the parameters to be completely accurate. Therefore, we propose that these parameters be fitted to a global data set that includes elastic scattering, many inelastic transitions with varying radial localization, and several different target nuclei. It is crucial to this procedure that the target densities be fully constrained by electron scattering. In an extended search procedure, the low density limit of the interaction might also be optimized. If this program produces a global parameterization that is independent of target and multipolarity, and whose parameters vary smoothly with energy, then the LDA description would be considered fully validated. The remaining task would then be a theoretical explanation of the values of these few parameters. We also note that such a parameterization will facilitate the study of structure problems and neutron densities.

ACKNOWLEDGMENTS

The work presented in this paper is the result of a long and fruitful collaboration which includes: W. Bertozzi, T. Buti, J. M. Finn, F. W. Hersman, C. Hyde, M. A. Kovash, B. Murdock, and B. Pugh (MIT); M. V. Hynes (LANL); B. E. Norum (U.Va.); A. D. Bacher, G. T. Emery, C. C. Foster, W. P. Jones, and D. W. Miller (IU); B. L. Berman (LLNL); W. G. Love (U.Ga.); J. A. Carr and F. Petrovich (FSU).

This work is supported in part by the U. S. Department of Energy under Contract No. DE-AC02-76ER03069.

REFERENCES

1. William Bertozzi and James J. Kelly in the Proceedings of the Conference on New Horizons in Electromagnetic Physics (to be published).
2. J. Kelly et al., Phys. Rev. Lett. 45, 2012 (1980).
3. J. Kelly, Ph.D. thesis (MIT, 1981) and to be published.
4. F. Petrovich et al., Phys. Rev. Lett. 22, 895 (1969); W. G. Love, Nucl. Phys. A312, 166 (1978).
5. T. N. Buti et al., MIT preprint.
6. H. Miska et al., Phys. Lett. 53B, 155 (1975).
7. J. C. Bergstrom et al., Phys. Rev. Lett. 24, 152 (1970).
8. H. Crannell, Phys. Rev. 148, 1107 (1966).
9. Y. Torizuka et al., Phys. Rev. Lett. 22, 544 (1969).
10. W. G. Love and M. A. Franey, Phys. Rev. C24, 1073 (1981).
11. J. P. Jeukenne, A. Lejune, and C. Mahaux, Phys. Rev. C15, 10 (1977); C16, 80 (1977).
12. F. A. Brieva and J. R. Rook, Nucl. Phys. A291, 299, 317 (1977); and A297, 206 (1978).
13. T. Hamada and D. Johnston, Nucl. Phys. 34, 382 (1962).
14. M. Lacombe et al., Phys. Rev. C21, 861 (1980).
15. H. V. von Geramb, "Table of Effective Density and Energy Dependent Interactions for Nucleons" (unpublished).
16. K. Nakano and H. V. von Geramb (unpublished); and H. V. von Geramb, these proceedings.
17. H. V. von Geramb, F. A. Brieva, and J. R. Rook, in Microscopic Optical Potentials, ed. by H. V. von Geramb (Springer-Verlag, Berlin, 1979), p. 104.
18. E. Clementel and C. Villi, Nuovo Cimento 2, 176 (1955).
19. P. Schwandt, these proceedings.
20. B. C. Clark, these proceedings.

DISCUSSION

Speth: If I understand correctly, the spin-orbit interaction is very important in your calculations, and you use experimental transition densities, taken from electron scattering. Might it not be importnat to include the transverse transition densities, which include some important spin effects in the nuclear structure?

Kelly: We have chosen states for which those transverse components are negligible.

Speth: But even small components can give a relatively large analyzing power effect.

Kelly: In this case they are on the 1% level. They don't affect the analyzing power, either.

Garvey: Are there high-q measurements of those in electron scattering?

174

<u>Kelly</u>: They are measured out to 3 fm^{-1} also, but these are collective states and the currents are less important than the charge.

<u>Scott</u>: Would you comment on how well the "consistent" calculations predict the elastic scattering analyzing power.

<u>Kelly</u>: The conclusion one would draw from the figure [Fig. 8, bottom] is that none of these things work very well for the analyzing power. The nuclear matter calculations tend to get the qualitative features of the oscillations, while the impulse approximation is totally out of phase. I want to stress that I consider it more important to study a first-order process, such as inelastic scattering, when you have a first-order theory. The trouble with giving too much weight to the elastic scattering is that it is essentially an infinite-order process so that you need very high accuracy to reproduce the asymptotic observables, whereas in the inelastic scattering if you make a first-order error in your theory you make a first-order error in your observables.

<u>Redish</u>: Have you looked at the elastic wave functions compared to what they look like in the phenomenolgial calculations?

<u>Kelly</u>: Yes, and they are not very different. Their differences are what you see in the inelastic calculations as the difference between the consistent and phenomenological distorted waves. It's not a very big effect for the inelastic scattering. It's just that they all tend to have the wrong slope as they get out of the nucleus, and they get out of phase, and you don't get the right scattering.

<u>Amado</u>: I don't understand that comment about first-order and many-order effects. It's certainly correct if distortions are zero, but since the observables in the inelastic process are being distorted in the same way as in the elastic process, I just don't understand it.

<u>Kelly</u>: I interpret that as a comment about how sensitive your are to those distortions in the inelastic scattering, and in fact you're not very sensitive to them, as far as the qualitative features go. They all have the same magnitude in the interior, they all give you the same absorption.

<u>Amado</u>: But if none of them fit the elastic analyzing power then I'm suspicious about the fit to the inelastic data.

<u>Kelly</u>: What I'm saying is, they don't have to fit the elastic scattering to give a sufficiently good description of the interior wave functions. You see that those negative analyzing powers [near 2.7 fm^{-1} in Fig. 8, bottom] are dominated by the central contribution to the inelastic amplitude, and no matter which distortion I used, if I had a highly repulsive central force at high momentum transfer I always get a deep negative analyzing power.

Unknown Questioner: Did you use in your calculations the prescription of Brieva and Rook, or some other method? They have used an approximate way of calculating the density-dependent effects.

Kelly: I have used that same form. I have used the effective interactions supplied by von Geramb, whose calculations with the Hamada-Johnston force are based on that approximation. With the Paris interaction he used a somewhat better approximation. Since we are not including effects of higher order, anyway, we cannot tell how important that approximation is.

Rawitscher: I have a question concerning exchange. You have implied that the properties specific to finite systems are of secondary importance. Now, when you put in exchange terms, which we know are very important (and produce cancellations) it seems to me that in calculating these exchange terms you are making use of specific shell-model wave functions. These shell-model wave functions should really be Hartree-Fock wave functions. They should change from one nucleus to the next nucleus, and they should be non-local. To what extent has this question been investigated?

Kelly: One has looked at the sensitivity of these local exchange calculations to the mixed density, whether you use the diagonal part or the off-diagonal part that depends on the Fermi momentum. It is really not very sensitive to that.

INELASTIC PROTON SCATTERING TO ONE-PARTICLE ONE-HOLE STATES

C. Olmer
Indiana University, Bloomington, IN 47405

INTRODUCTION

The study of intermediate-energy proton inelastic scattering to one-particle one-hole states has attracted considerable attention over the last few years. One primary purpose of these studies has been to exploit the structural simplicity of these excited states as a possible means of separating effects due to nuclear structure and those due to the reaction mechanism. Several excellent review talks have surveyed many of the current experimental and theoretical investigations.[1] This contribution will concentrate on one particular area of investigation, namely the energy dependence of proton inelastic scattering. In recent years, high-quality inelastic scattering data for several transitions have become available over a wide range of incident energy. In addition, there are now available several models of the effective nucleon-nucleon interaction and several prescriptions for making predictions of inelastic scattering. As a result, we are now in a position to examine in detail the energy dependence of the various effective interactions and to test the model predictions.

RESULTS AND DISCUSSION

This talk will focus on the excitation of three high-spin, particle-hole states in ^{28}Si: the 5^-, T=0 state at 9.70 MeV, the 6^-, T=0 state at 11.58 MeV and the 6^-, T=1 state at 14.35 MeV. These states have been studied at proton energies between 65 MeV and 800 MeV. Differential cross-section and analyzing-power data are available from Osaka for the 6^- states at 65 MeV[2]; both σ and A have been measured at IUCF for all three states at 80 MeV, 100 MeV, 135 MeV and 180 MeV[3,4]; and σ has been measured at LASL for all three states at 333 MeV, 500 MeV and 800 MeV.[5] A consistent analysis of all these data would be useful, but is not now available, and this paper will refer to only the IUCF data.

A region of the inelastic scattering spectrum measured at an incident energy of 180 MeV is displayed in Fig. 1. These high-spin states in ^{28}Si have been particularly useful in proton inelastic scattering studies since: (1) the states usually appear as strong, isolated peaks in the spectrum, (2) ^{28}Si is a self-conjugate nucleus so that differences between proton and neutron transition densities are not expected to be a problem, and (3) all three states have also been studied by pion inelastic scattering[6] and the 5^-, T=0 and 6^-, T=1 states have been studied by electron inelastic scattering[7], so that this is one of the few cases for which p-π-e complementary information exists.

The differential cross sections for the 5^-, T=0 state from 80 MeV to 180 MeV are displayed in Fig. 2. These distributions are all bell-shaped, and are peaked at similar large values of momentum

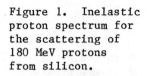

Figure 1. Inelastic proton spectrum for the scattering of 180 MeV protons from silicon.

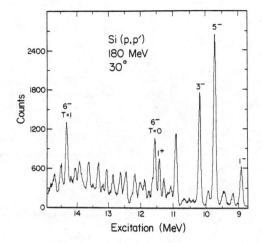

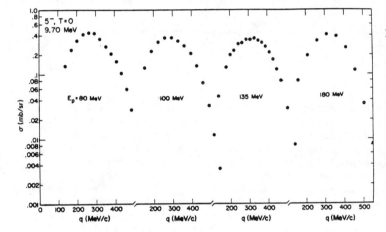

Figure 2. Momentum transfer dependence of the cross sections for proton inelastic excitation of the 5⁻, T=0 state at 9.70 MeV in ^{28}Si.

transfer. Some energy-dependent features are immediately apparent in the data, including variations of the maximum cross section σ_{max} with E_p and changes in the shape (i.e. width and asymmetry) of the peak over the indicated range in E_p. Several types of comparisons of model predictions with the data are thus possible: (1) the E_p-dependence of σ_{max}, (2) the E_p- and q-dependence of σ, and (3) the E_p- and q-dependence of A. A complete discussion of the results of these comparisons is not possible in the present talk, but will be presented elsewhere.[4]

The distorted-wave impulse approximation (DWIA) description of inelastic scattering to these states incorporates information concerning the nuclear wave functions of the states, the effective N-N interaction and p-^{28}Si elastic scattering distorted waves. The elastic scattering data measured at the various energies are displayed in Fig. 3. It is extremely important in a study of the energy dependence of inelastic scattering to measure both σ and A

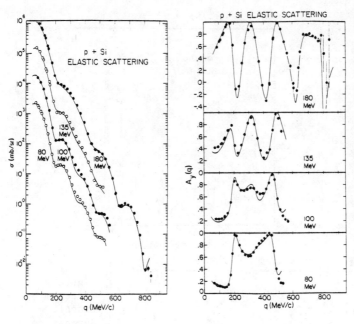

Figure 3. Momentum transfer dependence of the cross sections and analyzing powers for proton elastic scattering by silicon. The curves are the results of simultaneous optical-model fits to the cross section and analyzing power data at each energy.

for the elastic scattering over a wide range of both E_p and q. In the present study, a standard Woods-Saxon optical-model description of the elastic scattering was employed, and the resulting energy dependence of the deduced optical model parameters was investigated. By requiring a smooth dependence on E_p for these parameters (including the total reaction cross section), a significant reduction in the ambiguity of parameters was possible. The use of a non-standard optical potential (Woods-Saxon plus a squared Woods-Saxon term), which was found[8] to be important for reproducing proton elastic scattering from light nuclei, did not significantly affect the quality of the optical model fit at 180 MeV. This potential resulted in inelastic scattering analyzing-power predictions of similar shape and magnitude, and cross-section predictions of similar shape and slightly different magnitude as compared to the standard optical model treatment. A detailed description of the elastic scattering analysis will be presented elsewhere.[9]

The wave functions of the three states have been assumed to be of harmonic oscillator form, and the oscillator parameters derived from (e,e') studies[7] of the 5^-, T=0 and 6^-, T=1 transitions have been used (b=1.91 and b=1.74, respectively). The 6^-, T=0 state is assumed to have the same oscillator parameter as the 6^-, T=1 state. Both 6^- states are characterized by the $f_{7/2}d_{5/2}^{-1}$ configuration, whereas the 5^-, T=0 state is described by an RPA wave function[3], which has a dominant $f_{7/2}d_{3/2}^{-1}$ term.

Several different effective interactions have been considered in this study, and some of these interactions have been discussed in detail during this Workshop. These include:

    (1) the Love-Franey interaction[10] (LF), which has been separately derived at 100, 140 and 185 MeV,

    (2) The Geramb-Bauhoff interaction[11] (GB), which is a density-dependent interaction based on the Paris N-N interaction and is available at isolated energies (also the Hamada-Johnston[12] interaction (HJ), which is available for central and spin-orbit components only), and

    (3) the Picklesimer-Walker interaction[13] (PW), which explicitly incorporates both energy and momentum dependence.

The DWIA calculations employing the LF and PW interactions used the code DW81[14] and those employing the GB interaction used the code DWBA80[15]. All calculations treated the exchange term in an exact manner.

The simplest possible comparison of the DWIA predictions with the data involves the energy dependence of σ_{max}, since it ignores the dependence on the momentum transfer. The experimental values of σ_{max} for the 5^-, T=0 and 6^-, T=0 transitions are displayed in Fig. 4, together with predictions based on the various

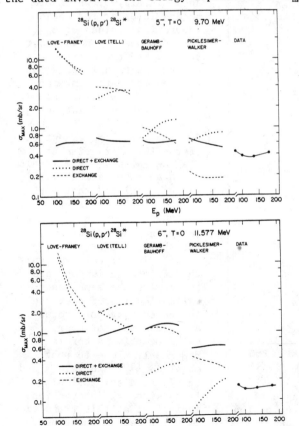

Figure 4. Energy dependence of the maximum cross sections for proton inelastic excitation of the 5^-, T=0 and 6^-, T=0 states in ^{28}Si. For the data, a smooth curve has been drawn through the four experimental measurements for each transition.

180

interactions. (An additional set of calculations is displayed, using an earlier form[16] of the Love interaction derived at 135 MeV. Also, note that the LF results are displayed only for energies of 100 MeV and above). The individual contributions of direct and exchange are displayed, and these vary dramatically for the various interactions. These terms interfere constructively for some interactions and destructively for others. For some interactions, very different direct and exchange terms combine to produce very similar predictions. Experimentally, the energy dependences of σ_{max} for these two transitions are very similar. Only the LF calculations seem to reproduce this observation, whereas the GB and PW calculations predict significantly different dependences on proton energy. Clearly an analysis over a broader range of energy is necessary in order to verify these conclusions.

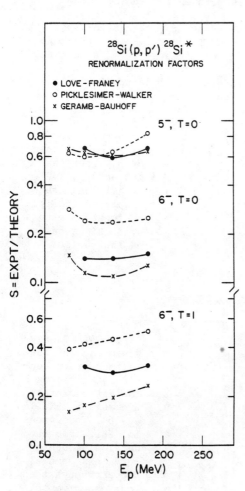

A further comparison can be made by examining the renormalization factors, shown in Fig. 5, which are needed in order for these theoretical values of σ_{max} to reproduce the data. For the 5^-, T=0 transition, all renormalization factors for all the indicated interactions have a similar value and a relatively flat energy dependence is seen for these factors (with the exception of the PW interaction at high energies). As a result of these observations, it may be concluded that both the LF and GB interactions reproduce reasonably well the energy dependence of σ_{max} for the 5^- transition. For the 6^-, T=1 transition, only the LF interaction appears to require the same renormalization factor over the indicated range of proton energies and, moreover, the value of this factor is remarkably close to that obtained [1,6,7] in (e,e') and (π,π') studies. Calculations with all interactions indicate that this transition is dominated by the tensor component of the interaction.

Figure 5. Energy dependence of the renormalization factors for proton inelastic excitation of the 5^-, T=0 6^-, T=0 and 6^-, T=1 states in ^{28}Si.

Thus it may be concluded that the increase in strength of the tensor component in this energy region (100-200 MeV) is reproduced reasonably well by the LF interaction, but that the energy dependence of the tensor component in both the GB and PW interactions does not appear to be quite correct.

Both the energy and momentum-transfer dependence of the 5⁻, T=0 cross section for the various interactions are displayed, together with the data, in Figs. 6-8. The calculations have been renormalized by the factors indicated in the appropriate figure captions. Calculations at 80 MeV for the LF interaction employed the 100 MeV force. The decomposition of the predicted cross section into central (C), spin-orbit (LS) and tensor (T) contributions is indicated for each of the various interactions. Calculations using the LF interaction, shown in Fig. 6, appear to

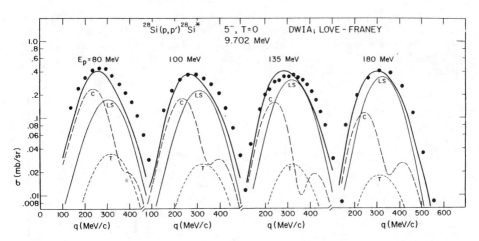

Figure 6. Momentum transfer dependence of the cross sections for proton inelastic excitation of the 5⁻, T=0 state at 9.70 MeV in ^{28}Si. The curves are DWIA calculations (multiplied by 0.67) using the Love-Franey effective interaction and optical potentials appropriate for the various energies.

reproduce the energy dependence of σ_{max} and q_{max} (the location of σ_{max}) reasonably well, but consistently fail to reproduce the width of the distributions. The predicted distributions are narrower than experimentally observed, and moreover, the calculations do not exhibit the observed increase in width of the distributions at the lower incident energies.

Calculations for the 5⁻ transition using the GB interaction, shown in Fig. 7, also appear to reproduce the energy dependence of σ_{max} and q_{max}, and here the calculations reproduce the experimental width of the distributions significantly better than was observed for the LF interaction. In particular, the shape of the distribution at 135 MeV is not exhibited in the LF calculation, but

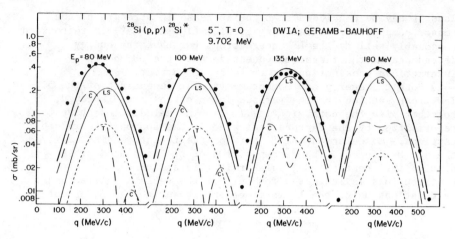

Figure 7. Momentum transfer dependence of the cross sections for
proton inelastic excitation of the 5^-, T=0 state at 9.70 MeV in
^{28}Si. The curves are DWIA calculations (multiplied by 0.67) using
the Geramb-Bauhoff effective interaction and optical potentials
appropriate for the various energies.

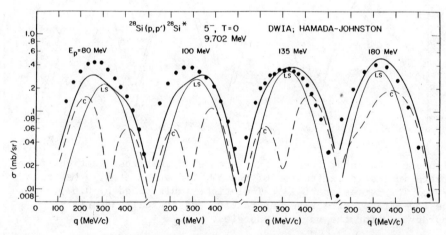

Figure 8. Momentum transfer dependence of the cross sections for
proton inelastic excitation of the 5^-, T=0 state at 9.70 MeV in
^{28}Si. The curves are DWIA calculations (multiplied by 0.67) using
the Hamada-Johnston effective interaction and optical potentials
appropriate for the various energies.

is well-reproduced in the GB calculation, where it results from the
large, second maximum of the central component (very weak in the LF
calculation) contribution to the large momentum transfer region.
This broadening of the distributions also results from calculations
using the HJ interaction, displayed in Fig. 8, although here the
energy dependence of σ_{max} is not reproduced. Both the GB and HJ

interactions are density dependent, and both exhibit a double-humped central contribution for proton energies between 100 and 180 MeV. Thus it appears that density dependence is important for a complete understanding of this transition.

The analyzing-power distributions for the 5^-, T=0 transition are shown in Fig. 9, together with predictions for the four interactions. The most positive value of A is observed to decrease from ~0.8 at 180 MeV to near zero at 80 MeV, and this general trend is apparent in calculations for all the interactions. Since A is sensitive to the interference of amplitudes, a comparison of the predicted and observed analyzing powers provides an independent test of the interaction, separate from that furnished by the cross

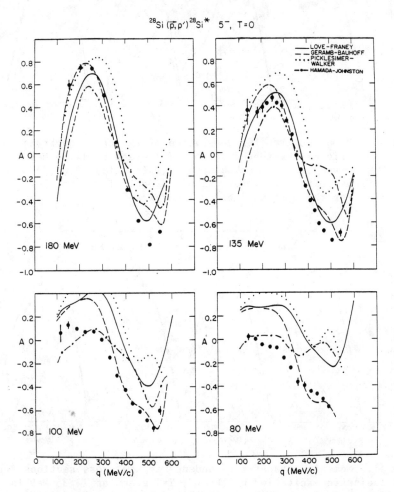

Figure 9. Momentum transfer dependence of the analyzing powers for proton inelastic excitation of the 5^-, T=0 state at 9.70 MeV in ^{28}Si.

184

section. Here, both the GB and LF calculations are in reasonable
agreement with the data, although the GB interaction appears to be
slightly preferred.

The cross-section distributions for the 6⁻, T=1 transition are
displayed in Figs. 10-12 with DWIA calculations using the LF, GB

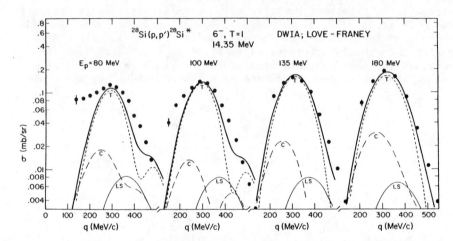

Figure 10. Momentum transfer dependence of the cross sections for
proton inelastic excitation of the 6⁻, T=1 state at 14.35 MeV in
^{28}Si. The curves are DWIA calculations (multiplied by 0.30) using
the Love-Franey effective interaction and optical potentials
appropriate for the various incident energies.

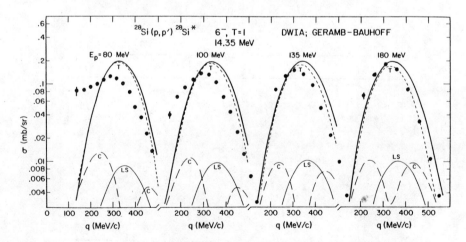

Figure 11. Momentum transfer dependence of the cross sections for
proton inelastic excitation of the 6⁻, T=1 state at 14.35 MeV in
^{28}Si. The curves are DWIA calculations (multiplied by 0.25) using
the Geramb-Bauhoff effective interaction and optical potentials
appropriate for the various incident energies.

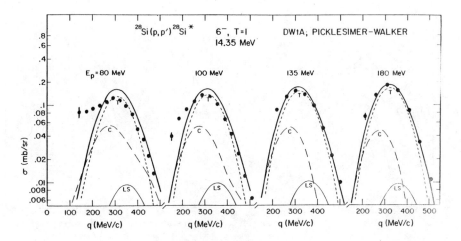

Figure 12. Momentum transfer dependence of the cross sections for proton inelastic excitation of the 6^-, T=1 state at 14.35 MeV in ^{28}Si. The curves are DWIA calculations (multiplied by 0.50) using the Picklesimer-Walker effective interaction and optical potentials appropriate for the various incident energies.

and PW interactions. As was noted earlier, all of these calculations indicate that this transition occurs predominantly through the tensor component of the interaction. The observed energy dependences of σ_{max} and q_{max} are reproduced reasonably well by calculations using the LF and PW interactions, but not for the GB interaction. Comparable agreement in shape is obtained for all three interactions at 135 MeV and 180 MeV. On the other hand, none of the calculations is able to reproduce the unusual behavior observed at low q for low incident energies; the origin of this behavior is not understood at present.

Displayed in Figs. 13 and 14 are the cross section distributions for the 6^-, T=0 transition and calculations using the LF and GB interactions. Note that for this transition, the tensor component is decreasing and the spin-orbit component is increasing as a function of increasing incident energy, but the details of this interchange are dependent on the specific interaction. In addition, the momentum-transfer dependence of the spin-orbit component for the two interactions is very similar, whereas that for the the tensor component is quite different, being peaked at smaller q for the GB interaction. For both interactions, a small change in the oscillator parameter (from the assumed 6^-, T=1 value) is necessary in order to reproduce the position and shape of the distribution at high incident energy. However, neither interaction, nor any other which has been examined in this study, can reproduce the shift of the peak to smaller q as the proton energy is reduced. The origin of this shift is not understood at present.

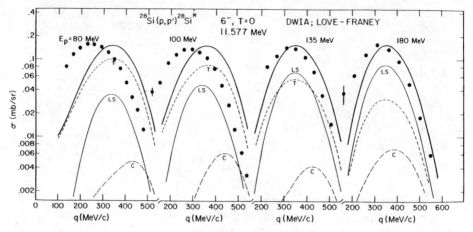

Figure 13. Momentum transfer dependence of the cross sections for proton inelastic excitation of the 6^-, T=0 state at 11.58 MeV in ^{28}Si. The curves are DWIA calculations (multiplied by 0.15) using the Love–Franey effective interaction and optical potentials appropriate for the various incident energies.

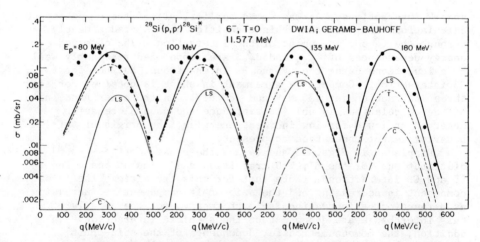

Figure 14. Momentum transfer dependence of the cross sections for proton inelastic excitation of the 6^-, T=0 state at 11.58 MeV in ^{28}Si. The curves are DWIA calculations (multiplied by 0.15) using the Geramb–Bauhoff effective interaction and optical potentials appropriate for the various incident energies.

The analyzing-power distributions for the 6^-, T=1 transition are shown in Fig. 15, together with calculations using the LF, GB, and PW interactions. The quality of the agreement with the experimental distributions is not very good here, as for the

6⁻, T=0 results. However, it should be noted that the analyzing powers for the 6⁻ states are sensitive to distortion effects and that the inclusion of density dependence in the calculations for these 6⁻ states seems to improve somewhat the quality of the agreement.

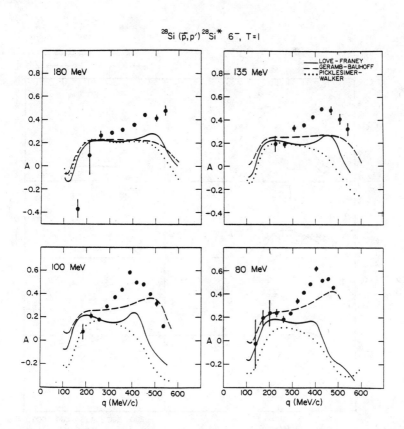

Figure 15. Momentum transfer dependence of the analyzing powers for proton inelastic excitation of the 6⁻, T=1 state at 14.35 MeV in ^{28}Si.

One area of recent interest concerns the difference between polarization and analyzing power, P–A. Preliminary theoretical calculations by several groups indicate that P–A should be sensitive to nonlocal effects in the NN effective interaction. Experimentally, P–A has been examined for only a limited number of transitions. Recent measurements for the 15.11 and 12.71 MeV 1⁺ states in ^{12}C indicate that P–A can be very different from zero, and DWIA calculations for the 15.11 MeV state are in qualitative agreement with the data.[17] Displayed in Fig. 16 are DWIA calculations using the LF and GB interactions for three states in ^{28}Si at incident energies of 80 and 180 MeV. For all states, large values of P-A are predicted at low incident energy and small to medium values of P-A are predicted at higher energies. (Note

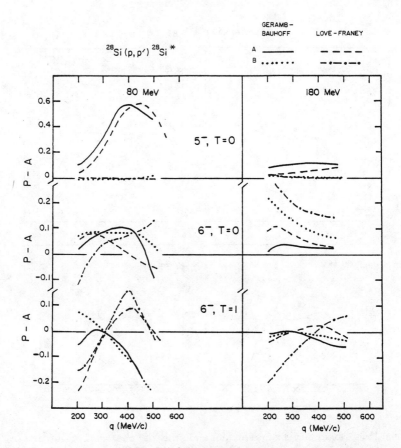

Figure 16. Momentum transfer dependence of P-A for proton inelastic excitation of the 5$^-$, T=0, 6$^-$, T=0 and 6$^-$, T=1 states in ^{28}Si. The curves labelled A and B are DWIA and PWIA calculations using the indicated effective interaction and the appropriate optical potentials.

that the region 300-350 MeV/c, where the cross section is largest, is of primary interest here). For the 5$^-$ state, the differences in P-A result from distortion effects, whereas for the 6$^-$ states, the influence of distortion on P-A is smaller. Measurable

sensitivities to the two interactions are predicted for these three
states. In conjunction with P-A measurements for lower
multipolarity states (e.g., the 1^+ levels in ^{12}C), measurements of
P-A for such high-spin states would provide a means of examining
the momentum-transfer dependence of nonlocal effects. Moreover, a
different manifestation of nonlocality may result, as compared to
the ^{12}C measurements, as a consequence of the very different
nuclear structure of the selected states in the two nuclei. The
theoretical situation for P-A is still uncertain and the origin of
large values of P-A is not fully understood. Additional work in
both experimental and theoretical areas is clearly needed.

SUMMARY

Detailed comparisons of several effective interactions with
inelastic scattering data is now possible over a broad range of
incident energy. Of fundamental importance in such an analysis is
a systematic, energy-dependent optical model analysis of the
elastic scattering. The energy dependence of σ_{max} was examined,
and a slight preference for the LF interaction was seen. The
shapes of the predicted cross section distributions were examined
and evidence was seen for the importance of density dependence,
although neither the GB nor the HJ interaction can exactly
reproduce the experimental results. A consideration of the energy
dependence of the analyzing power also revealed the importance of
incorporating density dependence in the effective interaction.

REFERENCES

1. A.D. Bacher, in Polarization Phenomena in Nuclear Physics--
1980, edited by G.G. Ohlsen et al. (AIP Conf. Proc. No. 69,
Am. Inst. of Phys., New York, 1981), p. 220; G.T. Emery, A.D.
Bacher, and C. Olmer, in Spin Excitations in Nuclei, ed. F.
Petrovich et al. (to be published by Plenum Press, New York,
1982); W.G. Love, M. Franey, and F. Petrovich in Spin
Excitations in Nuclei, ed. F. Petrovich et al. (to be
published by Plenum Press, New York, 1982); F. Petrovich and
W.G. Love, Proc. Int. Conf. on Nuclear Physics, Berkeley,
California, August 24-30, 1980 (in press).

2. K. Hosono, N. Matsuoka, K. Hatanaka, T. Saito, T. Noro, M.
Kondo, S. Kato, K. Okada, K. Ogino, and Y. Kadota, Phys. Rev.
C 26, 1440 (1982).

3. S. Yen, R.J. Sobie, T.E. Drake, A.D. Bacher, G.T. Emery, W.P.
Jones, D.W. Miller, C. Olmer, P. Schwandt, W.G. Love, and F.
Petrovich, Phys. Lett. 105B, 421 (1981); S.Yen, Ph.D. Thesis,
University of Toronto (1983).

4. C. Olmer, A.D. Bacher, G.T. Emery, W.P. Jones, D.W. Miller, P.
Schwandt, T.E. Drake, R.J. Sobie, and S. Yen, to be published.

190

5. N. Hintz, private communication.

6. C. Olmer, B. Zeidman, D.F. Geesaman, T-S. H. Lee, R.E. Segel,
    L.W. Swenson, R.L. Boudrie, G.S. Blanpied, H.A. Thiessen, C.L.
    Morris, and R.E. Anderson, Phys. Rev. Lett. 43, 612 (1979).

7. S. Yen, R.J. Sobie, H. Zarek, B.O. Pich, T.E. Drake, C.F.
    Williamson, S. Kowalski, and C.P. Sargent, Phys. Lett. 93B, 250
    (1980); S. Yen, to be published.

8. H.O. Meyer, P. Schwandt, G.L. Moake, and P.P. Singh, Phys. Rev.
    C 23, 616 (1981).

9. C. Olmer, A.D. Bacher, G.T. Emery, W.P. Jones, D.W. Miller, P.
    Schwandt, T.E. Drake, R.J. Sobie, and S. Yen, to be published.

10. W.G. Love and M.A. Franey, Phys. Rev. C 24, 1073 (1981).

11. K. Nakano and H.V. von Geramb, Paris Potential, A Table of
    Effective Density and Energy Dependent Interactions; Central
    Spin-Orbit and Tensor Potentials, preprint, Universitat Hamburg
    (1982); M. Lacombe, B. Loiseau, R. Vinh Mau, J. Cote, P. Pines,
    and R. de Journeil, Phys. Rev. C 23, 2405 (1981).

12. H.V. von Geramb, Table of Effective Density and Energy
    Dependent Interactions for Nucleons, Part A: Central
    Potential; W. Bauhoff and H.V. von Geramb, Part B: Spin-Orbit
    Potential, preprint, Universitat Hamburg (1980); T. Hamada and
    I.D. Johnston, Nucl. Phys. 34, 382 (1962).

13. A. Picklesimer and G. Walker, Phys. Rev. C 17, 237 (1978).

14. R. Schaeffer and J. Raynal (unpublished); modified by J.
    Comfort.

15. R. Schaeffer and J. Raynal (unpublished); modified by W.
    Bauhoff.

16. W.G. Love in The (p,n) Reaction and the Nucleon-Nucleon Force,
    edited by C. Goodman et. al. (Plenum, New York, 1980), p. 23.

17. T.A. Carey, J.M. Moss, S.J. Seestrom-Morris, A.D. Bacher, D.W.
    Miller, H. Nann, C. Olmer, P. Schwandt, E.J. Stephenson, and
    W.G. Love, Phys. Rev. Lett. 49, 266 (1982).

DISCUSSION

Schiffer: Can you do this sort of analysis for low-spin transitions, for example, for the 2^+ state, to get the low-momentum transfer properties?

Olmer: Yes. I haven't had the opportunity yet, but I will, and I think it will be very exciting.

Speth: Could you show the data and calculations for the 6^-, T=0 state?

Olmer: Any particular interaction? I feel like Baskin-Robbins.

Speth: Any one will do [e.g., Figs. 13,14]. If I understand this, it is a one-particle, one-hole form factor that you use. We have recently investigated the isoscalar structure and we get an important effect that may be relevant to the disagreement. You get a large contribution to the isoscalar transitions from the tensor exchange, and this means that you get contributions from the higher shells, in this case from the $3\hbar\omega$ shell. These components have contributions from larger momentum transfer. the interference of these terms with the $1\hbar\omega$ terms might be important.

Olmer: Can you produce a transition density for this state including those terms?

Speth: What we have done so far is ^{208}Pb for the 1^+ state, and there we get quite a bit of change.

Olmer: Such an effect might be relevant for the 12.7-MeV state in ^{12}C.

Redish: Are the harmonic oscillator form factors you use adequate? Does the electron scattering test them sufficiently well?

Olmer: The electron scattering data cover about the same range in momentum transfer.

COUPLING BETWEEN LOW-LYING NUCLEAR STATES

Munetake Ichimura
Institute of Physics, University of Tokyo
Komaba, Tokyo 153, Japan

ABSTRACT

Various treatments for nuclear direct reactions are briefly reviewed and some problems inherent to them are observed. Effects of coupling with deuteron breakup channels, discrete inelastic channels and rearrangement channels are discussed with respective examples such as ^{58}Ni(d,p)^{59}Ni(p$_{3/2}$), ^{24}Mg(p,d)^{23}Mg(1/2^{+}), and ^{208}Pb(p,t)^{206}Pb(3^{+}).

VARIOUS TREATMENTS FOR NUCLEAR DIRECT REACTIONS

I would like to start with a brief sketch of different kinds of treatments for nuclear direct reactions, and observe some topical problems each treatment involves. It must help us understand effects of coupling between low-lying states.

1) DWBA

Most commonly used method of analysis for nuclear direct reactions is distorted wave Born approximation (DWBA), which takes account of one-way coupling from the initial to the final channel, as is shown in Fig. 1.

It is noted that DWBA is not a uniquely defined treatment. If one classifies it by choice of distorting potentials, one may count up following variations.

a. Conventional DWBA

Usually the optical potentials, U_{opt}, are used as the distorting potentials both for the initial and final channels (the optical potentials of the excited channels are usually assumed to be the same as that of the corresponding elastic channel). This method is referred to as conventional DWBA.

b. Bare potential DWBA (BPDWBA)

The optical potential, U_{opt}, is used for the distorting potential of the initial channel, while for the distorting potential of the final channel, use is made of the bare potential, U_{bare}, the distorting potential used in coupled channel (CC) calculations. This type of DWBA was called asymmetric distorted wave approximation (ADWA) by Satchler[1]. Ascuitto et al.[2] recently revived it with successful analysis of heavy-ion inelastic scatterings and it was called bare potential DWBA by Kubo and Hodgson[3].

c. First order Born approximation of CC equations
    A sort of DWBA in which the distorting potentials are the
bare potentials for both the initial and final channels is nothing
but the first order Born approximation of CC equations.

d. Johnson-Soper's procedure (Adiabatic approximation)
    For deuteron involved reactions such as (d,p) and (p d),
Johnson et al.[4] proposed a sort of DWBA method in which effects of
deuteron breakup process are effectively included by an adiabatic
approximation to the CC equation. In its simplest version (zero-
range DWBA), the distorting potential for the deuteron channel is
taken to be the sum of the optical poentials for the proton and the
neutron with the half of the deuteron energy, $U_p + U_n$. This is so-
called Johnson-Soper's procedure.

e. Dirac phenomenology
    This will be discussed by Shepard[5] later.

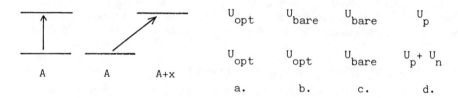

U_{opt}	U_{bare}	U_{bare}	U_p
U_{opt}	U_{opt}	U_{bare}	$U_p + U_n$
A   A   A+x
a.	b.	c.	d.

Fig. 1   Different kinds of DWBA

    Concerning with the title of this session, it is noted that
the effective interaction which should be used or which can be
extracted depends on the types of DWBA used.
    Concerning with the subject of my talk, one must note that
effects of coupling with low-lying states depends on the types of
DWBA, since the distorting potentials partialy take into account
the effect of the couplings according to which potentials being
used.
    For instance, Johnson-Soper's procedure approximately takes
into account the effects of the coupling with the deuteron continuum states on (d,p) reactions within the framework of DWBA. In
the analysis of $^{40}Ca(^{16}O,^{16}O')^{40}Ca(3^-;3.74MeV$ and $5^-;4.49MeV)$ done
by Ascuitto et al.[2], BPDWBA worked rather well even though the
strong coupling between the 3^- and the ground channels had been
known by CC calculation and failure of conventional DWBA had been
reported[6]. This was because the coupling effect was taken into
account rather well in the initial channel distorted wave through
the optical potential. The conventional DWBA erronously included
the coupling effect in the final channel. Ichimura and Kawai[7]
investigated the validity of BPDWBA for (d,p) reactions when the

d- and p-channels are strongly coupled and found that BPDWBA does not work well in this case contrary to the case of heavy ion inelastic scatterings. Detailed consideration for BPDWBA was given by Kubo and Hodgson[3], and Ichimura and Kawai[7].

One can also classifies DWBA by choice of effective interactions. In microscopic analysis of inelastic scattering, use of free nucleon-nucleon(N-N) t-matrix is called distorted wave impulse approximation (DWIA). Use of N-N G-matrix in nuclear medium with energy and density dependence was discussed by Kelly[8] just before. Sometimes the N-N G-matrix is also used in place of the n-p interaction V_{np} in (d,p) and (p,d) reactions[9,10]. However, contrary to the inelastic scattering, we[11] think that this procedure can not be justified for such transfer reactions since it overcounts V_{np} because the deuteron wave function has already taken into account the repetition of V_{np} while the G-matrix includes the free N-N t-matrix as the leading term. The effective interaction responsible for transfer should, of course, be energy and density dependent, but it may not be approximated by the N-N G-matrix used in the analysis of inelastic scatterings and calculation of optical potentials. An effective interaction for deuteron stripping reaction, which represents the Pauli effects, has been derived by Johnson, Austern and Lopes[12], who pointed out large difference in energy dependence between their effective interaction and the N-N G-matrix.

2) Two step approximation

The second order DWBA is also often used, and the processes dealt with this method is usually identified as "two-step process" (see Fig. 2). When transfer prosesses take place, it is known[13] that there are number of different forms of expressions for the two-step process, namely, prior-post, prior-prior, post-post and post-prior forms corresponding to the post and prior forms of DWBA. One must keep in mind that distinction between one- and two-step processes depends on the forms mentioned above and thus the number of steps is not well defined concept[13].

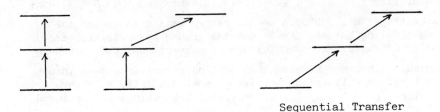

Sequential Transfer

Fig. 2   Some examples of two-step approximation

3) Coupled channel Born approximation (CCBA)

This approximation is illustrated in Fig.3. Strongly coupled channels are treated by CC method while weak but crucial process is treated by DWBA. This is often applied to transfer reactions, but scarecely used, to my knowledge, for inelastic scatterings as is shown in Fig. 3(b).

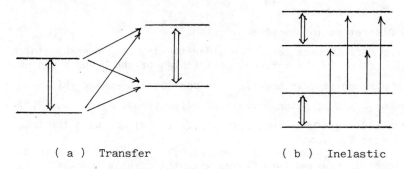

(a) Transfer   (b) Inelastic

Fig. 3   Some examples of CCBA calculations

4) Coupled channel (CC) analysis

When several channels including the initial and the final ones are strongly coupled (see Fig. 4) one should solve the full coupled channel equations. Let us discuss the coupling between inelastic channels and between rearrangement channels, separately.

continuum

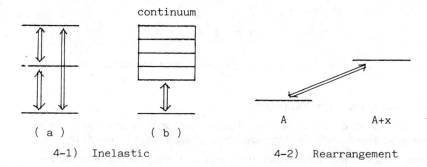

(a)   (b)   A   A+x

4-1) Inelastic   4-2) Rearrangement

Fig. 4   Some examples of CC calculations

4-1) Coupling with inelastic channels

 a. Coupling between discrete states

Strong coupling between discrete (bound) state channels is usually treated by standard coupled channel (CC) method[14].

 b. Coupling with continuum states

Recently much interest is growing up on the coupling with the

196

continuum states, especially with deuteron breakup states, d^*.

b-1) Coupled discretized continuum channels (CDCC) method[15]

In actual calculations one needs to approximate the continuum spectra by a finite number of discrete states. Such procedure is called coupled discretized continuum channels (CDCC) method[16]. Two methods of discretization is commonly used.

I. L^2- discretization[17]

II. Discretization by momentum-bins[16,18,19]

For definiteness let us consider the deuteron breakup states. In method I, the internal hamiltonian of the p and n system, h_{np}, is diagonalized in the finite dimentional L^2-space which is spanned by a set of given normalizable functions, $\hat{\phi}_n (n = 1,...,N)$. The obtained eigenfunctions, $\phi_i (i = 1,...,M < N)$ are used for the channel wave fuctions.

In method II, the relative momentum, k, between p and n in the range [0,K] is divided into N-bins with the size Δk and the averaged wave function

$$\phi_i = \frac{1}{\sqrt{\Delta k}} \int_{k_i - \Delta k/2}^{k_i + \Delta k/2} \phi(k) dk \qquad (1)$$

with energy $\varepsilon_i = \hbar^2 k^2/2\mu$ is used to specify the i-th discrete channel where $\phi(k)$ is the solution of the equation

$$h_{np}(k) = \frac{\hbar^2 k^2}{2\mu} \phi(k) \qquad (2)$$

The applicability of this method is extensively studied by Yahiro et al.[16].

b-2) Adiabatic approximation

Instead of discretizing d^* states, if one assumes all the d^* states are degenerate at the deuteron ground state energy, ε_d, then one can replace h_{np} by c-number ε_d and the coupled equations reduce to the simple form[19,20]

$$[E - \varepsilon_d - T_R - U_p(\vec{R} + \frac{\vec{r}}{2}) - U_n(\vec{R} - \frac{\vec{r}}{2})] \Psi(\vec{r},\vec{R}) = 0 \qquad (3)$$

This is an adiabatic approximation in the sense that the internal motion of p-n system relevant with the reaction is much slower than its center of mass motion. Johnson-Soper's procedure can be derived from this formalism by further approximations.

4-2) Coupling with rearrangement channels

There are essentially two ways of deriving the CC equations for the coupling between rearragement channels.

a. Channel coupling array (CCA) formalism

Rearrangement collisions are essentially a problem of many-body (at least 3-body) scattering theory, which involves formalistic difficulty. To manage it, someone thinks that one should rely on coupled integral equations with connected kernels, like Faddeev equation. A formalism based on the principle is CCA formalism[21], which gives the coupled integral equations for the component wave functions Ψ_μ's, where μ denotes the fragmentation and the total wave function Ψ is divided as

$$\Psi = \sum_\mu \Psi_\mu \tag{4}$$

In practice, one must, of course, introduce some approximations. For definiteness, let us hereafter take p + n + A(target) problem and consider only two fragmentations $\alpha = (p + n) + A$ and $\beta = (n + A) + p$ with setting $V_{pA} = 0$ [22]. In the bound state approximation (BSCCA) of Greben and Levin[21], the component wave functions are assumed to be

$$\Psi_\beta = \sum_i^N g_{\alpha i} \phi_{\alpha i} \quad \text{and} \quad \Psi_\beta = \sum_j^M g_{\beta j} \phi_{\beta j} \tag{5}$$

where $\phi_{\alpha i}$ and $\phi_{\beta j}$ are the bound state wave functions of p + n and n + A system, respectively (they could be discretized continuum states), the relative wave function $g_{\alpha i}$ and $g_{\beta j}$ are determined by the coupled equations (the differential form associated with the original coupled integral equations)

$$\langle \phi_{\alpha i} | E - H_\alpha | \Psi_\alpha \rangle = \langle \phi_{\alpha i} | V_{np} | \Psi_\beta \rangle$$
$$\langle \phi_{\beta j} | E - H_\beta | \Psi_\beta \rangle = \langle \phi_{\beta j} | V_{nA} | \Psi_\alpha \rangle \tag{6}$$

where $H_\alpha = T + V_{np}$ and $H_\beta = T + V_{nA}$, and T and V_{ab} are the total kinetic energy and the interaction between the particles a and b, respectively.

b. Coupled reaction channel (CRC) method

More commonly used method for numerical analysises is CRC[23], in which the total wave function is assumed to be

$$\Psi = \sum_i^N \chi_{\alpha i} \phi_{\alpha i} + \sum_j^M \chi_{\beta j} \phi_{\beta j} \tag{7}$$

and relative wave functions $\chi_{\alpha i}$ and $\chi_{\beta j}$ are determined by the coupled equations

$$<\phi_{\alpha i} | E - H | \Psi> = 0$$
$$<\phi_{\beta j} | E - H | \Psi> = 0 \qquad (8)$$

This method avoids the formal difficulty of many-body scattering theory by solving the Schrödinger equation within a truncated model space spanned by the bases

$$\phi_{\alpha i} \ \delta(\vec{r} - \vec{r}_\alpha), \ (i=1,\dots,N) \quad \text{and} \quad \phi_{\beta j} \ \delta(\vec{r} - \vec{r}_\beta), \ (j=1,\dots,M) \qquad (9)$$

where the point is that N and M are finite.

The important difference between BSCCA and CRC is that the non-orthogonality coupling terms appear in CRC but do not in BSCCA. This caused some controversies. Greben and Levin[21] maintained that the non-orthgonality terms in CRC is simply a theoretical artifact, while Austern[22] pointed out that BSCCA involves inconsistency within its formalism. The relation between CRC and BSCCA was lately clarified by Kawai, Ichimura and Austern[24]. Comparison with the exact calculations, CRC seems to be more favoured than BSCCA[25].

Now I come to the place to discuss the effects of coupling between low-lying states.

EFFECTS OF COUPLING WITH DEUTERON BREAKUP CHANNEL ON THE (d,p) REACTION

I first report an investigation for the effects of deuteron breakup channels on the reaction $^{58}Ni(d,p)^{59}Ni(1p_{3/2})$ at $E_d = 80.0$ MeV, which was carried out by Iseri, Yahiro and Nakano[26]. They analysed the reaction by CCBA, in which the deuteron ground state, d_0, and its breakup channels, d^*, are coupled as is shown in Fig. 5. The couplings are treated by CDCC with discretization by momentum -bins. Maximum momentum K is 1 fm^{-1} and number of bins N is 8. Angular momenta of breakup states are taken to be s- and d-waves.

In Fig. 5, the numerical result of CCBA is compared with those of the conventional DWBA, of the first order Born type DWBA (i.e. DWBA in which Watanabe potential (folding potential) is used in the d_0 channel), and of the adiabatic approximation for $d_0 \leftrightarrow d^*$ coupling part. One can clearly see appreciable effect of the coupling of the d^* channel. In Fig. 6 the cross sections of the process $d_0 \leftrightarrow d \longrightarrow p$ (process (1)) and of the process $d_0 \leftrightarrow d^* \longrightarrow p$ (process (2)) are separately shown. At backward angles the process (2) even exceeds the process (1). It is found that the adiabatic approximation well reproduces the CCBA result of the process (1) but not at

Fig.5 Differential
cross sections of
$^{58}Ni(d,p)^{59}Ni(p_{3/2})$

a.   CCBA
b.   CCBA (adiabatic)
c.   DWBA (conventional)
d.   DWBA (Watanabe pot.)

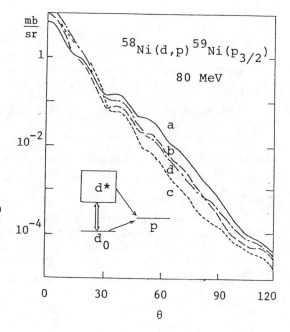

Fig.6 Differential
cross sections of
$^{58}Ni(d,p)^{59}Ni(p_{3/2})$

a.   CCBA [(1)+(2)]
e.   CCBA [(1)]
f.   CCBA [(2)]
g.   CCBA [(2)](adiabatic)

Note that the adiabatic
CCBA result for process (1)
is very close to line e,
and rather close to line c
of Fig.5.

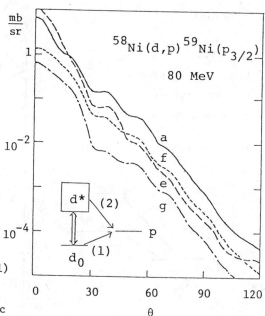

all that of the process (2). The conventional DWBA result is also
close to that of CCBA of the process (1), but the result of DWBA
with Watanabe potential is not. These are due to the fact that the
distorted wave of the elastic channel is rather well simulated by
the adiabatic approximation and by the optical model, though the
adiabatic approximation is very poor for the proceess (2).

EFFECTS OF COUPLING WITH DISCRETE INELASTIC CHANNELS

Here I woukd like to report a study of effects of the coupling of the discrete inelastic channels on the reaction ^{24}Mg(p,d)
^{23}Mg($1/2^+$;2.36MeV) at E_p = 95MeV[27,28], which is known as the famous
(or notorious) example of the failure of DWBA[28]. In the extensive
study of the reaction, Shepard, Rost and Kunz[28] concluded that CCBA
calculation did not help at all to explain the drastic discrepancy
between the result of conventional DWBA and the experimental data.
In CCBA calculation, they took into account the inelasic channels
of ^{24}Mg, (judging from the paper though not explicitly written).

Fig.7 Differential
cross sections of
^{24}Mg(p,d)^{23}Mg($1/2^+$)

a.   DWBA

b.   One + two step

c.   One − two step

d.   two step only

Experimental data from ref.27

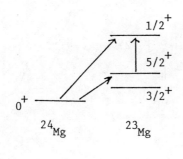

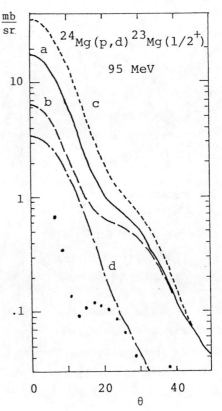

However, looking at the (p,d) cross sections[27] to various states of ^{23}Mg, one finds that the cross section to the $5/2^+$, 0.45 MeV state is much larger than those to the other levels. Therefore it seems worthwhile to estimate the process to the final $1/2^+$ state through the $5/2^+$ state.

Ichimura, Onishi and Yoshinada[29] performed the two-step approximation calculation for this process and found that its effect is very large though it could not eliminate the discrepancy. The results are shown in Fig. 7. At forward angles, reduction by almost factor 3 (or enhancement by factor 2, depending on the relative phase between the one- and two-step processes) are obtained by including the two-step process, while at larger angles its effect is very small in the cross sections. The spectroscopic facters are taken from the calculation of Chung and Wildenthal[30]. The form factor of the inelastic scattering ($5/2^+ \rightarrow 1/2^+$) is built by the collective model where the deformation parameter $_2$ is determined from the experimental E2 transition strength[31]. The optical potentials are taken from the global potential of Nadasen et al.[32] for the proton and that of Daehnick et al.[33] for the deuteron.

EFFECTS OF COUPLING WITH REARRANGEMENT CHANNELS

One of the typical examples which show the importance of the coupling with rearrangement channels is the sequential transfer of neutron in the (p,t) reactions to unnatural-parity states, since the one-step process is forbidden in the zero-range DWBA.

Here I would like to discuss an analysis by a two-step approximation for ^{208}Pb(p,t)^{206}Pb(3^+;1.34MeV) at E_p = 35MeV[34] done by Igarashi and Kubo[35]. They took account of the processes $p \rightarrow d^*$(s=0) $\rightarrow t$ and $p \rightarrow d^*$(s=1) $\rightarrow t$ as well as $p \rightarrow d_0 \rightarrow t$ process, where d_0, d^*(s=0) and d^*(s=1) are the ground state of deuteron, the singlet- and the triplet- states of unbound deuteron, respectively.

Recently Pinkston and Satchler[36] pointed out that taking only the $p \rightarrow d_0 \rightarrow t$ process introduces an artificial spin filter and the results could be misleading, and they stressed the importance to evaluate the effect of the $p \rightarrow d^*$(s=0) $\rightarrow t$ process. In fact, if the d_0 and d^*(s=0) states are degenerate and have the same spacial wave function (i.e. supermultiplet symmetry), the interference between the $p \rightarrow d_0^* \rightarrow t$ and $p \rightarrow d^*$(s=0) $\rightarrow t$ processes drastically reduces the two-step cross section of the transfer to the unnatural-parity states and exactly vanishes its analysing power. Thus the above reaction is also very interesting process as a good elucidative example for dependence of the choice of intermediate states in the two-step calculations.

Numerical results of Igarashi and Kubo [35] are shown in Fig. 8. The 3^+ state of ^{206}Pb is assumed to be of the pure configuration $(f_{5/2}^{-1}, p_{3/2}^{-1})$, and the wave functions of d_0, d^* and t are obtained by solving the Reid soft core potential. The continuum states are handled by discretization method by momentum-bins. The finite range calculation are performed both for the one- and two-step processes. Fig. 8. shows that the sequential transfer completely dominates and that the contribution from the $p \rightarrow d_0 \rightarrow t$ process much exceeds that from the $p \rightarrow d^*(s=0) \rightarrow t$ process but the interference effect between them is still appreciable. The contribution from the $p \rightarrow d^*(s=1) \rightarrow t$ is negligibly small.

Fig.8 Differential cross sections of ^{208}Pb(p,t)^{206}Pb(3^+)

a. total

b. $p \rightarrow d_0 \rightarrow t$

c. $p \rightarrow d^*(s=0) \rightarrow t$

d. $p \rightarrow d^*(s=1) \rightarrow t$

e. One step

Experimental data is taken from ref.34.

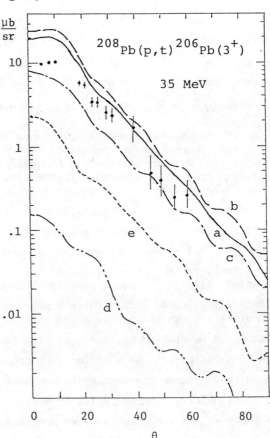

^{208}Pb(p,t)^{206}Pb(3^+)

35 MeV

As to effects of multistep processes at medium and high energies, in general, one may expect they decrease as the incident energy increases, but situation is not so simple and, to my knowledge, I do not have definite conclusions. For instance, in the excitation of higher members of rotational band, multistep processes are naturally essential[37]. A CRC calculation for the coupling between $p+^{12}C$ and $d+^{11}C$ channels at E_p = 185MeV showed its effects could be well simulated by the optical model[38]. Kawai et al.[39] reported that the effects of coupling between the p- and d-channels decrease more rapidly than those between the d_0 and d^* channels as the energy increase. Igarashi[40] reported that relative importance between the one-step $p \rightarrow t$ and two-step $p \rightarrow d_0 \rightarrow t$ processes does not change so much as the energy changes, and so on.

ACKNOWLEDGEMENTS

I would like to express sincere thanks to Y. Iseri, M. Yahiro and M. Nakano, and M. Igarashi and k-.I. Kubo for allowing me to use their results before publication. I also thank M. Igarashi, M. Kamimura, M. Kawai, N. Onishi, T. Terasawa, and K. Yasaki for stimulating and valuable discussions.

REFERENCES

1. G. R. Satchler, "Lecture in Theoretical Physics", vol.VIII, (ed. Kunz et al., The Univ. of Colorado Press, Boulder, 1966), p. 73.
2. R. J. Ascuitto, J. F. Petersen and E. A. Seglie, Phys. Rev. Letts. 41 , 1159 (1978).
3. K.-I. Kubo and P. E. Hodgson, Nucl. Phys. A366 , 320 (1981).
4. R. C. Johnson and P. J. R. Soper, Phys. Rev. C1 , 976 (1970). J. D. Harvey and R. C. Johnson, Phys. Rev. C3 . 636 (1971).
5. J. R. Shepard, Talk at this workshop. E. Rost, J. R. Shepard and D. Murdock, Phys. Rev. Letts 49 , 448 (1982).
6. K. E. Rehm et al., Phys. Rev. Letts. 40 , 1479 (1978).
7. M. Ichimura and M. Kawai, Prog. Theor. Phys. to be published.
8. J. Kelly, Talk at this workshop. J. Kelly et al., Phys. Rev. Letts. 45 , 2012 (1980).
9. B. Preedom, Phys. Rev. C5 , 587 (1972). S. Kosugi and T. Kosugi, Phys. Letts. 113B , 437 (1982).
10. P. J. van Hall, Eindhoven Univ. of Tech. preprint (1982). S. Kosugi and T. Kosugi, JAERI preprint (1982).
11. Discussions with Profs. T. Terasawa and K. Yazaki.
12. R. C. Johnson, N. Austern and M. H. Lopes, Phys. Rev. 26 , 348 (1982).
13. U. Gotz, M. Ichimura, R. A. Broglia and A. Winther, Phys. Rep. 16C , 115 (1975) and papers cited therein. M. Ichimura,

204

Proc . INS-IPCR International Sympo. Tokyo, (1975) p.547.

14. T. Tamura, Rev. Mod. Phys. $\underline{37}$, 679 (1965).

15. An excellent review was given by M. Kawai, Proc. the 3rd Varenna Conf. on Nuclear Reaction Mechanism. July, 1982.

16. M. Yahiro, M. Nakano, Y. Iseri and M. Kamimura, Prog. Theor. Phys. $\underline{67}$, 1467 (1982).

17. B. Anders and A. Lindner, Nucl. Phys. $\underline{A296}$, 77 (1978). M. Kawai, M. Kamimura and K. Takesako, Proc. 1978 INS International Sympo. Fukuoka (1978) p. 710, 711, 712.

18. G. H. Rawitcher, Phys. Rev. $\underline{C9}$, 2210 (1974). J. P. Farrel, Jr., C. M. Vincent and N. Austern, Ann. Phys. (N.Y.) $\underline{96}$, 333 (1976).

19. N. Austern, C. M. Vincent and J. P. Farrell, Jr., Ann. Phys. (N.Y.) $\underline{114}$, 43 (1978).

20. H. Amakawa, S. Yamaji, A. Mori and K. Yazaki, Phys. Letts. $\underline{82B}$, 13 (1979), H. Amakawa et al., Phys. Rev. $\underline{C23}$, 583 (1981) and papers cited therein.

21. J. M. Greben and F. S. Levin, Nucl. Phys. $\underline{A325}$, 145 (1979) and papeers cited therein.

22. N. Austern, Phys. Letts. $\underline{90B}$, 33 (1980).

23. G. H. Rawitcher, Phys. Letts. $\underline{21}$, 444 (1966). T. Ohmura, B. Imanishi, M. Ichimura and M. Kawai, Prog. Theor. Phys. $\underline{41}$, 391 (1969); $\underline{45}$, 347 (1970 , $\underline{44}$, 1242 (1970). W. R. Coker, T. Udagawa and H. H. Wolter, Phys. Rev. $\underline{C7}$, 1154 (1973).

24. M. Kawai, M. Ichimura and N. Austern, Z. Phys. $\underline{A303}$, 215 (1981).

25. J. Schwager, Ann. Phys. (N.Y.) $\underline{98}$, 14 (1976). Z. C. Kuruoglu and F. S. Levin, Brown University preprint,(1982).

26. Y. Iseri, M. Yahiro and M. Nakano, Kyushu Univ. preprint (1982) and private communicatiion. See also ref.15.

27. D. W. Miller et al., Phys. Rev. $\underline{C20}$. 2008 (1979).

28. J. R. Shepard, E. Rost and P. D. Kunz, Phys. Rev. $\underline{C25}$, 1127 (1982).

29. M. Ichimura, N. Onishi and K. Yoshinada, in progress.

30. W. Chung and B. H. Wildenthal, cited in ref.27.

31. R. Engmann et al., Nucl. Phys. $\underline{A171}$, 418 (1971).

32. A. Nadasen et al., Phys. Rev. $\underline{C23}$, 1023 (1981).

33. W. W. Daehnick et al., Phys. Rev. $\underline{C21}$, 2253 (1980).

34. W. A. Langford and J. B. McGrory, cited by N. B. de takacsy, Phys. Rev. Letts. $\underline{31}$, 1007(1973).

35. M. Igarashi and K.-I. Kubo, private communication. Partly published in Phys. Rev. $\underline{C25}$, 2144 (1982).

36. W. T. Pinkston and G. R. Satchler, Nucl. Phys. $\underline{A383}$, 61 (1982).

37. M. L. Barlett et al., Phys. Rev. $\underline{C22}$, 1168 (1980).

38. J. R. Comfort and B.C. Karp, Phys. Rev. $\underline{C21}$, 2162 (1980).

39 Ref. 15 and works cited therein.

40. M. Igarashi, private communication.

INFORMATION FROM QUASI-ELASTIC SCATTERING ON THE NUCLEON-NUCLEON INTERACTION IN NUCLEI

N. S. Chant
Department of Physics & Astronomy
University of Maryland, College Park, Maryland 20742

ABSTRACT

Methods of obtaining information on the nucleon-nucleon interaction in nuclei by means of quasifree knockout reactions are discussed. Examples of off-shell effects are shown and information on the radial region probed is presented. Various possible measurement strategies are considered as well as the effects of spin orbit distortions.

INTRODUCTION

As can be seen from the first order diagram shown in Fig. 1, the study of exclusive quasi-elastic scattering of nucleons from nucleons bound in nuclei can, at least in principle, yield information on both the momentum wave function of the struck particle and the nucleon-nucleon interaction responsible for its ejection. In Figs. 2 and 3, typical data are shown for nucleon knockout. Figure 2 shows 500 MeV ^{4}He(p,2p)^{3}H data from TRIUMF taken by Epstein, van Oers and co-workers.[1] Figure 3 shows typical data[2] for ^{40}Ca(p,2p)^{39}K taken at 100 MeV. Since the measurements cover a significant range of residual nucleus recoil momenta, these distributions largely reflect the struck nucleon momentum wave function and, while it is necessary to account for distortion effects, simple assumptions concerning the interaction appear to be sufficient. In contrast, few quasifree knockout experiments have been designed to specifically study the nucleon-nucleon interaction itself. This is partially due to experimental difficulties and partially due to limitations of available theoretical calculations. Since it is our hope that high quality data from the dual spectrometer system to be constructed at IUCF should remedy the first defect, and hopefully, stimulate greater theoretical activity, I should like to explore some ideas which may serve as useful orientation particularly for experimentalists who may be involved in this work.

Firstly, I should like to provide some background concerning current distorted wave impulse approximation calculations for quasifree knockout. Following this I should like to discuss some existing data and indicate the physics involved. Along the way I will mention a few measurements which could be improved or would become feasible using the spectrometers. Finally, I will comment on various relevant theoretical calculations.

206

THEORETICAL BACKGROUND

Although distorted wave impulse approximation calculations have been published for (p,2p) reactions using a local pseudo-potential, most current analyses of experimental data employ a factorized calculation[3] as outlined in Fig. 4, in which the two-body t-matrix is evaluated for the asymptotic kinematics. This, of course, is a half-shell t-matrix, since the struck nucleon is bound. Frequently, experimentalists make the additional approximation of replacing the half-shell t-matrix by a nearby on-shell value chosen according to some prescription such as equating the initial or final relative scattering energies. Nevertheless, according to the factorized DWIA it is the half-shell nucleon-nucleon cross section which is accessible to (p,2p) experiments.

Studies of the differences between on-shell and half-shell treatments of the nucleon-nucleon vertex in (p,2p) reactions have been reported by Redish et al.,[4] by Birrel et al.,[5] by Ioannides and Jackson[6] and by Miller and Thomas.[7] Quite significant off-shell effects are predicted for various (p,2p) experimental geometries even at incident energies of 300 MeV. In order to provide orientation for our ensuing discussion calculations from the work of Ioannides and Jackson, who use a Reid soft-core potential, are shown in Fig. 5 for $1p_{3/2}$ proton removal in the reaction $^{16}O(p,2p)^{15}N$ at 100 and 150 MeV. The variable θ_3 is the recoil angle of the residual nucleus. We will return to its significance later. The ordinate is simply the p-p cross section. For an incident energy of 150 MeV the ratio of P_{off}/P_{on} varies from 1.15 to 1.68 and it is clear that differences between the on-shell and half-shell calculations become significant well before the end of this range.

For the moment we will base our discussion on the factorized form of the theory and return to the possibility of eliminating the factorization approximation later. It must, of course, be emphasized that there are many other approximations and assumptions which we are taking for granted even assuming that higher order corrections to the DWIA are negligible. These include: the use of optical wave functions to represent all important multiple scattering effects of the target and residual nuclei; the use of a Woods-Saxon single particle wave function to represent the nuclear overlap; the assumption that the recoil mass is large enough to permit separation of the three-body final state; the neglect of Pauli effects on the scattering wave functions. This last correction has, in fact, been studied by Miller and Thomas[7] for a phenomenological separable non-local nucleon-nucleon potential. For 200 MeV (p,2p) data the effect was not very significant. It is, however, probably appropriate to regard this as an open question at the present time.

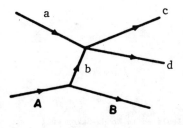

$$A(a,cd)B$$

PWIA

$$\frac{d^3\sigma}{d\Omega_c\,d\Omega_d\,dE_c} = K\,S_b\,|\phi(-\vec{p}_B)|^2\,\frac{d\sigma}{d\Omega}\Big|_{a+b}$$

K - kinematic factor

S_b - spectroscopic factor A→B+b

$\phi(-\vec{p}_B)$ - b-B momentum wave function

$\frac{d\sigma}{d\Omega}\Big|_{a+b}$ - half-shell a-b 2-body cross section for a+b → c+d

Fig. 1

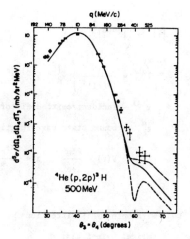

Fig. 2. Coplanar symmetric data; —— meson exchange corrected, DWIA with spin-orbit terms; --- same with no spin-orbit terms; -·-· Lim Eckart wave function.

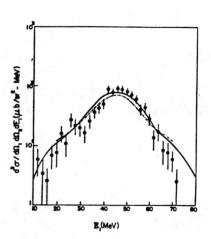

$\theta_1/\theta_2 = 41°/-41°$

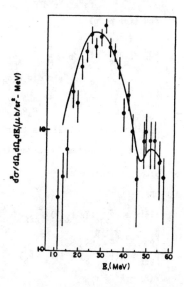

$\theta_1/\theta_2 = 52.2°/-29°$

Fig. 3.
Energy sharing distributions for ^{40}Ca(p,2p)^{39}K (1/2$^+$,2.52 MeV) Ep = 101.3 MeV; —— DWIA without spin orbit terms; -·-· DWIA with spin orbit terms.

DWIA

$$|\phi|^2 \to \sum_\Lambda |T^{\alpha L\Lambda}|^2$$

where

$$T^{\alpha L\Lambda} = \frac{1}{(2L+1)^{\frac{1}{2}}} \int \chi_d^{(-)^*} \chi_c^{(-)^*} \phi_{L\Lambda}^\alpha \chi_a^{(+)} d^3r$$

$\chi^{(\pm)}$ - incident/emitted distorted waves

$\phi_{L\Lambda}^\alpha$ - bound state wave function for b calculated in Woods-Saxon well

$$V(r) = \frac{V_0}{1+e^x} \; ; \quad x = \frac{r - r_0 A^{1/3}}{a}$$

$$(+V_{so} + V_{Coulomb})$$

DWIA with Spin-Orbit Distortions

Optical potentials:

$$V + iW \to V + iW + V_{so} \; \vec{\ell} \cdot \vec{s}$$

Distorted waves:

$$\chi^{(\pm)} \to \chi_{\rho\sigma}^{(\pm)} \qquad \rho, \sigma = \pm \tfrac{1}{2}$$

Hence:

$$T^{\alpha L\Lambda} \to T^{\alpha L\Lambda}_{\rho_a \sigma_a \rho_c \sigma_c \rho_d \sigma_d}$$

and (for p,2p)

$$\frac{d^3\sigma}{d\Omega_c d\Omega_d dE_c} = K' \, S_b \sum_{\substack{M \\ \rho_a \rho_c \rho_d}} \left| \sum_{\substack{\Lambda \rho_b \\ \sigma_a \sigma_c \sigma_d}} (L \Lambda \tfrac{1}{2} \rho_b | JM) \right.$$

$$\left. \times (2L+1)^{\frac{1}{2}} \, T^{\alpha L\Lambda}_{\rho_a \sigma_a \rho_c \sigma_c \rho_d \sigma_d} \langle \sigma_c \sigma_d | t | \sigma_a \rho_b \rangle \right|$$

$\langle | | \rangle$ - p+p $\frac{1}{2}$-shell t-matrix

Notice: factorization of amplitudes, not cross section.

For (p,2p), (p,pn), etc.

HALF OFF-SHELL

$$\langle |t| \rangle \equiv \langle \vec{P}_{on} | t(\epsilon(P_{on})) | \vec{P}_{off} \rangle \qquad \text{Fig. 4}$$

$$\vec{P}_{on} = \tfrac{1}{2}(\vec{P}_c - \vec{P}_d)$$

$$\vec{P}_{off} = \tfrac{1}{2}(\vec{P}_a - \vec{P}_b) = \tfrac{1}{2}(\vec{P}_a + \vec{P}_B)$$

$$\epsilon(P) = P^2 / 2 \, \mu_{cd}$$

INITIAL ENERGY PRESCRIPTION

$$\langle |t| \rangle \approx t(\epsilon(P_{off}), \theta(\vec{P}_{on}, \vec{P}_{off}))$$

FINAL ENERGY PRESCRIPTION

$$\langle |t| \rangle \approx t(\epsilon(P_{on}), \theta(\vec{P}_{on}, \vec{P}_{off}))$$

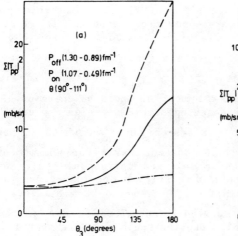
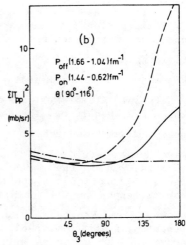

Fig. 5. Comparison of on-shell approximations with the half-off-shell prescription for the p-p matrix element $\Sigma|T_{pp}|^2$ in the fixed condition geometry with equal energies (FCG-A). The solid curve is the half-off-shell result, HSP, the dashed curve is the result for the final energy prescription (FEP) and the dash-dot curve is the result for the initial energy prescription (IEP). The range of values for the magnitudes of P_{off} and P_{on} and for their relative angle θ is indicated. The bound proton is removed from $1p_{3/2}$ state of ^{16}O with separation energy $E_s = 18.45$ MeV. The recoil momentum is $q_3 = 0.4$ fm^{-1}. The emitted protons have equal energies.

(a) Incident energy 100 MeV, P_{off}/P_{on} (1.21 → 1.81).
(b) Incident energy 150 MeV, P_{off}/P_{on} (1.15 → 1.68).

If one assumes the validity of the no-spin-orbit factorized DWIA expression for the three-body cross section, it is clear that one can determine the effective two-body pp cross section by dividing measured (p,2p) cross sections by $\sum_{\Lambda L} |T_L^{\Lambda}|^2$ and appropriate kinematic factors. Results of this type are shown in Fig. 6 for $^{40}Ca(p,2p)^{39}K(2.52$ MeV) at a number of incident energies. This is a $2s_{1/2}$ L=0 transition. The detected particle angles and energies have been selected subject to the constraint $P_B \approx 0$ which, in the plane wave limit, would lead to a constant value for the momentum wave function. Since the momentum selected corresponds to a maximum of the wave function for the L=0 transition considered, variations of the corresponding DWIA amplitude T_L^{Λ} tend to be minimized. The data shown, from Manitoba,[8] Maryland[9] and IUCF,[10] are plotted as a function of effective p-p scattering angle and have been arbitrarily normalized to optimize agreement with the on-shell free p-p cross sections which are shown as continuous lines. However, the corresponding spectroscopic factors are consistent with each other and with shell model predictions to within about 10% despite quite large changes in distortion effects with energy. (The ratio of DWIA/PWIA cross sections vary from ~1/100 at 45 MeV to ~1/4.5 at 150 MeV.) It is clear that, for the rather modest statistical accuracy of the data shown, the p-p cross sections extracted from the (p,2p) experiments are consistent with on-shell data. Notice also that corrections for spin dependent distortions are small. Clearly, in order to draw quantitative conclusions concerning the effective p-p interaction from data of this type improved statistical accuracy is desirable. With the dual spectrometer system under construction at IUCF data rates at least 100 times greater than achieved in the counter experiments shown should be possible. Thus, 10-fold reductions in the corresponding statistical errors can be expected.

Two additional comments on these data are in order. Firstly, according to the factorized DWIA approach, the measurement determines the half-shell p-p cross section. For the 150 MeV data shown, the value of P_{off}/P_{on} is only ~1.08 since the kinematics chosen actually minimize off-shell effects. Thus, measurements of this type serve to test our reaction calculations prior to studies employing geometries which enhance off-shell effects. Secondly, since the factorization approximation may or may not be valid, one should more properly regard any such measurement as determining an effective p-p cross section in the nuclear interior which is thus, at least in principle, fully off-shell. In order to understand better what this statement implies, we see in Fig. 7 histograms for the 150 MeV data of calculated contributions to the cross section as a function of nuclear radius compared with the nuclear density distribution. Very

Fig. 6(a). Factorization Test

Notes:

1) For an L=0 transition take data at various θ_c, θ_d such that $p_B = 0$.

2) Compute

$$Q(\theta^*) = \frac{d^3\sigma^{EXPT}}{d\Omega_c d\Omega_d dE_c} \bigg/ \left\{ K S_b \sum_\Lambda |T^{\alpha L \Lambda}|^2 \right\}$$

θ^* - two-body c.m. scattering angle.
{ } - varies slowly ($\propto$ K in PWIA)

3) No spin-orbit DWIA predicts $Q(\theta^*) = \frac{d\sigma}{d\Omega}\bigg|_{a+b}$

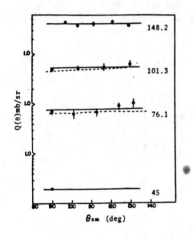

Fig. 6(b). $^{40}Ca(p,2p)^{39}K^*$ (2.52 MeV) L=0 factorization test at E_p = 45, 76.1, 101.3, and 148.2 MeV.

——— DWIA (no spin orbit) i.e., $d\sigma/d\Omega$ (pp)

---- DWIA (spin orbit distortions included).

212

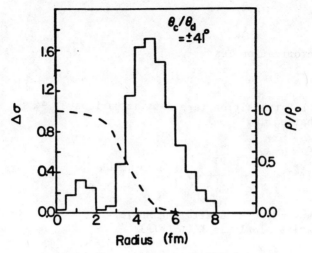

Fig. 7. Radial contributions to DWIA cross section for
^{40}Ca(p,2p)^{39}K(2.52 MeV,L=0) at 150 MeV. The detected
angles are ±41°. The detected energies are equal. The
broken curve is the ^{40}Ca density distribution.

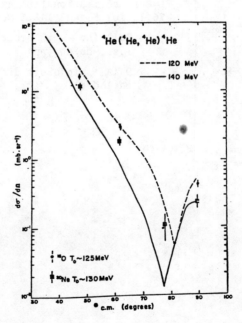

Fig. 8. Factorization test ^{16}O,^{20}Ne(α,2α) at 140 MeV, lines are
free cross section data.

roughly, the reaction takes place in a surface region having a density $\sim 10\%$ of central nuclear density or less.

As an aside, it is interesting to note that similar studies are possible for cluster knockout reactions[11,12] such as $(p,p\alpha)$ or $(\alpha,2\alpha)$. Results for $(\alpha,2\alpha)$ are shown in Fig. 8, where we see that the $\alpha+\alpha$ cross section extracted from $(\alpha,2\alpha)$ experiments is consistent with the angular dependence of the free cross section over almost 2 order to magnitude. My main point in including these data is as a reminder that many of the ideas discussed also apply to the interaction of clusters in or near the nuclear environment. For these data P_{off}/P_{on} is ~ 1.05 and the region of the nucleus where the scattering occurs has a density less than 0.1% of central nuclear density so that the good agreement between the $\alpha+\alpha$ and $(\alpha,2\alpha)$ data is no great surprise.

MEASUREMENT STRATEGIES

In order to examine systematically larger values of the ratio P_{off}/P_{on}, a variety of strategies are possible. One possibility[13] is to measure a coplanar symmetric angular distribution (CSAD) for equal emitted energies as a function of $\theta = \theta_1 = -\theta_2$. Typically, one might begin at angles such that $P_B \sim 0$ and P_{off}/P_{on} is small and take measurements at equal emitted proton energies for progressively smaller angles. This has the advantage that the effective p-p scattering angle is fixed at $90°$ which reduces spin dependent effects. However, both the nuclear recoil momentum P_B and the ratio P_{off}/P_{on} increase so that the shape of the distribution reflect both the struck nucleon momentum distribution and the variation of the nucleon-nucleon scattering. For the L=0 $^{40}Ca(p,2p)^{39}K(2.52$ MeV) reaction at 150 MeV the value of P_{off}/P_{on} reaches 1.49 at $\theta_1/\theta_2 = \pm 20°$ and 1.84 at $\pm 15°$. Measurements at such forward angles are prohibitively difficult using Germanium detectors but would become feasible with the dual spectrometers. In this connection it is worth noting that the minimum separation angle possible between the spectrometers is $\sim 30°$, making $\pm 15°$ the limit. It can be seen in Fig. 9 that, in addition to a surface contribution similar to the $\pm 41°$ case shown in Fig. 7, there is a $\sim 40\%$ contribution from radii below 2.5 fm where the density is roughly central density.

Various alternative strategies have been considered by Ioannides and Jackson.[14] One possibility (FCG-A) is to fix the magnitude of the recoil momentum P_B and to choose values of θ_1 and θ_2 in order to vary its angle from 0 to $180°$ for fixed detected particle energies. In a plane wave approximation this would eliminate variations due to the struck nucleon momentum wave

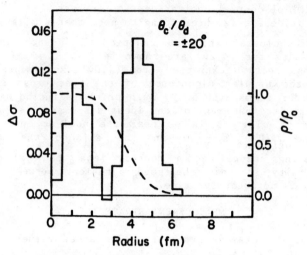

Fig. 9. Radial contributions to DWIA cross section for
^{40}Ca(p,2p) ^{39}K(2.52 MeV,L=0) at 150 MeV. The detected angles
are ±20°. The detected energies are equal. The broken curve
is the ^{40}Ca density distribution.

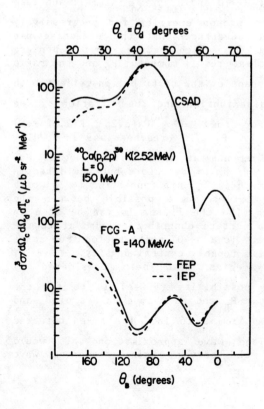

Fig. 10. Comparison of
CSAD and FCG-A predic-
tions for inital/final
energy prescriptions
(IEP/FEP). For FCG-A at
$\theta_B=0$, $\theta_c/\theta_d = \pm 56°$ and at
$\theta_B=180°$, $\theta_c/\theta_d = \pm 20°$.
Otherwise $|\theta_c| \neq |\theta_d|$.

function. The largest value of P_{off}/P_{on} reached occurs for $\theta_1 = -\theta_2$ at forward angles so that much the same surface localization considerations apply as shown for the coplanar symmetric angular distributions. A comparison of predicted cross sections for these two geometries, assuming two on-shell energy approximations to the p-p t-matrix, is shown in Fig. 10. These approximations, assuming either initial or final relative energy prescriptions, probably represent reasonable limits for half off-shell effects to be expected. One sees that in the FCG-A geometry there is still significant variation in the cross section arising from changing distortion effects even though the magnitude of the momentum transfer to the nucleus is fixed. Nevertheless, the procedure would still offer advantages if sensitivity to the various input parameters is, in fact, reduced. It may be worth pointing out that, even though the p-p scattering angle is close to $90°$, polarization analyzing powers are very large due mostly to distortion effects. This may prove useful. Owing to the symmetry of the CSAD analyzing powers are identically zero.

As far as I'm aware, no measurements using the FCG-A have been reported. CSAD results are shown[15] in Fig. 11 for $^{12}C(p,2p)^{11}B$-(ground state) at 100 MeV. This is, of course, a $1p_{3/2}$ transition. Shown are DWIA calculations using on-shell and half-shell calculated p-p cross sections. There are significant differences and neither agree with the data very well at forward angles. Similar problems occur for ^7Li which is shown in Fig. 12. Clearly, a careful remeasurement of some of the forward angle points would be very useful. However, a possible explanation of the discrepancy is a breakdown in the factorization approximation.

CORRECTIONS TO THE FACTORIZATION APPROXIMATION

The forward angle region, which is not well described by a factorized approximation for the 100 MeV ^7Li and ^{12}C data, is of course the interesting region where differences between on-shell and half-shell calculations are greatest. As we have already seen from the radial localization calculations, it is also a region where important contributions to the cross section arise from the nuclear interior. Clearly, the factorization approximation must be suspect. Indeed, it may well prove quite misleading to attempt to study off-shell effects and nuclear medium effects with a theory which, in essence, assumes the interaction takes place outside of the nucleus.

Non-factorized calculations, in which the p-p t-matrix is fully off-shell, were carried out a very long time ago by McCarthy and co-workers[16] using a rather crude local pseudo-potential. Their results are indicated by DWTA in Fig. 13. One sees significant differences with respect to the factorized DWIA result, particularly at forward angles. However, discrepancies between theory and data remain.

216

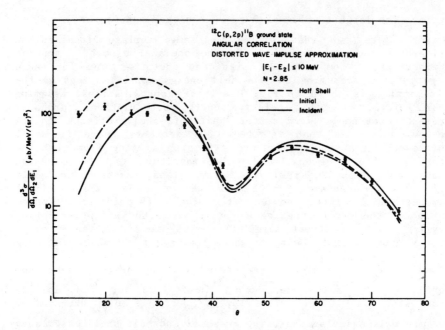

Fig. 11.   CSAD for ^{12}C(p,2p)^{11}B at 100 MeV.   The curves are DWIA
calculations.

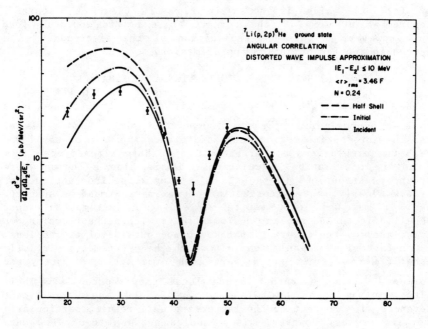

Fig. 12.   CSAD for ^{7}Li(p,2p)^{6}He at 100 MeV.   The curves are DWIA
calculations.

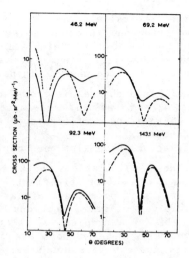

Fig. 13. Comparison of DWTA (full line) and DWIA (broken line) for (p,2p) on ^{12}C at various incident energies.

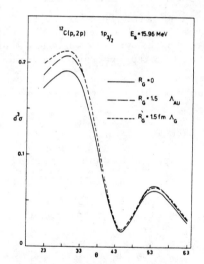

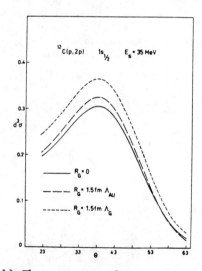

a) The cross section $d^3\sigma/d\Omega_1 d\Omega_2 dE$ for $1p_{3/2}$ knockout from ^{12}C at 160 MeV.

b) The cross section $d^3\sigma/d\Omega_1 d\Omega_2 dE$ for $1s_{1/2}$ knockout from ^{12}C at 160 MeV.

Fig. 14. Local energy approximation corrections to DWIA in di-proton approximation. Λ_G, method of Jackson, Λ_{Au}, method of Austern.

218

A first order treatment of fully off-shell effects was proposed by Redish.[17] In essence, the calculation determines momenta weighted by the distorted wave integrand in order to evaluate the off-shell p-p t-matrix. To my knowledge, this calculation has not yet been coded for the computer. However, rough estimates by Miller[7] for 200 MeV data at $\theta_1 = -\theta_2 = 30°$ yielded changes in polarization analyzing powers of less than 0.05 and changes in cross sections of less than 10%. These estimates were based on neglecting refraction and, I suspect, did not take into account effects due to imaginary parts of the three optical potentials.

Approximate calculations have also been outlined by Austern[18] and by Jackson.[19] These calculations both employ local energy approximations similar to earlier treatments for single nucleon transfer. Thus, derivatives of potentials are assumed to be small in comparison with derivatives of wave functions. Such approximations should thus work well at energies where the WKB method is satisfactory. Unfortunately, both authors are obliged to introduce additional assumptions. Austern does not treat the gradient of the bound state wave function properly while Jackson employs a di-proton approximation to describe the c.m. motion of the emitted nucleons. Results from Jackson's paper are shown in Fig. 14. For $1p_{3/2}$ knockout from ^{12}C at 160 MeV, corrections are of order 10-15% at forward angles using the Jackson calculation and are estimated to be somewhat smaller using an approximation to the Austern approach. For $1s_{1/2}$ knockout, assuming a 35 MeV binding energy, Jackson's correction is increased whereas Austern's is reduced which seems unphysical. A more careful evaluation of the Austern procedure is clearly of interest.

SPIN DEPENDENT EFFECTS

A number of (p,2p) and (p,pn) experiments using polarized incident protons have now been reported from TRIUMF[20,21] and IUCF.[22] In a factorized calculation, if spin-orbit distortions are negligible, the half-shell nucleon-nucleon cross section becomes

$$(\frac{d\sigma}{d\Omega})_{P_a} = (\frac{d\sigma}{d\Omega})_{unpol} (1 + (\vec{P}_a + \vec{P}_b)\cdot\vec{A} + \vec{P}_a\cdot\vec{P}_b C_{nn}) ,$$

where $\vec{P}_a$ is the incident beam polarization, $\vec{P}_b$ an effective struck nucleon polarization calculable in DWIA, and A and C_{nn} are the nucleon-nucleon analyzing power and spin correlation parameter, respectively. In principle, one can thus attempt to measure off-shell values for both A and C_{nn} since the reaction is providing us with a polarized proton target. In fact, for L=0 P_b=0. Thus, L=0

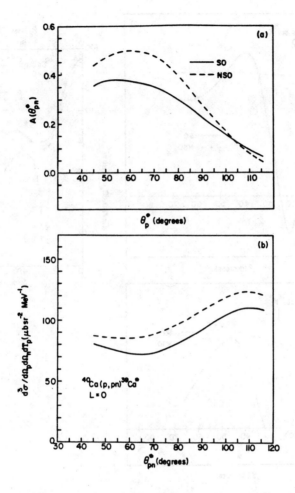

Fig. 15. Calculations for a quasifree angular distribution for
^{40}Ca(p,pn)^{39}Ca (2.47 MeV) at an incident energy of 150
MeV as a function of the effective p+n scattering angle.
The kinematics are chosen such that the residual nucleus
is at rest. (a) Polarization analyzing powers;
(b) Differential cross sections.

220

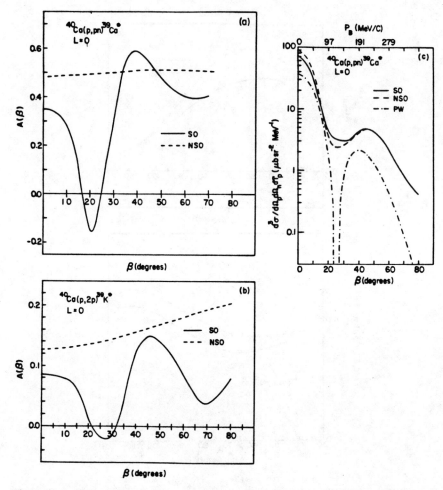

Fig. 16.   Calculations for L=0 nucleon knockout from   Ca at an
incident energy of 150 MeV using a noncoplanar geometry.
The detected particle angles are $\theta_c = 30°$, $\theta_d = 48.7°$ and

the detected energy of proton c is $T_c = 90.38$ MeV.

(a) Polarization analyzing power for $^{40}Ca(p,pn)^{39}Ca$
(2.47 MeV); (b) Polarization analyzing power for
$^{40}Ca(p,2p)^{34}K(2.52$ MeV); (c) Differential cross section
for $^{40}Ca(p,pn)^{39}Ca(2.47$ MeV).   The curve labeled PW is
the square of the struck neutron momentum wave function
in arbitrary units plotted as a function of P_B, the
recoil momentum of the residual nucleus.

221

analyzing powers should be identical to the free nucleon-nucleon values provided one is not too far off-shell. The most extensive studies using polarized incident protons are from the TRIUMF group. These 200 MeV data partially confirm the above analysis although in some angular regions one runs into problems which are unresolved. This is also true of the (p,pn) data from IUCF. As to corrections due to spin-orbit distortions, these are not very important for most of the TRIUMF data. There is some tendency to dilute the incident polarization, thus reducing the observed analyzing power, but not to introduce significant qualitative changes.

It is, however, dangerous to ignore spin-orbit distortions. In Fig. 15 a typical calculation[3] showing the dilution effect is shown. In contrast in Fig. 16 we see spectacular differences between spin-orbit and no-spin-orbit analyses which can arise.

CONCLUSIONS

1) New high precision (p,2p) and (p,pn) measurements, particularly at forward angles can potentially yield information on off-shell nucleon-nucleon matrix elements.

2) Surface localization calculations suggest a mix of half-shell and fully off-shell terms.

3) The kinematic flexibility of the reaction makes it possible to devise measurement strategies which minimize uncertainties from other effects.

4) Improved calculations which do not involve the factorization approximation are essential for meaningful analysis. As a first step use of the interactions developed for inelastic scattering studies would be instructive.

5) Methods such as the local energy approximation or the first order method of Redish may prove useful but need study. Redish's procedure may be very useful for experimental design in that one can estimate average values of the kinematic quantities. However, Miller has found situations where first order estimates may not suffice.

6) Polarized beams are certainly useful in unravelling the off-shell spin dependence but spin-orbit terms should be calculated. At present our ability to explain polarization analyzing power data is uncertain.

REFERENCES

1. M. B. Epstein et al., Phys. Rev. Lett. 44, 20 (1980); W.T.H. van Oers et al., Phys. Rev. C 25, 390 (1982).
2. C. Samanta, N. S. Chant, and P. G. Roos, to be published.
3. N. S. Chant and P. G. Roos, Phys. Rev. C 15, 57 (1977); N. S. Chant et al., Phys. Rev. Lett. 43, 495 (1979) and to be published.
4. E. F. Redish, G. J. Stephenson, Jr., and G. M. Lerner, Phys. Rev. C 2, 1665 (1970).
5. N. D. Birrel, I. E. McCarthy, and C. J. Nobel, Nucl. Phys. A271, 1665 (1976).

6. A. A. Ioannides and D. F. Jackson, Nucl. Phys. A308, 317 (1978).

7. C. A. Miller, Invited paper presented at 9th International Conference on the Few-Body Problem, Eugene, August 1980; C. A. Miller, in Common Problems in Low- and Medium-Energy Nuclear Physics, edited by B. Eastel, B. Goulard, and F. C. Khanna (Plenum Press, New York, 1979), p. 513.

8. K. H. Bray et al., Phys. Lett. B35, 41 (1971).

9. C. Samanta, Ph.D. Thesis, University of Maryland, 1981.

10. P. G. Roos et al., Phys. Rev. Lett. 40, 1439 (1978).

11. A. A. Cowley et al., Phys. Rev. C 15, 1650 (1977).

12. N. S. Chant et al., Phys. Rev. C 17, 8 (1978); C. W. Wang et al., Phys. Rev. C 21, 1705 (1980).

13. P. G. Roos, in Momentum Wave Functions-1976, edited by D. W. Devins (AIP Conference Proceedings No. 36, New York, 1977), p. 32.

14. A. A. Ioannides and D. F. Jackson, Nucl. Phys. A308, 305 (1978).

15. R. K. Bhowmik et al., Phys. Rev. C 13, 2105 (1976).

16. K. L. Lim and I. E. McCarthy, Nucl. Phys. 88, 433 (1966).

17. E. F. Redish, Phys. Rev. Lett. 31, 617 (1973).

18. N. Austern, Phys, Rev. Lett. 41, 1696 (1978).

19. D. F. Jackson, Physica Scripta 25, 514 (1982).

20. P. Kitching et al., Nucl. Phys. A340, 423 (1980).

21. L. Antonuk et al., Nucl. Phys., to be published.

22. J. Watson et al., Phys. Rev. C 26, 961 (1982) and to be published.

DISCUSSION

Austern: I wonder if I could ask you about the motivation for doing (p,2p). A long time ago we were told that it was going to be a magic way to get information about deep hole states of nuclei, and I think that has faded over the decades. Then the Maryland group began to tell us that it was a nice way to get information about the nucleon-nucleon vertex. But how distinctive is (p,2p) as a procedure for getting information about the nucleon-nucleon vertex? What can it give us that is not accessible from experiments that are in a less hostile environment?

Chant: I would say they are different environments, and that may prove to be useful. One of the features of the inelastic scattering studies is that the momentum transfer to the nucleon-nucleon vertex and the momentum transfer to the nucleus are coupled. That may be a useful feature or that may confuse the issue. In the (p,2p) studies, while there is an additional distorted wave coming into the picture, there is only a single-bound-state wave function in the problem, and one can hold the momentum transfer conjugate to that vertex constant, and look specifically at the nucleon-nucleon scattering. Secondly, one can do that for much lower momentum. The studies I indicated were for a fixed momentum transfer of 140 MeV/c, or q=0.7. As far as distortion is concerned, since the (p,2p) is much less mismatched, one is sensitive presumably to the elastic scattering predominantly at more forward angles, so one can hope for a little less sensitivity to the treatment of elastic scattering. I think those are all issues that have to be explored and maybe the bottom line will be that it really is easier to do the (p,p') experiment.

Redish: I'd like to make a comment relating what we've seen here to the G-matrix we saw earlier. One might imagine that the density dependence would be very important here because, as Nick showed, there are big contributions from the interior of the nucleus. I don't think this is actually the case. That's an interesting point in answer to Norman's question, because of the fact that we're measuring a different matrix element of the G-matrix. If you think in terms of a G-matrix for a finite nucleus, in the inelastic scattering you've got the projectile plane waves and then a bound state, initial and final, whereas in this case you've got a bound state initially but you have an outgoing continuum state. This means that the Q operator is projecting off states which are in a sense further away from the final state, and should have less of an effect. So what we may be able to do is make a continuous transition from a nearly free interaction to the G-matrix which has the density dependence in it.

Chant: Yes.

Kitching: Could you expand on the polarization measurements.

Chant: The point about the polarization measurements is basically that, in addition to the incident beam polarization, there is in fact an effective polarization of the struck nucleon due to localization effects. That's true for $\ell \neq 0$ transitions, and so the (p,2p) reaction provides us with a polarized proton target. That means that there are fewer games one can play. Basically, the study gives us information on the analyzing power and on the C_{nn} parameter. That expression is true in the factorized no-spin-orbit treatment, and works in a lot of cases. There are also a number of cases where one runs into problems, which may have to do with factorization. This does indicate some degree of richness with regard to spin structure. There are also special relationships between spin-orbit doublets. It is, however, important to take into account spin-orbit effects--sometimes they are small but sometimes they are spectacular effects.

Kitching; I would just like to remark on the data with polarized protons. In general one finds that at asymmetric angles the data don't agree with the theory, whereas they do at symmetric angles. One explanation of this could be refraction effects. A more intriguing possibility which Theo Maris has suggested is that if one puts the analyzing-power term to zero in the expression for the cross section you gave then one accentuates the C_{nn} term. This would require some kind of medium effect which would make the A term zero.

Chant: That also could be just a refraction effect, but yes, it could be some quenching in the medium, too.

EXCITATION OF M1 STRENGTH IN N=28 ISOTONES
BY INELASTIC PROTON SCATTERING AT 200 MeV*

N. Anantaraman, G.M. Crawley and A. Galonsky
National Superconducting Cyclotron Laboratory
Michigan State University, East Lansing, Michigan 48824

and

C. Djalali, N. Marty, M. Morlet, A. Willis, J.-C. Jourdain
Institut de Physique Nucléaire, Orsay, France

High-energy inelastic proton scattering at very forward angles has been found to selectively excite M1 states. We have used this reaction to study the M1 strength distribution in the N=28 isotones ^{48}Ca, ^{50}Ti, ^{51}V and ^{54}Fe. The measurements were performed with a beam of 201 MeV protons from the Orsay synchrocyclotron.

The nucleus ^{48}Ca has been a particularly interesting test case for the study of M1 transitions because of the simplicity of its structure. A strong sharp transition to a 1^- state at 10.21 MeV was first observed in (e,e').[1] In the present experiment, this state stands out very clearly above the background at very forward angles (see the spectrum at $4°$ shown in the top panel of the figure). Its angular distribution is very forward peaked and has been fitted with microscopic distorted wave Born approximation calculations. The measured shape is reproduced reasonably well. The ratio of experimental to predicted cross sections is 0.22 and 0.30, respectively, using a simple ($\nu f_{5/2}\nu f_{7/2}^{-1}$) and a more realistic full fp-shell wave function.

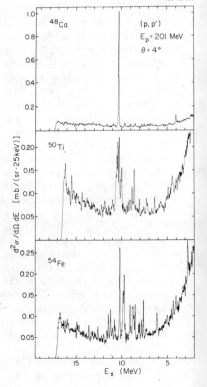

The effect of adding protons to the $f_{7/2}$ shell keeping a neutron number of 28 is shown in the lower two panels of the figure. The single peak in ^{48}Ca has become a more widely spread cluster of levels. These results agree with (e,e') observations for ^{50}Ti and ^{54}Fe,[2] but a broad feature seen in ^{51}V(p,p') is not seen in the (e,e') measurement.

*Work supported in part by NSF under grants PHY-78-22696 and INT-8116064

1. W. Steffen et al., Phys. Lett. 95B (1980) 23.
2. G. Eulenberg et al., Preprint (1982), and A. Richter, priv. comm.

IMPLICATIONS OF THE NON-LOCALITY IN THE EFFECTIVE N-N INTERACTION: NUCLEAR STRUCTURE AND SPIN OBSERVABLES*

W. G. Love

Department of Physics and Astronomy
University of Georgia
Athens, Ga 30602

J. R. Comfort

Physics Department, Arizona State University
Tempe, Az 85821

Some consequences of non-local spin-dependent terms in the effective N-N interaction are considered for nucleon-nucleus scattering. In particular, these terms are shown to probe the poorly understood spin $\otimes$ current modes of nuclear excitation. The simultaneous presence of this type of coupling together with appreciable spin $\otimes$ current transition densities can be observed experimentally[1] by measuring the differences between polarizations and analyzing powers.

In most of the presently used models of inelastic scattering this type of coupling arises[2] from knock-on exchange effects and although the size of these non-local terms are poorly understood, measurements of P-A offer the possibility of learning about them and the spin $\otimes$ current correlations which they probe. These nuclear structure correlations are untested by standard (e,e') measurements but do enter the (π,γ) reaction[3]; they also enter in β-decay where they are associated with "induced tensor" couplings[4].

Schematic calculations are given for $0^+ \rightarrow 1^+$ and $0^+ \rightarrow 0^-$ transitions to illustrate the role of spin $\otimes$ current couplings in calculations of P-A. DWIA calculations have been made for the $0^+ \rightarrow 1^+$ transition which support the more transparent schematic considerations.

*Work supported in part by the National Science Foundation.
[1]T. A. Carey, et al, Phys. Rev. Lett. 49, 266 (1982).
[2]W. G. Love, Part. Nucl. 3, 318 (1972).
[3]M. K. Singham and F. Tabakin, Phys. Rev. C21, 1039 (1980).
[4]B. R. Holstein, Phys. Rev. C4, 740 (1971).

MICROSCOPIC ANALYSIS OF ^{6}Li ($\vec{p}$,p)g.s. AND ^{6}Li ($\vec{p}$,p') (2.18 MeV)
ANALYZING POWERS AT 25, 35, AND 45 MeV*

C. H. Poppe, F. S. Dietrich and D. Rowley
Lawrence Livermore National Laboratory
University of California, Livermore, CA 94550

H. E. Conzett, D. Eversheim and C. Rioux
Lawrence Berkeley Laboratory
University of California, Berkeley, CA 94720

In the limit of the plane-wave Born approximation, analyzing powers
proton inelastic scattering result from interference between the central
spin-orbit parts of the effective nucleon-nucleon interaction. Because
real part of the effective spin-orbit interaction is generally much grea
than its imaginary part, one expects that the analyzing powers should exhi
a sensitivity to the imaginary part of the effective central interaction.
presence of distortion in the entrance and exit channels modifies t
conclusion somewhat, however for certain specific transitions this sensitiv
is still maintained.

To test this hypothesis we have measured analyzing powers for pro
scattering to the ground state (1^+; T=0) and first excited state (3^+; T
of ^{6}Li at incident energies of 25, 35 and 45 MeV using the polarized beam
the LBL 88 in. cyclotron. ^{6}Li was chosen because a recent study[1] of
Li isotopes described successfully various (p,p') and (p,n) cross sections
terms of a realistic effective force, however analyzing powers were
treated. In the present work, inelastic analyzing powers and cross secti
are compared to Born approximation calculations which use
density-dependent complex central interaction of Brieva and Rook and the r
spin-orbit and tensor interactions of Elliott.[2] Knockout excha
amplitudes are included in the calculation in the factorizat
approximation. Transistion densities are derived from the Cohen-Kur
p-shell wave functions suitably modified to reproduce measured B(E2
values. The effect of distortion on the calculations is studied for th
different optical potentials--two microscopic potentials based on the fold
model of Jeukenne, Lejeune and Mahaux and a phenomenological model deri
from previous fitting of elastic data. The predictions of these th
potentials for elastic analyzing powers are also compared to the pres
elastic data.

Agreement between the analyzing power data and the calculations is
very good for either the ground or first excited state, indicating s
serious deficiency in the model. Possible improvements will be discussed.

1. F. Petrovich, R. H. Howell, C. H. Poppe, S. M. Austin and G. M. Crawl
   Nucl. Phys. A383, 355 (1982).
2. G. Bertsch, J. Borysowicz, H. McManus and W. G. Love, Nucl. Phys. A2
   399 (1977).

*Work performed under the auspices of the U.S. Department of Energy
Lawrence Livermore National Laboratory under contract #W-7405-Eng-48.

TEST OF MICROSCOPIC THEORY WITH DATA FOR SIMPLE
TRANSITIONS IN THE ^{90}Zr(p,p') AND ^{89}Y(p,p') REACTIONS*

Alan Scott and W.G. Love
University of Georgia, Athens, GA 30602, USA

H.V. von Geramb
Theoretische Kernphysik, University of Hamburg, FRG

Proton transition densities for simple (p,p') transitions in
these nuclei have been 'calibrated' with (e,e') scattering results[1],
and two recent density-dependent (DD) N-N forces (Gl and G4) used in
different types of microscopic calculations at 159.5 MeV. The 8^+
state in ^{90}Zr is a pure proton state and the 6^+ state is dominantly
of the same configuration [in (e,e') scattering[1]], so proton
amplitudes were calculated with a $g^2_{9/2}$ density and Woods-Saxon
potential agreeing with electron scattering; first with exact
exchange and then using a pseudo potential for exchange amplitudes.
The computed asymmetries (A_y) had small differences; both
calculations are better fits to data than with a free N-N force, but
fits are still not fully satisfactory. Other calculations (ALN),
with direct input of (e,e') densities[2] (and neutron densities of the
same shape) and a pseudo potential for exchange, produce A_y shapes
differing from the exact calculations at larger angles; but much
larger differences in A_y shapes resulted from similar ALN calculations
with different DD forces (Gl and G4). Similar ALN calculations for
the 2^+_1 and 4^+_1 states in ^{90}Zr and simple transitions in ^{89}Y yield the
same conclusion; fits to cross sections are better with the Gl force,
A_y fits are better with the G4 force, and neither force is fully
satisfactory. Neutron strengths were extracted from these ALN fits.

These difficulties show A_y data are more sensitive tests of some
features and suggest improvements are needed in the neutron densities
and in the DD force. The influence of a 'wine-bottle' optical model
should also be investigated.

*Work supported in part by the National Science Foundation (U.S.), by
 BFMT contract 06HH726 (West Germany).

[1] J. Heisenberg, private communication.

[2] Program ALN, modification of Allworld program by J. Carr.

ENERGY DEPENDENCE OF THE RATIO OF ISOVECTOR EFFECTIVE INTERACTION STRENGTHS $|J_{\sigma\tau}/J_\tau|$ FROM 0° (p,n) CROSS SECTIONS

T.N. Taddeucci

Ohio University, Athens, Ohio 45701

ABSTRACT

Ratios of 0° (p,n) cross sections indicate that the spin-flip to non-spin-flip interaction strength ratio $|J_{\sigma\tau}/J_\tau|$ is approximately linear and target-mass independent for energies between 50 MeV and 200 MeV. To within a distortion-dependent factor of order unity, this latter ratio is well represented by $R(E_p) = E_p/(55.0 \pm 1.0 \text{ MeV})$.

The (p,n) reaction at intermediate energies has become an important tool for investigating spin-excitation strength distributions in nuclei. The circumstance responsible for this development is the dominance at energies larger than about 60 MeV of the isovector spin-flip effective interaction strength $V_{\sigma\tau}$ over the non-spin-flip interaction strength V_τ. The relative energy dependence of these two components of the effective interaction is dramatically illustrated by the ^{14}C(p,n) reaction. In Fig. 1 are shown ^{14}C(p,n) spectra obtained at a scattering angle of 0° at four different bombarding energies. The $0^+ \rightarrow 1^+$ transition to the 3.95-MeV state in ^{14}N is mediated by the $V_{\sigma\tau}$ component of the interaction while the $0^+ \rightarrow 0^+$ transition to the 2.31-MeV isobaric analog state

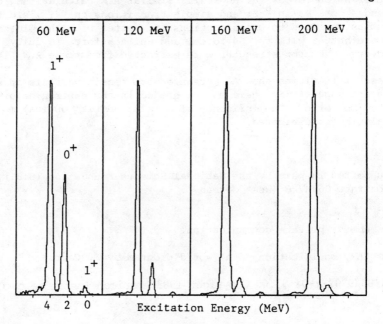

Fig. 1. Spectra for the ^{14}C(p,n) reaction at 0° for four energies.

0094-243X/83/970228-04 $3.00 Copyright 1983 American Institute of Physics

(IAS) is mediated by the V_τ component. At 60 MeV the IAS transition has 60% as much cross section as the transition to the 3.95-MeV state, but is only 5.5% as large at 200 MeV.

The relative energy dependence of $\Delta J^\pi = 1^+$ and $\Delta J^\pi = 0^+$ (p,n) transitions can be quantified in a systematic way be exploiting the proportionality between 0° (p,n) cross sections and the transition strengths for the analogous Gamow-Teller (GT) and Fermi (F) β decays. In the context of the DWIA, this proportionality takes the form

$$\sigma_\alpha(0°) \simeq K_\alpha(E_p)N_\alpha|J_\alpha|^2 B(\alpha), \tag{1}$$

where $K(E_p) = (E_i E_f/\pi^2)(k_f/k_i)$ is a kinematic factor, N_α is a distortion factor, J_α is the Fourier transform of the effective interaction at the momentum transfer q=0, $B(\alpha)$ is the β-decay transition strength, and the index α distinguishes between $\Delta J^\pi = 1^+$ transitions ($\alpha = \sigma\tau$ or GT) and $\Delta J^\pi = 0^+$ transitions ($\alpha = \tau$ or F). The β-decay strengths are related to measured ft values according to $B(F) + (1.25)^2 B(GT) = (6163.4 \text{ sec})/ft$, where $B(F) = N-Z$ for IAS transitions and is zero otherwise.

For even-A target nuclei with N-Z ≠ 0, the empirical quantity

$$[R(E_p)]^2 = [\sigma_{GT}(0°)/B(GT)K_{GT}(E_p)]/[\sigma_F(0°)/B(F)K_F(E_p)], \tag{2}$$

may be interpreted in terms of Eq. (1) as

$$R(E_p) \simeq |J_{\sigma\tau}/J_\tau|(N_{\sigma\tau}/N_\tau)^{1/2}. \tag{3}$$

A similar definition leading to the same interpretation may be made for odd-A targets.[1]

The quantity $R(E_p)$ as determined from (p,n) data on several odd-A and even-A targets is plotted in Fig. 2. This figure is taken from Ref. 1 and has been updated to include the results of recent measurements at 200 MeV. For energies larger than about 50 MeV, where the DWIA should be most valid, the data are well represented by the linear form

$$R(E_p) = E_p/(55.0 \pm 1.0 \text{ MeV}), \tag{4}$$

which is the straight line appearing in Fig. 2.

In Fig. 3 the experimentally-determined energy dependence of the ratio $|J_{\sigma\tau}/J_\tau|$ is compared to the results of a calculation by Brown, Speth, and Wambach (BSW)[2] and to values obtained from the t-matrix interactions of Love and Franey (LF)[3] and Picklesimer and Walker (PW).[4] The BSW curve incorporates a screening factor of $\gamma = 0.72$ that was omitted from their published calculations.[1,2,5] The experimental results are depicted as the shaded region. The width of this region approximately represents the energy and target-mass variation in the values of the distortion-factor ratio in Eq. (3). The t-matrix ratios are obtained by employing the asymptotic-energy approximation for the knock-on exchange contribution.[3] The high-energy trend of the data is poorly reproduced by the t-matrix

ratios. A recent fit to new phase-shift data, however, raises the
LF 210-MeV prediction from $|J_{\sigma\tau}/J_\tau| = 2.54$ to 3.05.

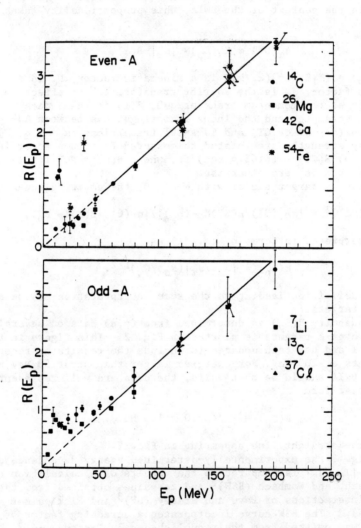

Fig. 2. The empirical quantity $R(E_p)$ as determined from 0° (p,n)
cross sections. The points for energies larger than 50 MeV are from
data obtained at IUCF. The lower-energy points are from (p,n) data
obtained at MSU, ORNL, LLL, Colorado, Tohoku, and Harwell by other
investigators.

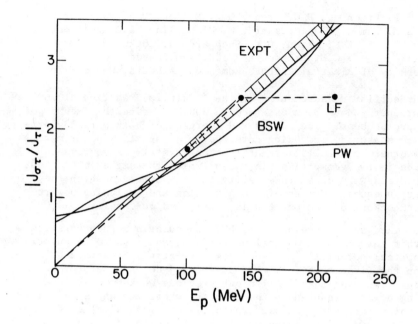

Fig. 3. Comparison of the experimentally-determined ratio $|J_{\sigma\tau}/J_{\tau}|$ to several predictions based on free N-N interaction studies.

The IUCF data presented here result from a collaboration of many workers: C.C. Foster, C. Gaarde, C.D. Goodman, C.A. Goulding, D.J. Horen, J.S. Larsen, T.G. Masterson, J. Rapaport, E. Sugarbaker, and T.P. Welch.

REFERENCES

1. T.N. Taddeucci et al., Phys. Rev. C 25, 1094 (1981).
2. G.E. Brown, J. Speth, and J. Wambach, Phys. Rev. Lett. 46, 1057 (1981).
3. W.G. Love and M.A. Franey, Phys. Rev. C 24, 1073 (1981).
4. A. Picklesimer and G.E. Walker, Phys. Rev. C 17, 237 (1978).
5. G.E. Brown and M. Rho, Nucl. Phys. A372, 397 (1981).
6. W.G. Love, private communication.

QUASI-ELASTIC SCATTERING OF POLARISED PROTONS AT 300 MeV*

P. Kitching, P.W. Green, C.A. Miller, D.A. Hutcheon,
A.N. James, W.J. McDonald, K. Michaelian, G.C. Neilson,
W.C. Olsen, D.M. Sheppard, J. Soukup,
G.M. Stinson and I.J. van Heerden
University of Alberta, TRIUMF, Edmonton, Alberta, Canada   T6G 2N5

Our earlier measurements[1] of the $^{40}Ca(\vec{p},2p)$ reaction at 200 MeV incident energy have been extended to 300 MeV using the TRIUMF polarised proton beam. The angles and energies of both outgoing protons were measured, one being detected in a magnetic spectrometer and the other in a NaI(Tℓ) scintillation crystal. Measurements were made in two geometries, in one of which the angles of the outgoing protons were close to equality and the other in which they were very unequal. Values of cross section and analysing powers for the knockout of protons for $1d_{\frac{3}{2}}$, $2s_{\frac{1}{2}}$, and $1d_{\frac{5}{2}}$ shells in ^{40}Ca were obtained and are compared to DWIA calculations incorporating spin orbit terms in the optical potentials, and also to calculations using optical potentials derived from a relativistic approach based on the Dirac equation. The expected j-dependence of the analysing power was seen. As in our previous work however, the analyzing powers measured in the geometry where the angles of outgoing protons are very unequal are consistent with a value of zero for the two body polarization parameter, P, while its value is expected to be ~0.2. Measurements for scattering from the $2s_{\frac{1}{2}}$ protons, which measure the value of P directly (if spin orbit distortions are neglected), show no such effect. We have not developed a satisfactory explanation for this behaviour. The possibility that refraction effects are responsible is being investigated.

*Supported by NSERC, Canada
1. L. Antonuk et al., Nucl. Phys. A370 (1981) 389

SESSION D
MACROSCOPIC CONCEPTS

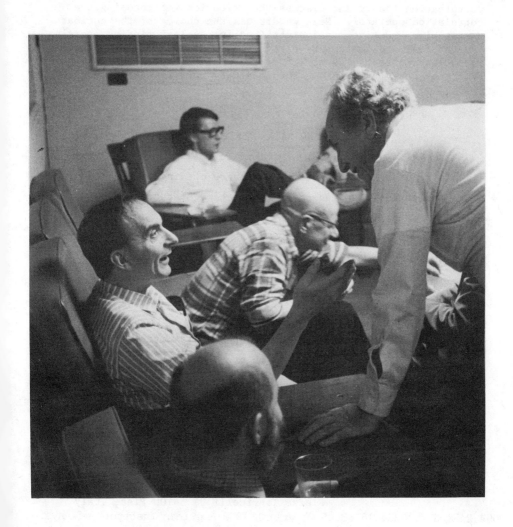

CORRELATED BASIS THEORY OF NUCLEON OPTICAL POTENTIAL IN NUCLEAR MATTER†

S. Fantoni*, B. L. Friman and V. R. Pandharipande
Department of Physics
University of Illinois at Urbana-Champaign
1110 W. Green Street
Urbana, Illinois 61801

I ABSTRACT

We give a brief and simple outline of correlated basis perturbation theory and discuss the criterion for choosing the correlation operator. Next we discuss the choice of the nuclear Hamiltonian and give results obtained for nuclear binding energies, and the real and immaginary parts of the nucleon optical potential in nuclear matter. The effect of the nonlocality of the real part of the optical potential on the imaginary part is also discussed.

II CORRELATED BASIS PERTURBATION THEORY

Correlated basis functions (CBF) are obtained by multiplying Fermi gas states with many-body correlation operators. The CBF state $|i\rangle$ is defined as

$$|i\rangle = |\Psi_i\rangle / \sqrt{\langle \Psi_i | \Psi_i \rangle} , \qquad (2.1)$$

$$|\Psi_i\rangle = G \, \Phi_i(n_i(k)). \qquad (2.2)$$

Here $n_i(k)$ are occupation numbers of the antisymmetric Fermi gas state Φ_i. The ground state $|0\rangle$ is obtained with

$$n_o(k<k_F) = 1 , \quad n_o(k>k_F) = 0 , \qquad (2.3)$$

---

† Work supported in part by NSF grants PHY 81-21399 and DMR 81-17182

* Present Address: Istituto di Fisica dell'Universita di Pisa
Piazza Torricelli 2, 56100 Pisa, Italy

and the particle, hole state $|\vec{p}\rangle$, $|\vec{h}\rangle$ etc. are obtained with

$$n_{\vec{p}}(k) = n_o(k) + \delta_{\vec{p}\vec{k}} \quad , \quad p > k_F; \tag{2.4}$$

$$n_{\vec{h}}(k) = n_o(k) - \delta_{\vec{h}\vec{k}} \quad , \quad h < k_F . \tag{2.5}$$

The correlation operator G is generally not unitary, thus the states $|i\rangle$ are not necessarily orthogonal. However we assume that they form a complete set.

When $G = 1$ the CBF theory reduces to the Brueckner-Bethe-Goldstone theory discussed by Mahaux[1] in this workshop. In the $G = 1$ basis the perturbation expansion is not convergent order-by-order, and we must do ladder sums etc. The Jastrow-CBF theory uses a product of pair correlation functions:

$$G = \prod_{a < b} f(r_{ab}) . \tag{2.6}$$

This theory was proposed by Feenberg and Clark[2], and has been applied to nuclear matter by Jackson et al[3]. The second order corrections in the Jastrow CBF theory are large, and the perturbation expansion is not obviously convergent.

In the present work we use

$$G = S\Pi F_{ab}, \tag{2.7}$$

where $S\Pi$ denotes a symmetrized product, and F_{ij} is a correlation operator

$$F_{ij} = \sum_{p=1,8} f^p(r_{ab}) O^p_{ab} , \tag{2.8}$$

$$O^{p=1,8}_{ab} = 1, \ (\tau_a \cdot \tau_b), \ (\sigma_a \cdot \sigma_b), \ (\tau_a \cdot \tau_b)(\sigma_a \cdot \sigma_b), \ S_{ab}, \ S_{ab}(\tau_a \cdot \tau_b)$$

$$\vec{L} \cdot \vec{S} , \ \vec{L} \cdot \vec{S} \ (\tau_a \cdot \tau_b) . \tag{2.9}$$

This correlation operator essentially describes the major part of the pair correlations induced by realistic nucleon-nucleon interactions. The Hamiltonian matrix H_{ij} is split up into two parts

$$H_{ij} = H_{o,ij} + H_{I,ij} \ . \tag{2.10}$$

Here $H_{o,ij}$ contains only the diagonal elements of H

$$H_{o,ij} = \langle i|H|j \rangle \ \delta_{ij} \equiv E_i^v \ \delta_{ij} \ , \tag{2.11}$$

where E_i^v are variational energies of the state $|i\rangle$. The perturbation $H_{I,ij}$ has only non-diagonal matrix elements

$$H_{I,ij} = (1-\delta_{ij}) \ \langle i|H-E|j \rangle \ . \tag{2.12}$$

The energy E, in the matrix elements of H_I, is due to the non-orthogonality of the CBF. In calculating second order corrections to E_i^v we must set $E = E_i^v$ (see ref. 4,5).

In order to understand the convergence criterion of CBF theory we expand the state $|i\rangle$ using the eigenstates $|\tilde{i}\rangle$ of the Hamiltonian

$$H|\tilde{i}\rangle = E_i |\tilde{i}\rangle \ , \tag{2.13}$$

$$|i\rangle = |\tilde{i}\rangle + \sum_j \alpha_{ij} |\tilde{j}\rangle \ . \tag{2.14}$$

The 0th order ground-state energy is

$$E_o^v = E_o + \sum_{i \neq o} \alpha_{oi}^2 \ (E_i - E_o) \ . \tag{2.15}$$

It is always above the true E_o, the error being of order α^2. The second-order contribution to E_o is

$$E_o^{(2)} = \sum_{i \neq o} \frac{|\langle i|(H-E_o^v)|o\rangle|^2}{E_o^v - E_i^v}$$

$$= - \sum_{i \neq o} \alpha_{oi}^2 (E_i - E_o) + \alpha^3 \text{ and higher terms.} \qquad (2.16)$$

Thus the error in $E_o^v + E_o^{(2)}$ is of order α^3. If the CBF states $|i\rangle$ are close to the true states $|\tilde{i}\rangle$ the α's are small and the convergence is good.

The best way to choose the $_{ij}$ may not be to minimize the E_o^v. The problem is that E_o^v is not sensitive to large admixtures of low lying states having $E_i \sim E_o$. A better way appears to be to minimize $\kappa^{(2)}$ defined as

$$\kappa^{(2)} = \sum_{i \neq o} \frac{|\langle i|(H-E_o^v)|o\rangle|^2}{(E_o^v - E_i^v)^2}$$

$$= \sum_{i \neq o} \alpha_{oi}^2 + \alpha^3 \text{ and higher terms.} \qquad (2.17)$$

It provides an unbiased measure of the magnitudes of the α's, and it is related to the κ of Brueckner's theory. In our calculation $E_o^{(2)}$ has the smallest magnitude when $\kappa^{(2)}$ is minimized.

III MODEL OF THE NUCLEAR HAMILTONIAN

It now appears that a reasonable nuclear Hamiltonian has the form

$$H = - \frac{\hbar^2}{2m} \sum_i \nabla_i^2 + \sum_{i<j} v_{ij} + \sum_{i<j<k} (v_{ijk}^{2\pi} + v_{ijk}^R) . \qquad (3.1)$$

Here v_{ij} is a realistic two-body interaction that fits the two-body scattering data. It can be written in the form

$$v_{ij} = \sum_{p=1,14} v^p(r) \, O_{ij}^p . \qquad (3.2)$$

The operators $O^{p=1,8}$ are given by eq. (2.9), and in the Urbana v_{14} interaction[6] used in this work the $O^{p=9,14}$ are taken as

$$L^2, \; L^2(\tau_1 \cdot \tau_2), \; L^2(\sigma_1 \cdot \sigma_2), \; L^2(\sigma_1 \cdot \sigma_2)(\tau_1 \cdot \tau_2), \; (L \cdot S)^2 \text{ and } (L \cdot S)^2(\tau_1 \cdot \tau_2).$$

$$(3.3)$$

These operators are chosen so that the scattering data can be fitted with rather weak $v^{9-14}(r)$. The total contribution of $\sum\limits_{p=9,14} v^P(r) \; 0^P_{ij}$ to the binding energy of nuclear matter is only 2.4 MeV per nucleon. The Paris interaction[7] has terms $(\nabla^2 \, v^x_{T,S}(r) + v^x_{T,S}(r)\nabla^2)$ in each spin-isospin (T,S) channel, inplace of the four L^2 terms of the Urbana v_{14}. These $v^x_{TS}(r)$ terms of the Paris interaction give a very large contribution ($\sim$ - 40 MeV per nucleon at $k_F = 1.6 \text{ fm}^{-1}$) which is hard to calculate accurately. The Paris and Urbana v_{14} interactions give quite similar phase shifts[6].

The $V^{2\pi}_{ijk}$ is the Fujita-Miyazawa two-pion-exchange three-nucleon interaction. Its parameters are taken from Model V of ref. 8). It is $\sim$30% more attractive than the Tucson model[8] of $V^{2\pi}_{ijk}$, and is necessary to obtain reasonable binding energies for ^{3}H and ^{4}He nuclei. The V^R_{ijk} is a phenomenological representation of the multi-pion exchange three-nucleon interaction. It is necessary to obtain a reasonable saturation density for nuclear matter. Its parameters are also taken from model V of ref. 8). The Hamiltonian (3.1) with Urbana v_{14} two-nucleon interaction and the model V three-nucleon interaction gives a fair description of the binding energies and radii of ^{3}H, ^{3}He and ^{4}He nuclei.

The second order CBF calculations are done with an approximate Hamiltonian in which the $V^{2\pi}_{ijk}$ is neglected and V^R_{ijk} approximated by a density-dependent two-nucleon interaction.

IV GROUND STATE CALCULATIONS

It is necessary to perform ground-state calculations to set up the correlated basis. The $f^P(r)$ are taken as solutions of approximate "Euler-Lagrange" equations[9] with constraints

$$f^p(r > d_p)^{\cdot} = \delta_{p1} \; . \qquad (4.1)$$

The 0th order E_o^v can be calculated quite accurately using chain summation techniques[8,9]. Fig. 1 shows E_o^v at k_F = 1.13, 1.33 and 1.53 fm^{-1} calculated as a function of d_t = the range of the tensor

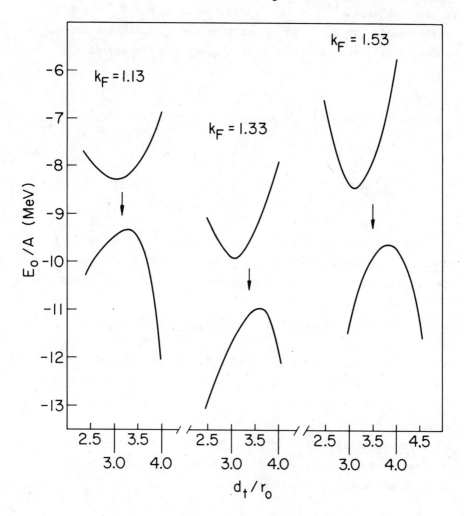

Fig. 1. The top curves give E_o^v while the bottom curves give $E_o^v + E_o^{(2)}$ as functions of d_t/r_o. The arrows mark the values of $d_t = d_{tp}$ at which $\kappa^{(2)}$ is minimum. Note that d_{tp} is a little larger than d_{tv} at which E_o^v has its minimum.

(S_{12} and $S_{12}\tau_1 \cdot \tau_2$) correlations. These calculations use the approximate Hamiltonian. The lower curves of fig. 1 show $E_o^v + E_o^{(2)}$ as a function of d_t, and the arrows mark the values $d_t = d_{tp}$ at which $\kappa^{(2)}$ is minimized. It is obvious that the convergence of the perturbation expansion becomes poor as one goes away from d_{tp}. Near d_{tp} however, the $\kappa^{(2)}$ is very small (table I) and $E_o^{(2)}$ is also fairly small. Thus we can expect that the perturbation expansion converges well for this choice of the correlation operator.

Table I

$E_o^{(2)}$ and $\kappa^{(2)}$ in nuclear matter at $k_F = 1.33 \text{fm}^{-1}$

d_t/r_o	$E_o^{(2)}/A$	$\kappa^{(2)}$
2.5	-3.8	0.032
3.0	-1.8	0.015
3.5	-1.7	0.011
4.0	-4.0	0.025

The 0th order energy $E_o^v(k_F, d_{tv})$ obtained[8] with the <u>full Hamiltonian</u> is shown in fig. 2 (d_{tv} denotes the d_t at which E_o^v has its minimum). The second order correction to $E_o^v(k_F, d_{tv})$ defined by

$$\Delta E_2(k_F) = E_o^v(k_F, d_{tp}) + E_o^{(2)}(k_F, d_{tp}) - E_o^v(k_F, d_{tv}), \qquad (4.2)$$

is estimated from the curves in fig. 1 (calculated with the approximate Hamiltonian). The dashed curve of fig. 2 shows $E_o^v(k_F, d_{tv})$ (full H) + $\Delta E_2(k_F)$ (approx. H), and the empirical saturation curve, obtained with $E_{o,eq} = -16\text{MeV}$, $k_{F,eq} = 1.33 \text{ fm}^{-1}$ and incompressibility K = 250 MeV, is shown for comparison. The differences between the empirical and the $(E_o^v + \Delta E_2)$ are not significant except at low densities.

V REAL PART OF THE OPTICAL POTENTIAL

Let $E_p(E_h)$ be the real part of the energy of the one particle (hole) state with momentum p(h). The single-particle energy is defined as

$$e(p) = E_p - E_o, \qquad (5.1)$$
$$e(h) = E_o - E_h, \qquad (5.2)$$

and the real part of the optical potential is then given by

$$e(k) = \frac{\hbar^2}{2m} k^2 + U(\rho, e), \qquad (5.3)$$

where ρ is the density of nuclear matter, and k can be either $p(>k_F)$ or $h(<k_F)$. Equation (5.3) gives us the wavelength of a particle of

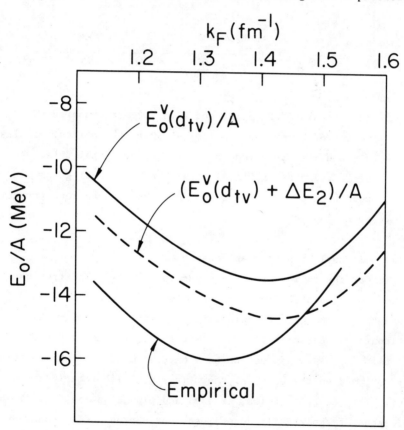

Fig. 2. The calculated saturation curves of nuclear matter in 0th and $0^{th} + 2^{nd}$ order CBF theory are compared with the empirical saturation curve.

242

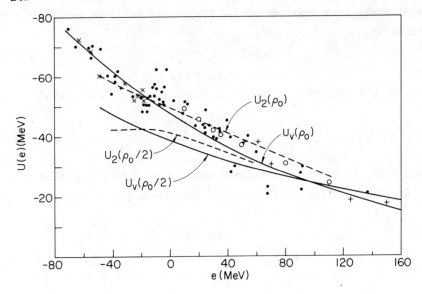

Fig. 3. The calculated $U_v(\rho_0,e)$ and $U_2(\rho_0,e)$ are compared with
depths of Woods-Saxon wells (shown by the points, circles,
x's and +'s) required to fit the experimental data. The
$U(\rho_0/2,e)$ crosses $U(\rho,e)$ at ~100 MeV indicating the
possibility of wine-bottle shaped nucleon-nucleus potentials
at $e \gtrsim$ 150 MeV.

energy e in nuclear matter at density ρ.

In CBF theory we obtain an expansion for U. The first term
$U_v(\rho,e)$ is calculated from the variational (i.e. 0th order)
E_o^v and E_k^v . The $U_v(\rho,k)$ calculated[10] with the approximate
Hamiltonian is shown in fig. 3. In ref. 10) the contribution of
$V_{ijk}^{2\pi}$ to $U_v(\rho_0,e)$ is assumed to be - 6MeV; a more realistic estimate
is -3MeV, and so the $U_v(\rho_0,e)$ shown in fig. 3 is probably off by
~3MeV. The points, circles, crosses and plus signs in fig. 3) show
the depths of empirically determined Woods-Saxon nucleon-nucleus
potentials from various compilations given in ref. 10). These
should be compared with the $U(\rho_0,e)$.

The $U_2(\rho,e)$ is calculated[5] with E_o and E_k correct up to second order, using a realistic estimate of the $V_{ijk}^{2\pi}$ contribution. The difference between $U_v(\rho,e)$ and $U_2(\rho,e)$ is rather small; the $U_2(\rho,e)$ is in marginally better agreement with the empirical well depths in the -60 to $+60$ MeV energy range.

The effective mass $m^*(e)$ is given by

$$\frac{m^*(\rho,e)}{m} = 1 - \frac{\partial U(\rho,e)}{\partial e} ; \qquad (5.4)$$

it determines the velocity of particles and density of states in nuclear matter. Both these quantities are important for calculating the imaginary part of the optical potential. The effective mass $m^*(\rho,e)$ obtained from $U_v(\rho,e)$ and $U_2(\rho,e)$ are shown in fig. 4 by the curves labeled $m_v^*(e)$ and $m_2^*(e)$ respectively; $m_2^*(e)$ has a peak in the region of $e \sim e(k_F)$. Evidence for such a peak was first pointed out by Brown et al[11], and it has been discussed extensively in the literature, particularly by Mahaux and collaborators[12]. Mahaux factorizes the effective mass into a k-mass $\tilde{m}$ and e-mass $\bar{m}$, so that $m^* = \tilde{m}\,\bar{m}\,m$. The CBF calculations are done "on the energy shell", and hence it is not possible to extract the $\tilde{m}$ and $\bar{m}$ from m_v^* and m_2^*. Numerically we find that the variational m_v^*/m is close to Mahaux's $\tilde{m}$, while the ratio m_2^*/m_v^* is close to his $\bar{m}$.

VI IMAGINARY PART OF THE OPTICAL POTENTIAL

The single-particle CBF states $|p\rangle$ decay into two-particle one-hole states $|p'p''h\rangle$ in second-order. Their life-time $\tau(p)$ is given by the golden rule

$$\frac{1}{\tau(p)} = \frac{2\pi}{\hbar} \int \frac{d^3h}{(2\pi)^3} \int \sin\theta\,d\theta\,d\phi\, |\langle p|(H-E_p^v)| \, p'p''h\rangle|^2 \rho(e(p) + e(h)), \qquad (6.1)$$

where $\rho(e(p) + e(h))$ is the density of final states at

$$e(p') + e(p'') = e(p) + e(h). \qquad (6.2)$$

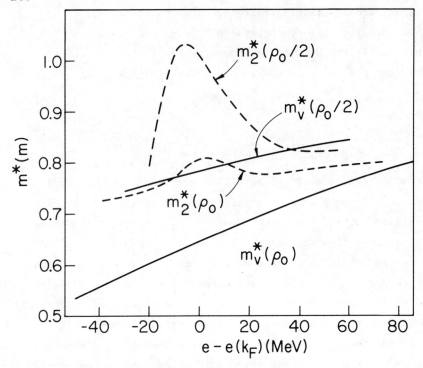

Fig. 4. The calculated effective mass $m^*(\rho,e)$ in nuclear matter in 0th and (0th + 2nd) order CBF theory.

The scattering responsible for this decay is illustrated in fig. 5. This lifetime can be associated with an imaginary part $W(p)$ of the single-particle energy

$$W(p) = 2\hbar\tau(p). \qquad (6.3)$$

We note that in 0th (and 1st) order CBF theory $W(p) = 0$. In the second order calculation of $W(p)$ we should use the 0th order $e_v(k)$ in eq. (6.2), and to calculate the density of final states, given by[13]

$$\rho(e(p) + e(h)) = m_v^*\left(\sqrt{\tfrac{1}{2}(p'^2 + p''^2)} \right) \frac{|\vec{p}' - \vec{p}''|}{4\hbar^2(2\pi)^3}. \qquad (6.4)$$

The finite lifetime implies that a beam of particles travelling

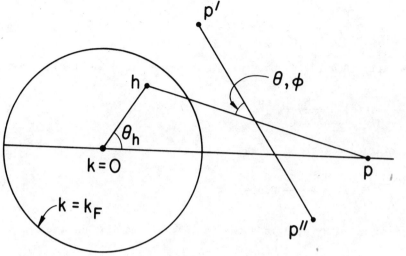

Fig. 5. A typical scattering leading to the decay of one particle
     states in nuclear matter.

with velocity $v(p)$ in nuclear matter is attenuated by a factor e
over a distance λ (called the mean-free path)

$$\lambda = \tau(p)\ v(p) = \frac{\hbar}{2W(p)}\ \frac{\hbar p}{m^*(p)}\ . \qquad (6.5)$$

Here $v(p) = \hbar p/m^*(p)$ is the velocity of the particles in nuclear
matter.

In the optical model elastic scattering is analyzed with a
Schrödinger equation using the bare mass m

$$\left[-\frac{\hbar^2}{m}\ \nabla^2 - U(e) - iW_o(e)\right]\ \Psi(r) = e\Psi(r), \qquad (6.6)$$

where $U(e) + iW_o(e)$ is the optical potential. The plane-wave
solutions of this equation are

$$\Psi(r) = \exp\left[i(p + \frac{i}{2\lambda})\ 2\right], \qquad (6.7)$$

$$p^2 = \frac{1}{4\lambda^2} + \frac{2m}{\hbar^2}(e + U_o(e)) \; , \tag{6.8}$$

(generally $1/(4\lambda^2) \ll p^2$, so this term can be neglected), and

$$\lambda(e) = \frac{\hbar}{2W_o(e)} \frac{\hbar p}{m} \; . \tag{6.9}$$

Comparing equations (6.5) and (6.9) for λ we obtain

$$W_o(e) = \frac{m^*(p)}{m} W(p(e)) \; . \tag{6.10}$$

The calculated $W_o(p,e)$ is shown in fig. 6. At low energies

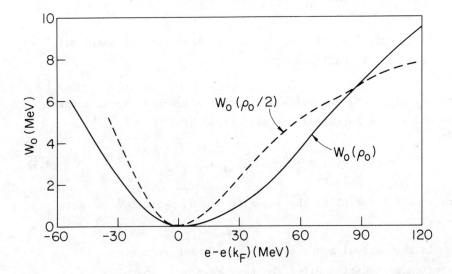

Fig. 6. The calculated $W_o(\rho_o)$ and $W_o(\rho_o/2)$ in nuclear matter.

$W_o(\rho_o/2,e) > W_o(\rho_o/e)$ indicating the well known enhanced surface absorption.

The effect of the factor m^*/m, which only recently was noticed independently by Negele and Yazaki[14] and by us[15], is to weaken $W_o(\rho_o,e)$ by a factor $\sim.7$. Calculations of $W_o(e)$ using the impulse approximation[16] also use m instead of m^* in the density of states (6.4). Thus these calulations are off by a factor $(m^*/m)^2$.

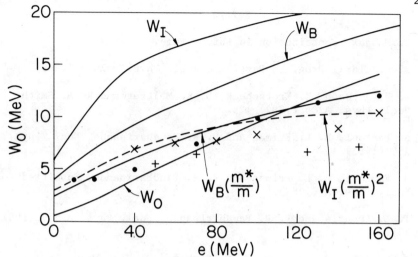

Fig. 7. The impulse approximation W_I, and the corrected $W_I(m*/m)^2$, the Brueckner theory W_B, and the corrected $W_B(m*/m)$ and the present W_O are compared with empirical Woods–Saxon strengths shown by •, + and x, in the "standard" model. The • are from the compilation by Bohr and Mottelson,[18] and the x and + are from the analysis of proton scattering off ^{208}Pb and ^{40}Ca respectively by the Indiana group.[19]

In fig. 7 we show, using an average value of m* =.7m, the results of older impulse approximation estimates[16] W_I and Brueckner theory estimates[17] W_B of W_O. We see that on correcting W_I by $(m*/m)^2$ and W_B by $(m*/m)$ we obtain a crude agreement between all estimates of $W_O(e)$. The theoretical $W_O(e)$ is a little larger than the strengths of Woods–Saxon potentials used in the so called "standard" optical model, but much smaller than that obtained in optical models with a "wine-bottle" shaped real potential[17].

References

1) C. Mahaux, contribution in this volume.

2) J. W. Clark, Prog. in Part. and Nucl. Phys. vol. 2 (1969).

3) A. D. Jackson, E. Krotscheck, D. E. Meltzer and R. A. Smith, Nucl. Phys. A386 (1982) 125.

4) S. Fantoni, B. L. Friman and V. R. Pandharipande, Nucl. Phys. A 386 (1982) 1.

5) S. Fantoni, B. L. Friman and V. R. Pandharipande, Submitted to Nucl. Phys. A (1982).

6) I. E. Lagaris and V. R. Pandharipande, Nucl. Phys. A 359 (1981) 331.

7) M. Lacombe, B. Loiseau, J. M. Richard, R. Vinh Mau, J. Côté, P. Pires and R. deTourreil, Phys. Rev. C21 (1980) 861.

8) J. Carlson, V. R. Pandharipande and R. B. Wiringa, submitted to Nucl. Phys. (1982).

9) I. E. Lagaris and V. R. Pandharipande, Nucl. Phys. A 359 (1981) 349.

10) B. Friedman and V. R. Pandharipande, Phys. Lett. 100B (1981) 205.

11) G. E. Brown, J. H. Gunn and P. Gould, Nucl. Phys. 46 (1963) 598.

12) R. Sartor and C. Mahaux, Phys. Rev. C 21 (1980) 2613; J. P. Jeukenne, A. Lejeune and C. Mahaux, Phys. Reports 25C (1976) 83.

13) E. Krotscheck and R. A. Smith, Phys. Lett. 100B (1981) 1.

14) J. W. Negele and K. Yazaki, Phys. Rev. Lett. 47 (1981) 71.

15) S. Fantoni, B. L. Friman and V. R. Pandharipande, Phys. Lett. 104B (1981) 89.

16) J. Dabrowski and A. Sobiczewski, Phys. Lett. 5 (1963) 87.

17) P. Schwandt, in this volume.

18) A. Bohr and B. Mottelson, Nuclear Structure, Vol. I (W. A. Benjamin, New York, 1969.)

19) A. Nadasen, P. Schwandt, P. P. Singh, W. W. Jacobs, A. D. Bacher, P. T. Debevec, M. D. Kaitchuck and J. T. Meek, Phys. Rev. C 23 (1981) 1023.

A PRAGMATIC APPROACH TO THE CONTINUUM SPECTRUM
IN THE QUASIFREE SCATTERING

G. Ciangaru
Department of Physics and Astronomy
University of Maryland, College Park, MD 20742

ABSTRACT

An expression for the continuum background in the quasifree scattering is given as the convolution of a number of rescattering steps of the quasifree particles on the residual nucleus. It is assumed that the rescattering probability can be expressed in terms of the inelastic scattering cross section of the quasifree particles. Comparison with the $^{58}Ni(p,2p)^{57}Co$ data at $E_0 = 198$ MeV supports the model prediction.

During the recent years a considerable attention has been given to the study of the continuum spectrum in the inclusive reactions. In contrast, still very little is known about the continuum spectrum in the exclusive (a,ab) quasifree (QF) reactions. So far, it was customary in the analyses of the knockout reactions to draw continuum background lines in an arbitrary manner. However, a Monte Carlo simulation showed[1] that the contribution from the background to the momentum distribution can be structured and therefore there is a distinct possibility of misinterpreting it as being part of the quasifree spectrum. The present contribution introduces a method which for the first time allows a quantitative description of the continuum spectrum in the quasifree scattering.

Our model is based on the following simple physical picture illustrated by the inset in Fig. 1 for a (p,2p) scattering. We postulate that the coincidence continuum spectrum results from the rescattering of the QF particles on the spectator (S) part of the target nucleus. Thus, in the case of the (p,2p) scattering for example, the process of forming the continuum is initiated by a QF collision which excites a two particle - one hole (2p-1h) doorway state having both protons unbound. Upon elevating above the Fermi level particles from different ν orbits, there are several such doorway states possible. The doorway states can decay via two routes: into the $p_1 + p_2 + S$ open channel, thus feeding the QF particle loci, or into more complicated p-h configurations involving a multiple scattering of the two protons on the S nucleus, thus feeding the continuum PEQ spectrum. Notice that, in contrast with Rogers and Saylor,[3] we are concerned here with the particles which, as a result of rescattering, are removed from the QF kinematic loci.

The above ideas can be implemented by extending the formalism of the Feshbach, Kerman, and Koonin[4] statistical theory of the multi-step direct two-body reactions. Thus, we assume that the energy averaged coincidence continuum cross section can be expressed as a convolution of the QF cross section with two chains of subsequent rescattering probabilities, one for each particle. We showed[5] that, in the case of the (p,2p) scattering, it is physically reasonable to further express the rescattering chains in terms of the probability that the two protons from the QF step, with energies E_1' and E_2' at angles Θ_1' and Θ_2', will inelastically scatter to the final angles Θ_1 and Θ_2 with energies E_1 and E_2. We assume that these probabilities are related to the cross section for inelastic (IN) scattering on the S nucleus by the expression $d^2\sigma^{IN}/d(\Omega_i-\Omega_i')dE_i(E_i';\theta_i,E_i)/(2\pi\sigma_{TOT}^{IN}(E_i))$, where $\sigma_{TOT}^{IN}(E_i')$ is the θ_i- and E_i-integrated inelastic scattering cross section of proton i with incident energy E_i' and $\theta_i = |\Theta_i-\Theta_i'|$ is the inelastic scattering angle.

In order to expedite the test of these theoretical ideas we consider here the case when the rescattering of one of the QF particles can be neglected. The (e,e'p) scattering is an example where this should be an excellent approximation. For such situations, the energy averaged coincidence continuum cross section is given by the expression

$$\left\langle \frac{d^4\sigma}{d\Omega_1 dE_1 d\Omega_2 dE_2} (\Theta_1,<E_1>,\Theta_2,E_2) \right\rangle$$

$$= \int d\Theta_2' \sin\Theta_2' \sum_\nu \frac{d^3\sigma_\nu^{QF}}{d\Omega_1 dE_1 d\Omega_2'} (\Theta_1,<E_1>,\Theta_2',E_2') \sigma_{TOT}^{IN}(E_2')^{-1}$$

$$\cdot \frac{d^2\sigma^{IN}}{d(\Omega_2-\Omega_2')dE_2} (E_2';\theta_2,E_2) \, ,$$

where the QF cross sections, corresponding to the initial instant when particles are not yet lost from the QF kinematic loci, assume distorted waves for the incident 0 particle and the 1 particle and plane waves for the 2 particle.

For the purpose of this contribution we consider the average yield in a 40 MeV wide slice through the two-dimensional energy (p,2p) spectra centered at $<E_1> = 130$ MeV, for two angles $\Theta_1 = -12°$ and $-30°$ and several positive angles Θ_2. The coincidence data is provided by our recent[2] coplanar measurement of the $^{58}Ni(p,2p)^{57}Co$ reaction using 198 MeV protons from IUCF, and the inelastic scattering cross sections are provided by the available experimental data for $E_2' = 62$ MeV protons on ^{54}Fe.[6] Assuming that the rescatter of the p_1 proton with $E_1 > E_0/2$ is mostly at small angles, we will neglect the contribution to the continuum at positive angles Θ_2 coming from the QF protons initially emitted at negative angles Θ_1'. Thus, the conditions which we have chosen will allow us to test the above simplified procedure for calculating the continuum spectrum in the (p,2p) scattering. An obvious prerequisite for solving the convolution integral is a correct evaluation of the doorway stage σ_ν^{QF} cross section. Since this is not an observable,

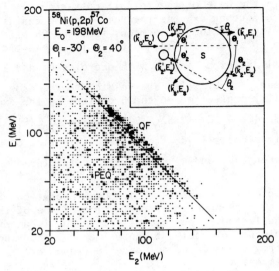

Fig. 1. Two-dimensional energy spectrum above the particle evaporation (equilibrium) region.[2] The diagonal line marks approximately the separation between the three-body QF region and the pre-equilibrium (PEQ) continuum region. The inset presents schematically our model of the mechanism of the concidence continuum.

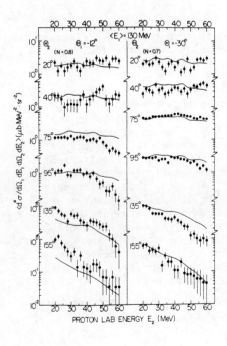

Fig. 2. A calculation (full line curves) of the continuum spectrum from the $^{58}Ni(p,2p)^{57}Co$ reaction at $E_0 = 198$ MeV. The data corresponds to the average yield in a slice through the two-dimensional energy spectra as explained in the text. The agreement in magnitude is reflected by the ratios $N = \sigma_{exp}/\sigma_{theor}$.

we derive the occupancy probabilities of the ν orbits from the comparison of a DWIA calculation[2] with the data corresponding to the particles detected on the QF loci shown in Fig. 1.   The convolution integral considering that most of the continuum particles come from the $\nu = 0f_{7/2}$ and $1s_{1/2}$ shell model orbits is compared with the data in Fig. 2.   The remarkable agreement which is obtained argues for further pursuing the ideas introduced here.

References

1.   K. Nakamura et al., Nucl. Phys. A268, 381 (1976).
2.   G. Ciangaru et al., Phys. Rev., to be published.
3.   J. G. Rogers and D. P. Saylor, Phys. Rev. C 6, 734 (1972).
4.   H. Feshbach, A. K. Kerman, and S. Koonin, Ann. Phys. 125, 429 (1980).
5.   G. Ciangaru et al., Phys. Rev. Lett., submitted for publication.
6.   F. E. Bertrand and P. W. Peel, ORNL Report No. 4469 (1970).

254

A GLOBAL STUDY OF THE p + ^{27}Al REACTION AT 180 MeV[*]

S-H. Zhou[+], K. Kwiatkowski, T.E. Ward, and V.E. Viola, Jr.
Indiana University, Bloomington, IN 47405

H. Breuer, A. Gokmen, and A.C. Mignerey
University of Maryland, College Park, MD 20742

G.J. Mathews
Lawrence Livermore National Laboratory, Livermore, CA 94550

ABSTRACT

A global study of the mass, energy and angular distributions
of all products formed in collisions of 180-MeV protons with ^{27}Al
is reported. These data are compared with semiempirical and
intranuclear-cascade-plus-evaporation calculations, as well as data
from recent fission-fragment angular correlation studies with
intermediate-energy light-ions incident on heavy target nuclei. It
is found that there is substantial evidence for enhanced energy
deposition in nucleon-nucleus collisions relative to the
predictions of the intranuclear cascade calculation.

EXPERIMENTAL PROCEDURES

In order to derive a broader undertanding of the salient
mechanisms which characterize nucleon-nucleus collisions at
intermediate energies, a global study of the p + ^{27}Al reaction has
been performed. The term global, as applied to this experiment,
implies the measurement of complete energy spectra, angular
distributions and isobaric cross sections for all fragments
produced in the reaction.

In the measurements reported here we have employed a
channel-plate-fast-timing detector with a silicon semiconductor
detector telescope to detect with discrete mass resolution all
reaction products from the 180-MeV p + ^{27}Al reaction. The studies
were carried out at the Indiana University Cyclotron facility where
a 200-400 nA beam of 180-MeV protons was used to bombard a 100
μg/cm^2 self-supporting Al target. Fragments were detected in a
164-cm diameter scattering chamber. The detector system consisted
of a channel-plate (CP) fast-timing start detector (located 13 cm
from the target) and a triple silicon semiconductor detector
telescope containing elements of thickness 50 μm, 500 μm and 5mm,
respectively. The 50 μm detector, placed 67 cm behind the CP
device, served as a timing stop detector. With this system a
timing resolution of <140 ps was achieved, yielding a mass
resolution of 0.4 - 0.8u (FWHM), depending on fragment mass and

[*]Work supported by the U.S. National Science Foundation, Department
of Energy, and Office of Naval Research.

[+]Permanent address: Instiute of Atomic Energy, Academica Sinica,
Beijing, PRC.

energy. A sample mass spectrum is shown in Fig. 1. This system provided complete definition of the energy spectra for all fragments with E/A > 0.05 MeV/u which stopped in the 50 μm detector. Complete spectra for energetic He, Li and Be nuclei which penetrated the 50 μm detector were determined from the triple silicon-detector telescope; hydrogen spectra for E < 35 MeV were similarly recorded; these will be reported in a more complete paper.

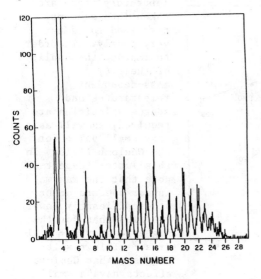

Figure 1. Mass spectrum from fragments with energies above 10 MeV observed at 20 deg in the reaction 180-MeV p + ^{27}Al.

Fragment energies were corrected for energy losses in the target, a 30-μg/cm^2 carbon foil on the CP detector and the 50-μg/cm^2 Au layer covering the 50 μm detector. All differential cross section values, $d^3\sigma/d\Omega dE dA$, were extrapolated systematically to zero energy to include the missing yield below the electronic cutoff energy, using a spectrum shape of the form P(E)αE$^{1/2}$ exp(-E/T). By the use of very thin foils neither the energy correction (<0.2 MeV maximum) nor the low-energy cross-section extrapolations (≤ 10 percent) are significant for A < 24. Because the recoil energies for a significant fraction of the yield for simple reactions such as (p,p') and (p,pn) fall below our lower detection limit, isobaric cross-sections for A = 26 and 27 used in these discussions were taken from the in-beam gamma ray studies of Ref. 1. Our data for A = 25 may be in error by as much as 50 percent due to similar effects, although our value is larger than that of Ref. 1.

DISCUSSION OF RESULTS

In Fig. 2 representative energy spectra for A = 7, 16 and 22 fragments at angles of 20, 40, and 70 deg are shown. The spectra are characterized by a broad peak at low energies followed by an exponential decrease with increasing energy, extending up to relatively high energies compared to kinematic predictions of cascade models.[2-4] These spectra exhibit many similarities with data from 2-GeV proton bombardments of ^{27}Al (Ref. 5). The slopes of the exponential tails become systematically steeper as a function of both increasing fragment mass and angle, suggesting the fragments are emitted from a hot, moving source.[5,6] However, attempts to fit all spectra with a uniform model using a spectral shape P(E)αE$^{1/2}$ exp[-E/T], where Coulomb effects, source velocity

256

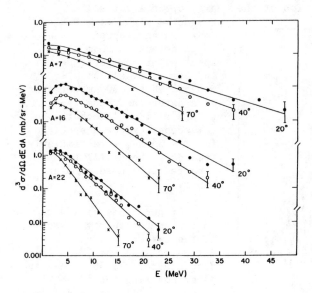

Figure 2. Fragment energy distributions, $d^3\sigma/d\Omega dEdA$, for fragments with A = 7, 16 and 22 observed at angles of 20, 40, and 70 in the 180-MeV p + ^{27}Al reaction. Lines are eye-guiders.

and temperature, and transformation to the laboratory frame are taken into account, described the data very poorly. In order to improve the quality of the fit, mass-dependent temperatures and source velocities are required, as well as a very small value for the Coulomb field. In this latter respect we note that the maxima in our spectra occur at significantly lower energies than for the higher energy p + ^{27}Al data.[5] This result suggests that Coulomb effects have a small influence on the emission mechanism in our case, perhaps indicative that n-fragment breakup processes (n > 3) exhibit considerable strength in these interactions.

It should be noted that rather substantial yields of relatively energetic A = 12 and 16 fragments are observed at 90 degrees. Recent measurements[7] have shown that these fragments continue to be observed as far back as 170 deg and furthermore, instead of having complementary heavy partners, these fragments are associated primarily with H and He ions. This result again suggests that a significant fraction of events in these reactions involve large energy deposition collisions followed by multifragment breakup of the residual nucleus. Hence, for such a process a model based on statistical emission of fragments from an excited moving source is not appropriate.

In Fig. 3 the mass distribution of the reaction products with A ≥ 6 is shown. The fragment yields are spread broadly over the entire range of possible products, with a large contribution from relatively light fragments. Peak yields are found to be in the A = 20-22 region. Under the assumption that nucleon removal is correlated with excitation energy, the results again imply that relatively large energy deposition processes account for a substantial fraction of the reaction cross section. It is also noted that among the lighter fragments the 4n nuclei with A = 12

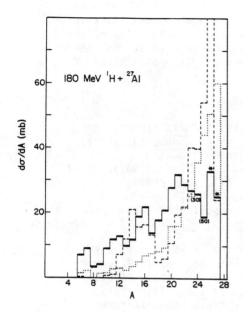

Figure 3. Isobaric cross sections for fragments from the 180-MeV proton + ^{27}Al reaction; ———, experimental data from this work; * - data from Ref. 1, ----- intranuclear cascade/evaporation code of Ref. 3-4, and ····· semiempirical calculation of Ref. 8.

and 16 have substantially enhanced yields, presumably due to nuclear Q-value effects.

Also shown in Fig. 3 are yield predictions based on intranuclear cascade/evaporation calculations[3,4] and the semiempirical estimates of Silberberg and Tsao.[8] The latter estimates clearly underestimate the probability for high energy deposition events leading to light fragment production. This shortcoming may significantly influence current calculations of cosmic-ray abundance anomalies[9], which depend on these calculated cross sections.

The intranuclear cascade calculations also over-predict the yields of products near the target; this is due to relatively low excitation energies of residual nuclei found in simple quasi-free scattering processes. The cascade calculation also exhibits a peak near A ~ 14 which arises from compound-nucleus-like products with high excitation energies. The experimental results suggest a much broader distribution of energy deposition in which the cascade events are substantially more effective in transferring linear momentum and excitation energy than predicted by the model. Specifically, the cascade overpredicts the probability for few-nucleon removal processes (A = 23-27) by about 30 percent, while failing to account for this same amount of cross section for lighter fragments (A = 6-22). Thus, it would appear from these comparisons that in addition to the basic nucleon-nucleon aspects of these collisions, as represented by the cascade code, some additional mechanism for energy dissipation must also be present. This conclusion is consistent with that of Ref. 10 which suggests that per nucleon, proton-induced reactions are more effective agents of linear momentum and energy transfer than previously believed.

Finally, we obtain a total reaction cross ssection for these data that ranges from σ_R = 370 mb, assuming all fragments with 6 ≤ A ≤ 9 have heavy partners, to σ_R = 394 mb assuming that all fragments with A $\geq$ 6 have no heavy partners. Recent experiments[7] favor the latter assumption. Within the uncertainties of the data these values are consistent with elastic scattering analyses which give σ_R = 410 mb[11] and compare favorably with calculated values[12,13] for σ_R.

REFERENCES

1. O. Artun, et al, Phys. Rev. Lett. 35, 773 (1975).
2. V.S. Barashenkov, et al, Nucl. Phys. A1867, 531 (1972).
3. K. Chen, et al, Phys. Rev. 166, 949 (1968).
4. G.J. Mathews, et al, Phys. Rev. C25, 2181, (1982).
5. G. Westfall, et al, Phys. Rev. C17, 1368 (1978).
6. A. Goldhaber, Phys. Rev. C17, 2243 (1978).
7. M. Walker, Indiana University, unpublished data.
8. R. Silberberg and C.H. Tsao, Ap. J. 25, 315 (1973) and private communication.
9. R.A. Mewalt, et al, Ap. J. 235, L95 (1980); M.E. Weidenbeck and D.E. Greiner, Phys. Rev. Lett. 46, 682 (1981); M.E. Weidenbeck and D.E. Greiner, Ap. J. 247, L119 (1981).
10. F. St. Laurent, et al, Phys. Lett. 110B, 372 (1982).
11. C. Olmer, Indiana University, private communication.
12. N.J. DeGiacomo, et al, Phys. Rev. Lett. 45, 527 (1980).
13. P. Karol, Phys. Rev. C11, 1203 (1975).

SESSION E
RELATIVISTIC APPROACHES

DIRAC PHENOMENOLOGY AND THE NUCLEAR OPTICAL MODEL*

by

B. C. Clark and S. Hama
Department of Physics
The Ohio State University
Columbus, Ohio 43210 U.S.A.

and

R. L. Mercer
International Business Machines
Watson Research Center
Yorktown Heights, New York 10598 U.S.A.

Abstract

A Dirac equation based optical model treatment of proton-nucleus scattering is discussed. Results obtained using a relativistic optical model consisting of Lorentz scalar and Lorentz four-vector components are given. The parameters in the model have been varied to obtain high quality fits to elastic differential cross section, analyzing power, and spin rotation function measurements. Features of these Dirac optical model potentials are compared with various theoretical relativistic calculations.

Interest in relativistic aspects of nuclear structure has increased dramatically over the past few years. There are now extensive programs considering various aspects of the relativistic quantum field theory of nuclei (see Refs. 1-9 and the citations to earlier work contained in those papers). In many cases these treatments have had remarkable success in calculating single particle properties of nuclei. In parallel with the development of these relativistic treatments of the nuclear many-body problem has been the development of a relativistic optical model treatment of the nuclear scattering problem [10-12]. This model, based on the Dirac equation as the relevant wave equation, has been used to describe elastic scattering observables throughout the energy range of interest in nuclear physics [12], and has recently been applied to various inelastic nuclear reactions [13-15]. In this work we describe some results of the relativistic optical model.

A fundamental characteristic of the Dirac equation approach is that the Lorentz character of the potentials must be specified. The most general local, time independent Dirac equation contains five tensor types; scalar, pseudoscalar, vector, axial vector and tensor. In this case [16] the Dirac equation is ($\hbar = c = 1$)

$$\left\{ \vec{\alpha} \cdot \vec{p} + \beta \left[m + U_s(\vec{r}) + \gamma^\mu U_{v_\mu}(\vec{r}) + \gamma^5 U_{ps}(\vec{r}) + \gamma^\mu \gamma^5 U_{a_\mu}(\vec{r}) \right. \right.$$

$$\left. \left. + \sigma^{\mu\nu} U_{t_{\mu\nu}}(\vec{r}) \right] \right\} \psi(\vec{r}) = E \, \psi(\vec{r}) \ . \tag{1}$$

The Dirac optical model used in this work employs two potentials: one, $U_s(r)$, transforms like a Lorentz scalar and the other, $U_0(r)$, transforms like the time-like component of a Lorentz four-vector. The Dirac equation with these potentials is

$$\left\{\vec{\alpha}\cdot\vec{p} + \beta\left[m + U_s(r)\right] + \left[U_o(r) + V_c(r)\right]\right\}\Psi(\vec{r}) = E\,\Psi(\vec{r})\ ,$$ (2)

where $V_c(r)$ is the Coulomb potential determined from the empirical nuclear charge density, m the nucleon mass and E the nucleon total energy in the c.m. frame. Equation (2) is solved using partial wave decomposition and the complete four component wave functions are obtained [17].

A two component reduction of Eq.(1) yields an equation for the upper two components of the wave function which contains central, spin-orbit, and Darwin terms. If the pseudoscalar and axial vector terms are included, other more complicated terms arise. Table I summarizes the various contributions. The pseudoscalar and axial vector terms are ordinarily omitted because of parity arguments, leaving scalar, vector and tensor potentials. In Appendix I, we give the reduction of the Dirac equation to second order form including these interactions. It is interesting that the space-like part of the vector potential does not explicitly appear in the second order equation. More generally it can be shown that when the spatial portion of the vector potential in Eq.(is spherically symmetric it has no effect on elastic scattering [18]. Examination of Eq.(7) and Eqs.(12-14) in Appendix I indicates that, at a minimum, one should consider U_o and U_s or U_o and U_t or U_s and U_t in order that the required central and spin-orbit potentials may be obtained. We choose U_o and U_s because these potential types appear in various relativistic mean field theories [1,2,4-9,19,20] and are the largest terms in relativistic Brueckner-Hartree-Fock calculations [3,21] of the optical potential.

A number of macroscopic features of the Dirac equation optical model treatment using $U_o(r)$ and $U_s(r)$ potentials are clear from Eqs.(7,12-14) in the Appendix. They are:

1) The central potential has explicit energy dependence.

2) The central potential has a non-local Darwin term.

3) Non-linear terms involving U_o, U_s and $A(r)$ are present.

4) The spin-orbit potential occurs naturally.

5) The spin-orbit and central potentials are constrained by the choice of U_o and U_s.

6) A "Coulomb correction" term, $U_o V_c$, appears (it is complex if U_o is complex).

7) The real and imaginary parts of the optical potential are mixed through the $U_o^2(r)$, $U_s^2(r)$ and $A(r)$ terms.

These considerations lead, in the case of the large repulsive potential $U_o(r)$ and large attractive potential $U_s(r)$ usually found, to central and spin-orbit potentials of reasonable size. In addition, the real central potential exhibits a radial dependence which changes with energy in a manner similar to that of nonrelativistic microscopic calculations of the real central potential [22-27]. Similar effects are found from the analysis of large angle elastic scattering data [28-29].

Table I

Contributions of various tensor types to the second order Dirac Equation.

Tensor Type	Central	Spin-Orbit	Darwin
U_s	Yes	Yes	Yes
U_v^o	Yes	Yes	Yes
U_v^r	Yes	No	Yes
U_t	Yes	Yes	No
U_{ps}	Yes	No	No
U_a	Yes	Yes	Yes

We now discuss an application of Dirac phenomenology to the analysis of p - ^{40}Ca elastic scattering over a wide range of energies. First we base the analysis on the simple Dirac-Hartree or folding model described in Ref. 11. The complex potentials are written

$$U_o(r) = V_o f_o(r) + i W_o g_o(r) \tag{3}$$

$$U_s(r) = V_s f_s(r) + i W_s g_s(r) . \tag{4}$$

If the shape functions are chosen to be two-parameter Fermi-functions $[1 + \exp(r - c)/z]^{-1}$, this model contains 12 adjustable parameters, as does the standard phenomenological Schrödinger equation model. The folding prescription fixes the geometries of $f_o(r)$ and $f_s(r)$. In this earlier work we found that with these geometries good fits to experiment, see Fig. 1, can be obtained over a wide range of energies [11,12]. The strengths V_o and V_s are energy dependent and the ratio R_R defined by

$$R_R = \frac{J_o}{J_s} = \frac{\int V_o(r) d\vec{r}}{\int V_s(r) d\vec{r}} \tag{5}$$

varies linearly with energy as shown in Fig. 2. This behaviour is similar to that observed in the relativistic Hartree-Fock calculations of Ref. 19. The energy variation of the volume integrals of the central and spin-orbit effective potentials (see Appendix I) shown in Fig. 3 is smooth, with the real central volume integral exhibiting a logarithmic energy dependence. The behaviour of the real central effective potential shown in Fig. 4 shows the development of a "wine-bottle" shape in the transition energy region and the persistence of a small attractive potential in the nuclear surface region, even at 800 MeV.

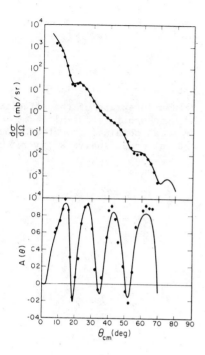

Fig. 1. Elastic cross sections and analyzing powers for ^{40}Ca at 181 MeV. The calculation employs the Dirac phenomenology given in Ref. 11.

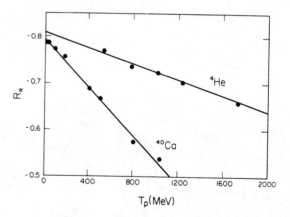

Fig. 2. The ratio R_R obtained from analyses of p - ^{40}Ca and p - ^{4}He from Ref. 12.

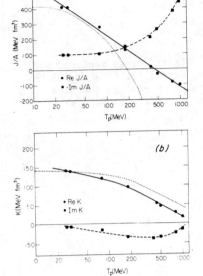

Fig. 3. Volume integrals of the effective central and spin-orbit potentials as a function of T_p as obtained from the Dirac equation optical model analysis given in Ref. 12.

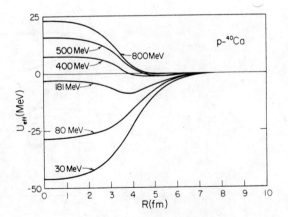

Fig. 4. Energy dependence of the effective central potential for p - ^{40}Ca given in Ref. 12.

Next we consider improvements in the phenomenology obtained by using the relativistic mean field calculations of Serot and Horowitz [6]. These calculations, which contain only four parameters, give good results for the charge distributions and single particle energies of closed shell nuclei. The imaginary parts of the optical potentials were again taken to be two-parameter fermi shapes except at energies below 50 MeV, where surface peaked absorption was used. The results for p - ^{40}Ca at 26 MeV are shown in Figs. 5 and 6, and quite reasonable agreement with experiment is obtained. The relativistic Hartree potentials of Ref. 6 are, of course, energy independent; however, a simple energy dependence may be introduced by the following assumption regarding the real potentials,

$$U_o = C_1 U_s^{Stanford} \quad ,$$

$$U_s = C_2 U_s^{Stanford} \quad .$$

The parameters C_1 and C_2 are varied in addition to the parameters determining the imaginary parts of the optical potentials. We find quite good agreement with both cross section and analyzing power data at 181 and 500 MeV. The results are shown in Figs. 7-10. The predicted spin rotation function, $Q(\theta)$, based on the fit to $\sigma(\theta)$ and $A(\theta)$ at 500 MeV shown in Fig. 11 is in good agreement with the recent LAMPF measurement [30].

Two approaches for further development of the Dirac equation based optical model are now underway. First, we plan to redo the analysis at other energies and for other target nuclei using the Stanford potentials as well as those obtained by the Boguta [1]. Second, we are currently reanalyzing the p - ^{40}Ca data without restricting the geometries of the real optical potentials

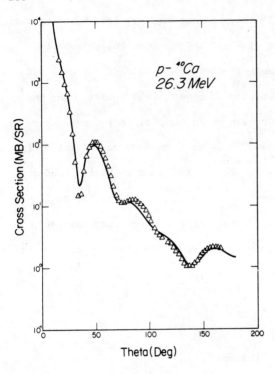

Fig. 5. Calculated elastic p - ^{40}C
cross section values at 26.3 MeV.
$\sigma(\theta)$ data are from Ref. 38. The re
optical potentials described in Ref
are used.

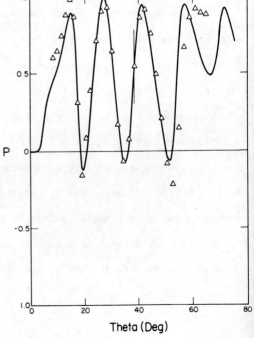

Fig. 6. Calculated elastic p - ^{40}Ca
polarization values at 26.3 MeV. The
P(θ) data are from Ref. 39. The real
optical potentials described in Ref. 6
are used.

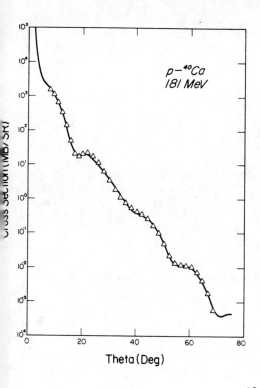

$p - {}^{40}Ca$
$181\ MeV$

Fig. 7. Calculated elastic p - ^{40}Ca
cross section values at 181 MeV. The
data are from Ref. 11. The real optical
potentials of Ref. 6 are used with
$C_1 = 0.61$ and $C_2 = 0.64$, see text.

Calculated elastic p - ^{40}Ca
ing power values at 181 MeV. The
e from Ref. 11. The real optical
als of Ref. 6 are used with
61 and $C_2 = 0.64$, see text.

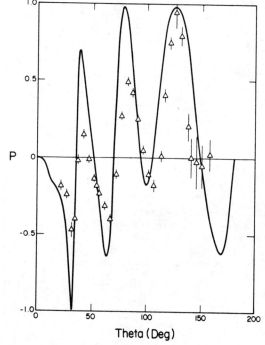

270

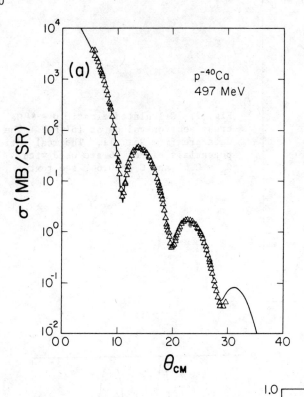

(a)

$p-^{40}Ca$
497 MeV

$\sigma\,(MB/SR)$

θ_{CM}

Fig. 9. Calculated elastic p - cross sections at 497.5 MeV. Th are from Ref. 31. The real opti potentials of Ref. 6 are used wi $C_1 = 0.49$ and $C_2 = 0.59$, see te

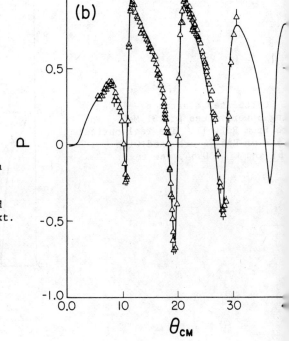

(b)

P

θ_{CM}

Fig. 10. Calculated elastic p - ^{40}Ca analyzing powers at 497.5 MeV. The data are from Ref. 31. The real optical potentials of Ref. 6 are used with $C_1 = 0.49$ and $C_2 = 0.59$, see text.

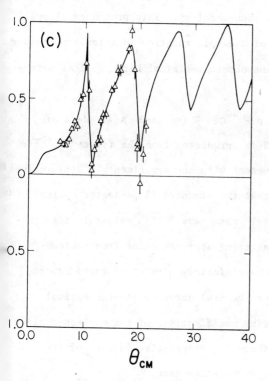

Fig. 11. Calculated values of the
spin rotation function predicted from
the $\sigma(\theta)$ data of Ref. 30. The $Q(\theta)$
data are from Ref. 29.

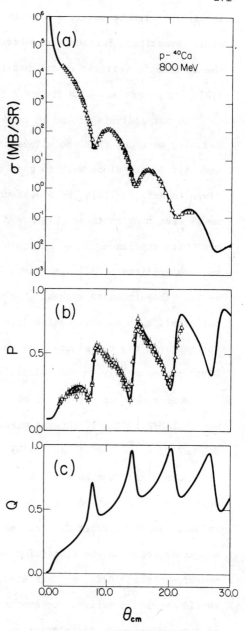

Fig. 12. Calculated values of 800 MeV
p - ^{40}Ca cross sections (a), analyzing
powers (b), and the spin rotation
function (c) predicted from the fit to
the $\sigma(\theta)$ and $A(\theta)$ data of Ref. 32.

272

in order to remove possible bias in the phenomenology. This procedure allows

us to investigate further the question of the spin rotation function, reconsider

the systematic features of the model, and obtain empirical Dirac optical poten-

tials for comparison with theory.

In our analysis of the 497.5 McV p - ^{40}Ca $\sigma(\theta)$ and A(θ) data of

Ref. 31, we found that the values of Q(θ) predicted from the fit to the $\sigma(\theta)$

and A(θ) data alone were in good agreement with the experimental Q(θ) values

given in Ref. 30 [10]. We also found that the standard 12-parameter optical

model treatment of these $\sigma(\theta)$ and A(θ) data gave Q(θ) values in disagree-

ment with experiment. The multiple scattering approach using free nucleon-

nucleon amplitudes [32] also gives an unsatisfactory account of these experi-

ments. These results suggest that a more general geometry for the optical

potential, such as would arise from medium modifications of the nucleon-nucleon

interaction, may be required. We attribute the comparative success of the

Dirac phenomenology to the relativistic formulation used.

Measurement of Q(θ) for 800 MeV protons on ^{40}Ca are currently under-

way at LAMPF. Fig. 12 shows a prediction of Q(θ) based on our fit to the

800 MeV p - ^{40}Ca $\sigma(\theta)$ and A(θ) data of Ref. 33. Alterations in the poten-

tial parameters from this fit to $\sigma(\theta)$ and A(θ) which increase the χ^2 per

degree-of-freedom by one produce very little change in the predicted Q(θ)

values. As the largest differences between Schrödinger and Dirac phenomenology

appear to occur in the transition energy region from 200 MeV to 600 MeV,

measurements of Q(θ) in this energy region for a number of target nuclei

should provide a test of our approach as well of nonrelativistic treatments

of the nucleon-nucleus interaction.

The first results of our reanalysis of the p - ^{40}Ca data from 26 to

800 MeV show that most of the systematic features obtained earlier are unchange

The volume integrals of the effective central potential shown in Fig. 13 behave as before. The rms radii of the real part of the individual vector and scalar potentials, shown by dots in Fig. 14, are within $\pm$ 5% of the values obtained in the folding model. The volume integral of the effective spin-orbit potential shows a deviation from our previous results at low energies, $T_p < 50$ MeV. This behaviour, shown in Fig. 15, is probably due to the change in the parameterization of the imaginary potentials from volume to surface form. We are investigating this feature using different parameterizations for the imaginary potentials. Figure 16 shows the energy variation in R_R from the present analysis and again this variation appears to be linear. However, the point at 26 MeV is in sharp disagreement with the trend of the rest of the analysis. In view of the results for $T_p < 50$ MeV, we cannot, at this time, consider the potential obtained below this energy to be reliable. We are analyzing other data in this energy region using several alternate treatments of the imaginary potentials. The consistency of our results at higher energies with results from our previous analyses leads us to believe that the Dirac potentials in this energy region are much better determined. We are, however, planning investigations of uniqueness and model dependence throughout the entire energy range.

With the provisos mentioned above, we give in Figs. 17 and 18 our results for the strengths of the real and imaginary scalar and vector potentials at the center of the ^{40}Ca nucleus. Figure 17 compares the real strengths, $V_0(o)$ and $V_s(o)$, with the results of two relativistic mean field treatments in the Hartree approximation [1,6] and with the relativistic Brueckner-Hartree-Fock calculation of Ref. 21. The energy independent relativistic Hartree calculations, as expected, do not agree with the phenomenological results, which exhibit considerable energy dependence. The trend of the RBHF calculations

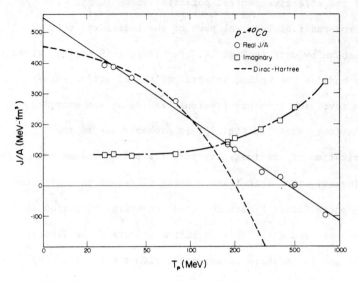

Fig. 13. Calculated values of the volume integrals of the real effective central (open circles) and imaginary effective central (open squares) potentials. The solid line is a least squares fit to the open circles. The long dashed line is to guide the eye. The dashed line gives the Dirac Hartree values using the potentials of Ref. 6.

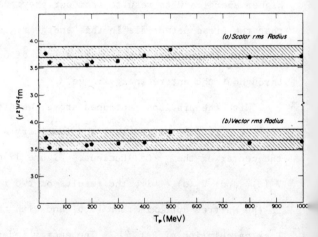

Fig. 14. Calculated values of the rms radi of the real U_S and real U_0 optical potentia The solid lines in each case are the values using the folding model of Ref. 11. The ba show ± 5% limits.

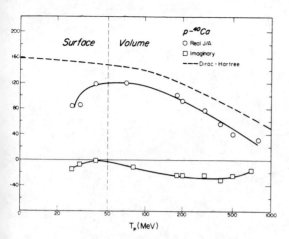

Fig. 15. Calculated values of the volume integrals of the real effective spin-orbit (open circles) and imaginary effective spin-orbit (open squares) potentials. The dashed line gives the Dirac Hartree values from the potentials of Ref. 6.

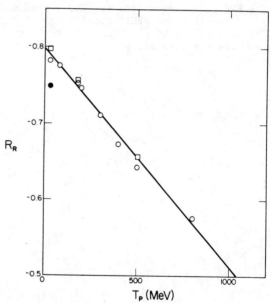

Fig. 16. Calculated values for the ratio R_R obtained from the 12-parameter fits to p - ⁴⁰Ca elastic data described in the text (open circles). The open squares are the results of the analyses of 26, 181, and 500 MeV p - ⁴⁰Ca elastic data using the Stanford potentials, see text. The shaded circle is the 26 MeV result.

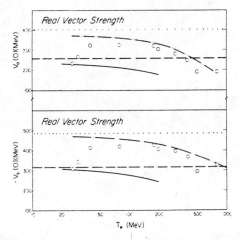

Fig. 17. Values of the real scalar V_S and V_O
potentials at r = 0 determined from the
12-parameter analysis of p - ^{40}Ca data. The
upper dotted lines are the values from the RMF
calculation of Ref. 11. The dashed lines are
the RMF values from Ref. 1. The curved solid
lines are from the RBHF calculations of Ref. 21.
The dash-dot lines give the results of the fixed
geometry analysis of Ref. 12.

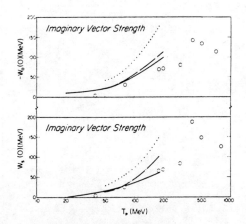

Fig. 18. Values of the imaginary scalar W_S
and vector V_O potentials at r = 0 determined
from the 12-parameter analysis of p - ^{40}Ca
data described in the text. The dotted line
are the calculations of Ref. 35, the dash-do
lines are the calculations of Ref. 34 and t
smooth curves are the calculations of Ref. 2

exhibit energy dependence which seems to follow the trend of the analysis. However, the potential strengths are smaller than the phenomenological results. There could be a number of reasons for this discrepancy and, clearly, possible model dependence of the phenomenology is an important question. Results for the imaginary parts of the scalar, $W_s(o)$, and vector, $W_o(o)$, potential strengths at the nuclear interior are given in Fig. 18. The results are compared with three recent theoretical calculations of the imaginary parts of the relativistic optical potential. The calculation of Horowitz [34], shown by the dot-dashed lines in Fig. 18, uses the Walecka model while that of Celenza, Pong and Shakin [21], shown by the solid lines, employs a relativistic Brueckner-Hartree-Fock approach. The dotted lines in Fig. 18 are calculations of Jaminon [35] which are based on a different interpretation of Horowitz's results. These theoretical treatments result in a positive W_s and a negative W_o potential. The magnitude of both potential types increase with energy. The theoretical calculations are in reasonable agreement with the phenomenological results. These first direct comparisons of relativistic optical models with the results of Dirac phenomenology are encouraging and point up one of the advantages of the treatment employed in this work.

The results of Dirac phenomenology as applied to a nuclear optical model treatment of elastic scattering show promise, but considerable work is still to be done. Questions of model dependence and uniqueness must be answered. Applications to other targets and other energies using, where possible, input from relativistic nuclear structure calculations is clearly desirable. The inclusion of other potential types in Dirac phenomenology is under investigation [36]. Consideration of other probes such as neutrons or antiprotons is underway [36]. In the latter case questions of G-parity can be expected to play an important role in the determinations of model potentials. Application to reactions other than elastic scattering is progressing [13-15,37] and the results are encouraging.

278

In this paper we have discussed only one aspect of Dirac phenomenology. A survey of a number of other applications has been given by Jaminon and Mahaux [9].

In conclusion, it appears that Dirac phenomenology can play a useful role in the relativistic description of nuclei.

We thank J. Boguta, E. D. Cooper, C. Horowitz, M. Jaminon, C. Mahaux, B. Mulligan, L. Ray, B. Serot, C. Shakin and H. Sherif for helpful correspondence and discussions. The Stanford R.M.F. Hartree potentials were provided by B. Serot and we thank him for letting us use them in our calculations. This work was supported by NSF Grant No. PHY-8107397.

References

1. J. Boguta, Nucl. Phys. $\underline{A372}$, 386 (1981); Phys. Lett. $\underline{106B}$, 241, 245, 250 (1981).

2. A. Bouyssy, Nucl. Phys. $\underline{A381}$, 445 (1982).

3. L. S. Celenza, W. S. Pong and C. M. Shakin, Phys. Rev. C$\underline{25}$, 3115 (1982); and B.C.I.N.T. preprint 81/121/111.

4. J. M. Eizenberg, Nucl. Phys. $\underline{A355}$, 269 (1981).

5. C. J. Horowitz and J. D. Walecka, Nucl. Phys. $\underline{A364}$, 429 (1981).

6. C. J. Horowitz and Brian D. Serot, Nucl. Phys. $\underline{A368}$, 503 (1981); and ITP-718 preprint 6/82.

7. M. Jaminon and C. Mahaux, Conference proceedings of the Conference on New Horizons in Electromagnetic Physics, University of Virginia, April, 1982.

8. I. Lovas, Nucl. Phys. $\underline{A367}$, 509 (1981).

9. J. V. Noble, Nucl. Phys. $\underline{A368}$, 447 (1981).

10. B. C. Clark, R. L. Mercer and P. Schwandt, Phys. Lett. in press

11. L. G. Arnold, B. C. Clark, R. L. Mercer and P. Schwandt, Phys. Rev. C$\underline{23}$, 1949 (1981), and references therein.

12. L. G. Arnold, B. C. Clark, E. D. Cooper, H. S. Sherif, D. A. Hutcheon, P. Kitching, J. M. Cameron, R. P. Liljestrand, R. N. MacDonald, W. J. McDonald, C. A. Miller, G. C. Neilson, D. M. Sheppard, G. M. Stinson, D. K. McDaniels, J. R. Tinsley, R. L. Mercer, L. W. Swensen, P. Schwandt and C. E. Stronach, Phys. Rev. C$\underline{25}$, 936 (1982).

13. E. D. Cooper and H. S. Sherif, Phys. Rev. Lett. $\underline{47}$, 818 (1981).

14. J. R. Shepard, E. Rost, and D. Murdock, Phys. Rev. Lett. $\underline{49}$, 14 (1982); E. Rost, J. R. Shepard, and D. Murdock, Phys. Rev. Lett. $\underline{49}$, 448 (1982).

15. B. C. Clark, S. Hama, E. Sugarbaker and R. L. Mercer, OSU TR-82-222, (1982)

16. L. D. Miller, Ann. Phys. (N.Y.) $\underline{91}$, 40 (1975).

17. R. L. Mercer, Phys. Rev. C$\underline{15}$, 1786 (1977).

18. B. C. Clark, S. Hama and R. L. Mercer, in preparation.

19. J. D. Walecka, Ann. Phys. $\underline{83}$ 491 (1974).

20. M. Jaminon and C. Mahaux, Phys. Rev. C$\underline{24}$, 1353 (1981).

21. L. S. Celenza, W. S. Pong and C. M. Shakin, B.C.I.N.T. preprint 82/082/11(
    and C. M. Shakin, invited talk at the IUCF Conference on The Interaction
    Between Medium Energy Nucleons and Nuclei (1982).

22. J.-P. Jeukenne, A. Lejeune and C. Mahaux, Phys. Rev. C$\underline{16}$, 80 (1977);
    Phys. Rep. $\underline{25C}$, 83 (1976).

23. C. Mahaux, in Microscopic Optical Potentials, edited by H. V. von Geramb
    (Springer, New York, 1979), p. 1.

24. F. A. Brieva and J. R. Rook, Nucl. Phys. $\underline{A291}$, 299 (1977); $\underline{A291}$, 317
    (1977); $\underline{A297}$, 206 (1978); $\underline{A307}$, 493 (1978).

25. W. Bauhoff, H. V. von Geramb, and G. Pálla (unpublished).

26. J. Kelly, W. Bertozzi, T. N. Buti, F. W. Hersman, C. Hyde, M. V. Hynes,
    B. Norum, F. N. Rad, A. D. Bacher, G. G. Emery, C. C. Foster, W. P. Jones
    D. W. Miller, B. L. Berman, W. G. Love, and F. Petrovich, Phys. Rev.
    Lett. $\underline{45}$, 2012 (1980).

27. B. Friedman and V. R. Panharipande, Phys. Lett. $\underline{100B}$, 205 (1981).

28. L. R. B. Elton, Nucl. Phys. $\underline{89}$, 69 (1966).

29. H.-O. Meyer, P. Schwandt, G. L. Moake, and P. P. Singh, Phys. Rev. C$\underline{23}$,
    616 (1981).

30. A. Rahbar, B. Aas, E. Bleszynski, M. Bleszynski, M. Haji-Saeid, G. J. Ig(
    F. Irom, G. Pauletta, A. T. M. Wang, J. B. McClelland, J. F. Amann,
    T. A. Carey, W. D. Cornelius, M. Barlett, G. W. Hoffmann, C. Glashausser,
    S. Nanda, and M. M. Gazzaly, Phys. Rev. Lett. $\underline{47}$, 1811 (1981).

31. G. W. Hoffmann, L. Ray, M. L. Barlett, R. Ferguson, J. McGill, E. C. Mil(
    Kamal K. Seth, D. Barlow, M. Bosko, S. Iverson, M. Kaletka, A. Saha, and
    D. Smith, Phys. Rev. Lett. $\underline{47}$, 1436 (1981).

32. M. L. Barlett, G. W. Hoffmann, and L. Ray, Bull. Am. Phys. Soc. $\underline{27}$, 729 (1982) and L. Ray, private communication, 1982.

33. L. Ray, G. W. Hoffmann, M. Barlett, J. McGill, J. Amann, G. Adams, G. Pauletta, M. Gazzaly, and G. S. Blanpied, Phys. Rev. C$\underline{23}$, 828 (1981).

34. C. J. Horowitz, Phys. Lett. $\underline{117B}$, 153 (1982).

35. M. Jaminon, contribution to the IUCF Conference on The Interactions Between Medium Energy Nucleons and Nuclei, (1982), and preprint.

36. B. C. Clark, S. Hama and R. L. Mercer, Proceedings of the International Conference on Nuclear Structure, Aug. 30-Sept. 3, 1982, Amsterdam, p. 23.

37. J. M. Shepard, invited talk at the IUCF Conference on The Interactions Between Medium Energy Nucleons and Nuclei (1982).

38. K. H. Bray, K. S. Jayaraman, G. A. Moss, W. T. H. van Oers, D. O. Wells, and Y. I. Wu, Nucl. Phys. $\underline{A167}$, 57 (1971).

39. D. L. Watson, J. Lowe, J. C. Dore, R. M. Craig, and D. J. Baugh, Nucl. Phys. $\underline{A92}$, 193 (1967).

APPENDIX I

In this appendix we obtain the Schrödinger equivalent potential for a Dirac equation treatment which contains Lorentz scalar, four-vector and tensor potentials. The Dirac equation in this case may be written

$$\left\{ \vec{\alpha}\cdot\vec{p} + \beta\left[m+U_s(\vec{r})\right] + \beta\gamma^\mu U_{v_\mu}(\vec{r}) + \beta\sigma^{\mu\nu}U_{t_{\mu\nu}}(\vec{r}) \right\}\Psi(\vec{r}) = E\,\Psi(\vec{r}) . \tag{1}$$

The notation is that of Miller [45]. For rotationally symmetric potentials $U_s(\vec{r}) = U_s(r)$, $\gamma^\mu U_{v_\mu}(\vec{r}) = \gamma^o U_v^o(r) - \vec{\gamma}\cdot\vec{U}_v^r(r) = \gamma^o U_v^o(r) - \vec{\gamma}\cdot\hat{r}\,U_v^r(r)$, and $\sigma^{\mu\nu}U_{t_{\mu\nu}}(\vec{r}) = -\gamma^o\vec{\gamma}\cdot\hat{r}\,U_t^r(r) = \beta i\vec{\alpha}\cdot\hat{r}\beta\,U_t(r)$ so that Eq.(1) becomes

$$\left\{ \vec{\alpha}\cdot\vec{p} + \beta(m+U_s) - (E - U_v^o - V_c) - \beta\vec{\gamma}\cdot\hat{r}\,U_v^r + i\vec{\alpha}\cdot\hat{r}\beta\,U_t \right\}\Psi(\vec{r}) = 0 \tag{2}$$

where V_c is the Coulomb potential. Equation (2) may be written as two coupled equations for the upper (ψ_U) and lower (ψ_L) components of Ψ. Solving for ψ_L in terms of ψ_U gives

$$\psi_L = \frac{1}{(E+m)A(r)}\left[\vec{\sigma}\cdot\vec{p} - (\vec{\sigma}\cdot\hat{r})U_v^r + i(\vec{\sigma}\cdot\hat{r})U_t \right]\psi_U \tag{3}$$

where

$$A(r) = (m+U_s+E - U_v^o - V_c)/(E+m) . \tag{4}$$

The equation for $\psi_U(\vec{r})$ is

$$\left[(E - U_v^o - V_c)^2 - (m+U_s)^2 - Q(r) \right]\psi_U(\vec{r}) = 0 , \tag{5}$$

where

$$Q(r) = A(r)\left[\vec{\sigma}\cdot\vec{p} - (\vec{\sigma}\cdot\hat{r})U_v^r - i(\vec{\sigma}\cdot\hat{r})U_t\right]\frac{1}{A(r)}$$

$$\times \left[\vec{\sigma}\cdot\vec{p} - (\vec{\sigma}\cdot\hat{r})U_v^r + i(\vec{\sigma}\cdot\hat{r})U_t\right] \quad . \tag{6}$$

Carrying out the indicated algebra gives

$$\left\{\nabla^2 + (E - U_v^o - V_c)^2 - (m + U_s)^2 - U_v^{r^2} - U_t^2\right.$$

$$+ \left[\frac{1}{rA}\frac{\partial A}{\partial r} - 2\frac{U_t}{r}\right](\vec{\sigma}\cdot\vec{L}) - \frac{2}{r}\left(iU_v^r + U_t\right)$$

$$+ \frac{1}{A}\frac{\partial A}{\partial r}\left(iU_v^r + U_t\right) + \frac{1}{r}(\vec{r}\cdot\vec{p})\left(U_v^r - iU_t\right)$$

$$\left. - \left[i\frac{1}{rA}\frac{\partial A}{\partial r} - 2\frac{U_v^r}{r}\right](\vec{r}\cdot\vec{p})\right\}\psi_U = 0 \quad . \tag{7}$$

From Eq.(7) we see that the tensor term contributes both to the spin-orbit term and the central terms. The 3-vector part of U_v^r contributes to both the Darwin and the central terms. To remove the first derivative terms, we let

$$\psi_U(\vec{r}) = K(r)\phi(\vec{r}) \tag{8}$$

with $K(r) \to 1$ as $r \to \infty$. Direct substitution of (8) into (7) gives

$$\frac{\partial}{\partial r}K(r) = \frac{1}{2}\left[\frac{1}{A}\frac{\partial A}{\partial r} + 2iU_v^r\right]K(r) \tag{9}$$

or

$$K(r) = A^{\frac{1}{2}} \exp \int iU_v^r(r)dr \quad . \tag{10}$$

Using Eq. (9), we may write the Schrödinger equivalent equation

$$\left\{ \nabla^2 + (E - U_v^0 - V_c)^2 - (m + U_s)^2 - U_t^2 + \frac{U_t}{A}\frac{\partial A}{\partial r} \right.$$

$$- 2\frac{U_t}{r} - \frac{\partial}{\partial r}U_t - \frac{3}{4}\frac{1}{A^2}\left(\frac{\partial A}{\partial r}\right)^2 + \frac{1}{2r^2 A}\frac{\partial}{\partial r}\left(r^2\frac{\partial A}{\partial r}\right)$$

$$\left. + \left(\frac{1}{rA}\frac{\partial A}{\partial r} - 2\frac{U_t}{r}\right)(\vec{\sigma}\cdot\vec{L}) \right\} \phi(\vec{r}) = 0 \ . \tag{11}$$

Notice that the 3-vector part of the vector potential does not appear in (11). The tensor potential contributes in a complicated way to the central potential and also contributes to the spin-orbit term. In addition there are cross terms between U_t and derivatives of U_v^0, V_c and U_s. One may then define Schrödinger equivalent central, spin-orbit and Darwin potentials given by

$$U_{eff} = \frac{1}{2E}\left[2EU_0 + 2mU_s - U_0^2 + U_s^2 - 2V_c U_0 + U_t^2 \right.$$

$$+ \left(-\frac{U_t}{A}\left(\frac{\partial A}{\partial r}\right) + 2\frac{U_t}{r} + \frac{\partial U_t}{\partial r} \right)$$

$$\left. + \left(-\frac{1}{2r^2 A}\frac{\partial}{\partial r}\left(r^2\frac{\partial A}{\partial r}\right) + \frac{3}{4A^2}\left(\frac{\partial A}{\partial r}\right)^2 \right) \right] \ , \tag{12}$$

$$U_{Darwin} = \frac{1}{2E}\left[-\frac{1}{2r^2 A}\frac{\partial}{\partial r}\left(r^2\frac{\partial A}{\partial r}\right) + \frac{3}{4A^2}\left(\frac{\partial A}{\partial r}\right)^2 \right] \tag{13}$$

$$U_{so} = \frac{1}{2E}\left[-\frac{1}{rA}\left(\frac{\partial A}{\partial r}\right) + 2\frac{U_t}{r} \right] \tag{14}$$

for a Schrödinger equation given by

$$\left[p^2 + 2E(U_{eff} + U_{so}\vec{\sigma}\cdot\vec{L}) \right]\phi(\vec{r}) = \left[(E - V_c)^2 - m^2 \right]\phi(\vec{r}) \ . \tag{15}$$

DISCUSSION

Rawitscher: Why doesn't the mean field theory give energy dependent real potentials? After all, the nucleon-nucleon potential is known to be energy dependent.

Clark: Those are Hartree calculations which give no energy dependence. Hartree-Fock calculations would give energy dependent potentials.

Walecka: Can I make a comment?. The potentials which you put into a Dirac equation are energy independent in a Hartree calculation. When you reduce it to a Schrödinger equation you get a linear energy dependence. Your equivalent Schrödinger potential is energy dependent. There is an additional energy dependence in the fits to the data because you actually have to scale the Dirac potentials .

Rawitscher: I was not worried about that additional energy dependence. I understand that you get an automatic energy dependence when you go from the Dirac equation to the Schrödinger equation, but when you think of the nucleons acting with the nucleus and think of the nucleus as a relativistic object which has e.g., mesons in it, why doesn't that interaction, due to relativistic nucleons in the nucleus, become energy dependent?

Walecka: It will, eventually. Carl will comment on this later.

MacFarlane: Two questions. One is that you referred by initials I didn't recognise to what I assume were relativistic One-Boson Exchange Potentials, what was the reference on that? Secondly, how big are the small components in your Dirac Spinors at 26 MeV?

Clark: I haven't looked at the lower components at 26 MeV. Certainly, for bound states, they can be around 15 to 20 percent of the upper components.

Shakin: I'd like to make a comment. It's very deceptive to call these Hartree calculations. It's nothing to do with the kind of Hartree calculation we do in the case of atoms because the forces here are much too strong for the Hartree approximation to be applicable. These are calculations which look like Hartree calculations but have adjusted parameters.

MacFarlane: What was the reference to the One-Boson Exchange Potentials ?

Shakin: There's been a large body of work by people in Bonn, Hollinde and Machleit, there are two Review articles in Phys. Reports that summarise their work very well I can give you the references later.

Hwang: Is there any theoretical grounds for adjusting the constants c_1 and c_2 to fit the data?

Clark: If you look at the Relativistic Hartee Fock calculations you can see there is certainly an energy dependence, although it's not linear.

Hwang: So you introduce another energy dependence?

Clark: Yes.

Serot: I want to answer the question on small components. I only know the results for the bound states near the Fermi surface, the valence particles, where the maximum amplitude of the smaller components is about 20 percent that of the larger components; upon squaring, this becomes about 5 percent.

Walecka: You can see the energy dependence quite clearly from the viewgraph the last speaker showed, would the speaker put the graph up please?

Clark: These are the strenghts of the extracted vector and scalar potentials in the interior of the nucleus. May I remind you again that I feel our phenomenology has failed at the lower energies. Previously we had quite good fits to the data when the energy dependence of the Schrodinger equivalent potential was constant. It was only when we got up above 200 MeV that we started to see a great deal of energy variation in the vector and scalar potentials that fit the data.

Walecka: Karl's calculations, which he's going to talk about, give this additional energy dependence.

Miller: Can you give some idea of the narrowness of the χ^2 minimum? Specifically are we talking about a broad flat minimum so that if you, e.g., didn't put in all this energy dependence it wouldn't appreciably spoil the fits?

Clark: You can't fit the data without the energy dependence, that's certain.

Miller: How bad is it? That's the question.

Clark: You mean you would like to see the χ^2 per degree of freedom if one didn't adjust the strengths of the potentials?

Miller: If it didn't go through all the data, so what? How seriously can you say that this is an extracted variation?

Clark: These are results of twelve-parameter fits, trying to determine the optical potential in that way. When we keep the geometry fixed e. g., using the Stanford potential, we reduce the number of parameters that are varied, the change in the real potentials, even in the most severe case, (which is around 500 MeV where the potentials are essentially cancelling each other out) the change between their strength parameters and the ones that have been extracted from the phenomenology is 20 percent. That's in the strength of the real potential. The imaginary parts are closer to each other than that. At 200 MeV the strength of both real and imaginary parts are within that, and, as a matter of fact the imaginary potentials that I get with 12 parameters at 100 MeV have the strengths of -71 MeV and +66 MeV; the ones from the Stanford potentials are -71 MeV and +59 MeV, and those, I think, are very good values and I must mention that the χ^2 per degree of freedom has changed by more than 1 between those two fits, even though when you see the fits to the data from the Stanford potentials they look very fine.

Brown: Do any of the models have any mechanism to give a real scalar part that increases in magnitude up to 100 MeV.

Clark: Let me say again that these three points (the ones obtained for energies lower than 40 MeV) are an artifact of the fact that we have allowed the subroutine to search freely for a minimum.

DIRAC PHENOMENOLOGY FOR NUCLEAR REACTIONS

J. R. Shepard, University of Colorado, Boulder, CO   80309

ABSTRACT

The application of Dirac phenomenology to nuclear reactions is discussed. The large vector and scalar potentials necessary to describe the nucleon-nucleus spin-orbit interaction are found to enhance the small components of nuclear wavefunctions. This is found to have a potentially large effect on some electron- and nucleon-induced inelastic scattering amplitudes.

Dirac phenomenology has provided a very satisfying, economical description of nucleon-nucleus elastic scattering.[1] For example, the magnitude of the spin-orbit interaction can be simply related both to relativistic nucleon-nucleon potentials[2] and to the binding energy and saturation density of infinite nuclear matter[3] as well as properties of finite nuclei.[4] The energy dependences of, and geometry differences between, the central and spin-orbit potentials can also be accounted for. This approach has been recently extended to deuteron-nucleus elastic scattering.[5] For the most part, the observables calculated in these studies--e.g. phase shifts, binding energies and total nucleon densities--are sensitive only to gross features of the wavefunctions. Reaction calculations, on the other hand, are generally more sensitive to details of the wavefunctions. It is therefore of interest to test further the wavefunctions of Dirac phenomenology by using them in such calculations.

Until very recently, the only calculations of this type to be published were for the (p, π^+) reaction[6] which is a fundamentally relativistic process. However, such calculations--though they provide a successful description of the data in many cases--cannot be considered as tests of the Dirac wavefunctions since there are so many unanswered questions about the nature of the reaction mechanism. Calculations of more conventional nuclear reactions should constitute more readily interpretable tests. We at the University of Colorado have recently done preliminary relativistic (p,d) and (p,n) calculations.[7] In these calculations, the usual Schroedinger continuum and bound state wavefunctions were replaced by the large components of solutions to the Dirac equation containing phenomenological scalar and 4th component vector potentials adjusted to reproduce elastic scattering data. The Dirac solutions differ from their non-relativistic counterparts in that a damping factor is present as a consequence of the Darwin term[8] which appears when the Dirac equation is manipulated to resemble a Schroedinger equation. This term has the same origin as the spin orbit term and is correspondingly large. It has little influence on elastic scattering.[1] The resulting damping factor is

$$\sqrt{1 - \frac{V_v(r) - V_s(r)}{E + m}}$$

where V_v and V_s are the vector and scalar potentials and E and m are the projectile total energy and mass. This factor is unity at large distances, but roughly 0.75 at the center of a nucleus. This represents an appreciable reduction of flux in the nuclear interior. The overall effect in reaction calculations is further increased since such a factor is present for <u>each</u> distorted wave. Figs. 1 and 2 show that sizeable modifications of calculated observables can result when the Dirac wavefunctions are used.

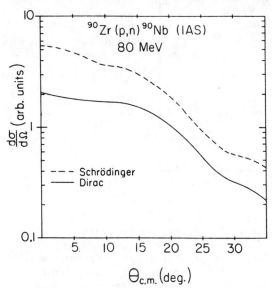

Fig. 1. Comparison of Schrödinger and Dirac calculations for the reaction $^{90}Zr(p,n)^{90}Nb$ (isobaric analog state) at T_p=80 MeV.

What do we learn from such preliminary calculations? Given their shortcomings--the most important being that non-relativstic interactions have been used and small components ignored--we are in fact only able to draw the general conclusion that relativistic effects are likely to be appreciable. This provides motivation to begin the non-trivial task of doing better calculations.

The formulation of a consistent (within the confines of Dirac phenomenology) relativistic treatment of nuclear reactions has begun, but specific calculations are not yet available. Nevertheless, some general features of such an approach can already be anticipated. For instance, the small components of the bound and continuum wavefunctions--which have been ignored in our calculations to date--will of course contribute to reaction amplitudes. For pseudo-scalar coupling between interacting nucleons, for example, <u>only</u> large component to small component couplings are present. Since the magnitudes of the small components depend strongly on

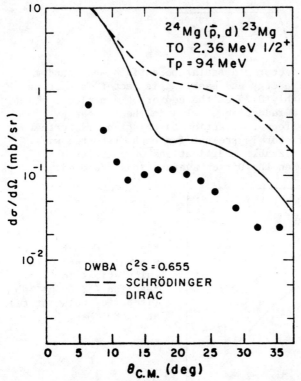

Fig. 2a. Zero-range
DWBA calculations
for the reaction
^{24}Mg(p,d)^{23}Mg (2.36
MeV, 1/2$^+$) at T_p=
94 MeV using
Schrödinger and
Dirac wavefunctions
are compared with
data.

kinetic energy, appreciable energy dependence in the calculations
can be expected. The magnitudes of the small components are also
influenced by the strengths of the strong vector and scalar
potentials which are at the heart of present Dirac phenomenology.
Consequently those interactions which involve small components are
likely to depend strongly on the local values of the vector and
scalar potentials. These effects can be illustrated by considering
specific reactions and we begin with electron scattering.
   Electron scattering probes the nuclear current,[9]

$$J^{\mu}_{fi} = \overline{\Psi}_f \left[\gamma^{\mu} F_1(q^2) + \frac{i\kappa\sigma^{\mu\nu}q_{\nu}}{2m} F_2(q^2) \right] \Psi_i \qquad (1)$$

where Ψ refers to the nucleon spinor, the F's are the nucleon form
factors, κ is the appropriate nucleon anomalous magnetic moment, m
the nuclear mass, q the momentum transfer and γ^{μ} and $\sigma^{\mu\nu}$ are Dirac
matrices. This is a relativistic object and usually a non-relativ-
istic reduction is made. The process begins by assuming that F, κ,
and m have their free space values. The large component, u, of the
nucleon spinor is then identified with the non-relativistic
wavefunction, φ , and the small component, w, is assumed to be
related to the large component by the free Dirac equation,

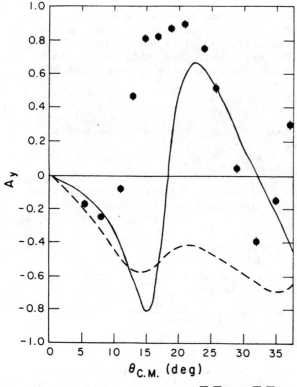

Fig. 2b. Zero-range DWBA calculations for the reaction ^{24}Mg(p,d)^{23}Mg (2.36 MeV, 1/2$^+$) at T_p= 94 MeV using Schrödinger and Dirac wavefunctions are compared with data.

$$w = \frac{\overline{\sigma \cdot p}}{E+m} u \rightarrow \frac{\overline{\sigma \cdot p}}{E+m} \varphi \qquad (2)$$

where E, m, and p are the free space nucleon energy, mass and momentum, respectively.

Of course, a non-relativistic reduction is not necessary when bound state wavefunctions from Dirac phenomenology are used. The current of Eq. 1 can be evaluated as it stands. It is, nevertheless, useful to make a reduction in order to compare with the non-relativistic approach. In doing this, we find that, at the very least, Dirac phenomenology suggests that the use of the free space values for E, m, and p for bound nucleons is inappropriate since the strong vector and scalar fields of the nucleus modify these quantities as follows (in the limit of infinite nuclear matter):

$$E \rightarrow \mathcal{E} \equiv E - V_v \qquad (3a)$$
$$m \rightarrow \mathcal{m} \equiv m + V_s \qquad (3b)$$
$$p \rightarrow \rho = \sqrt{\mathcal{E}^2 - \mathcal{m}^2} \qquad (3c)$$

Beyond this, there are indications that the anomalous magnetic moments and the nucleon form factors can also be appreciably modified by the nuclear medium.[10]

For the moment, let us examine the consequence of medium modification of the kinematical quantitites only. The effect on the magnetization density (second term, Eq. 2) is obvious since it contains a factor $1/m \rightarrow 1/\mathcal{M}$. For $V_s \simeq -450$ MeV, this factor alone suggests an increase of $\sim$ x 2. The convection current terms are also modified. Consider

$$\vec{J}_c \equiv \bar{\Psi}_f \vec{\gamma} \, \Psi_i$$

where the form factor has been dropped. Performing the implied matrix multiplication gives

$$\vec{J}_c \equiv u_f^* \, \vec{\sigma} \, w_i + w_f^* \, \vec{\sigma} \, u_i$$

in terms of large and small components. Thus small components enter at lowest order. To calculate them in a way consistent with Dirac phenomenlogy, we use

$$w = \frac{\vec{\sigma} \cdot \vec{p}}{\mathcal{E} + \mathcal{M}} \, u, \tag{4}$$

rather than Eq. 2. The importance of this difference can be estimated qualitatively by examining the ratio

$$\left(\frac{\vec{p}}{\mathcal{E} + \mathcal{M}} \right) \div \left(\frac{p}{E + m} \right) = \sqrt{\left(\frac{T - V_c}{T} \right) \cdot \left(\frac{E + m}{E + m - V_v + V_s} \right)} \tag{5}$$

where $V_c = V_v + V_s$ and corresponds to an effective central potential. The first factor on the right-hand-side of Eq. 5 is therefore the usual binding energy effect. The second factor comes from Dirac phenomenology and is appreciably different from unity for the same reason that the nucleon-nucleus spin orbit potential is large, i.e. $V_v - V_s \simeq 800$ MeV is very large relative to the nucleon rest mass. For $E = m$, the second factor on the r.h.s. of Eq. 5 is found to be ~ 1.3. Thus, the strong nuclear vector and scalar potentials have the effect of increasing the magnitude of the small components of bound nucleons by up to 30%. Consequently, for electron scattering, operators containing factors of $p/(E+m)$ will be increased by up to 30% due to effects of the nuclear medium.

The above considerations are quite schematic and the large effects discussed may be masked or cancelled by others which have been ignored. For instance, Noble[10] has suggested that the nucleon form factor "softens" as effective mass decreases which would tend to reduce the kinematic effects discussed above. In any case, the time appears to be ripe for testing the above speculations by performing explicit electron scattering calculations using bound state wavefunctions from Dirac phenomenology.

Nucleon induced inelastic scattering can also be examined within the framework of Dirac phenomenology. The simplest approach is to use a relativistic impulse approximation. The free nucleon-nucleon invariant amplitude can be parameterized as[11]

$$(2ip)^{-1}f_{NN} = A + \bar{\sigma}_1 \cdot \bar{\sigma}_2 B + i(\bar{\sigma}_1 + \bar{\sigma}_2) \cdot \bar{q} \times \hat{z} \, C + \bar{\sigma}_1 \cdot \hat{q} \bar{\sigma}_2 \cdot \hat{q} \, D$$
$$+ \bar{\sigma}_1 \cdot \hat{z} \bar{\sigma}_2 \cdot \hat{z} \, E \tag{6a}$$

$$= \bar{\Psi}_{1f} \bar{\Psi}_{2f} \sum_j F_j \Psi_{1i} \Psi_{2i} \tag{6b}$$

where the first expression is the usual (local) non-relativistic parametrization while the second is a relativistic one where:

$$\Psi = \sqrt{\frac{E+m}{2m}} \begin{pmatrix} 1 \\ \frac{\bar{\sigma} \cdot \bar{p}}{E+m} \end{pmatrix} \chi_{Pauli} \, e^{-ip \cdot x}$$ is a free nucleon Dirac spinor, and

$$\sum_j F_j = F_S + \gamma_1 \cdot \gamma_2 F_V + \gamma_1^5 \gamma_2^5 F_P + \gamma_1^5 \gamma_2^5 \gamma_1 \cdot \gamma_2 F_A + \sigma_1^{\mu\nu} \sigma_{2\mu\nu} F_T \tag{7}$$

is a phenomenological invariant interaction constructed to give the invariant amplitude in Born approximation. In this sense, $\sum_j F_j$ is a relativistic analogue of effective interactions such as those of Love and Franey.[12] The simplest form of a relativistic impulse approximation consists of replacing the plane wave spinors in Eq. 6b with the relevant bound state and continuum wavefunctions from Dirac phenomenology. Although such an amplitude would be evaluated without further approximation, it is again useful to make reductions in order to compare the usual non-relativistic formulations.

We see, for example, an explicit energy dependence in the effective interaction. Consider the NN spin orbit amplitude due to F_S and F_V:

$$f_{NN}^{so} = \frac{i(\bar{\sigma}_1 + \bar{\sigma}_2) \cdot \bar{q} \times \hat{z}}{4\sqrt{2}} \cdot \frac{p}{E} \cdot (3F_V(q) - F_S(q)), \tag{8}$$

there being an explicit energy dependence through $p/E=\beta$. Perhaps this kind of energy dependence will be found to explain the bulk of that observed phenomenologically in the same way that Dirac phenomenology does for nucleon-nucleus elastic scattering.[1]

When reduced to a non-relativistic form, an apparent medium dependence in the effective interaction also is found, much as it was for electron scattering except that now the projectile is also subject to the strong nuclear potentials. Again, this medium effect is a consequence of $V_V - V_S$ being large compared to the nucleon mass as is required to describe the nucleon-nucleus spin orbit potential. Since these large potentials increase the small components of the the nucleon wavefunctions, those interactions appearing in the non-relativistic reduction which originate with the small components can be appreciably affected. An example of such an interaction is the spin orbit amplitude of Eq. 8. Fig. 3 shows f_{pp}^{so} calculated using Eq. 8 and assuming F_V and F_S are given by the Born terms of OBE potentials[13] arising from σ and ω exchange. We then have

294

$$F_v \rightarrow \frac{g_\omega^2}{4\pi} \cdot \frac{1}{q^2+m_\omega^2} \cdot F^2(q^2) \qquad (9)$$

$$F_s \rightarrow \frac{-g_\sigma^2}{4\pi} \cdot \frac{1}{q^2+m_\sigma^2} \cdot F^2(q^2) \qquad (9)$$

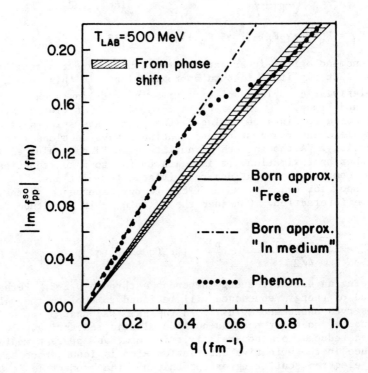

Fig. 3. The Born approximation pp spin-orbit amplitude for T_{lab}= 500 MeV is compared with the range of phenomenological amplitudes determined from phase shifts. The solid line is the free space amplitude while the dashed line uses kinematical quantities reflecting the presence of the nuclear vector and scalar potentials. The dotted curve is the phenemenological amplitude of Hoffmann et al.[15] determined by fitting $\vec{p}$-nucleus elastic scattering data.

where both direct and exchange contributions must be considered. As Fig. 3 shows, the free space Born approximation is in excellent agreement with the phenomenological amplitude extracted from phase shift analyses.[11]

The nuclear vector and scalar potentials alter E, m, and p for the nucleon. For V_v= +253 MeV and V_s= -437 MeV,[1],[14] E=1054 MeV → 800 MeV, m = 935 MeV → 498 MeV, and p = 483 MeV/c → 626 MeV/c. Making these substitutions in Eqs. 8 and 9, the dashed curve of Fig. 3 results. Note that for q < 0.6 fm^{-1} the increased medium-modified amplitude agrees qualitatively with the phenomenological amplitude found necessary to fit $\vec{p}$ + nucleus elastic scattering at T_p=500 MeV.[15] (Qualitative uncertainty principle arguments suggest that the mean-field concept may not be appropriate for q > 1/R, R = internucleon separation, and that f→f_{free} in this region as also found in Ref. 15.) It should be emphasized that this apparent medium dependence appears only when a non-relativistic reduction is made and that the relativistic interactions of Eq. 6b are assumed to be those appropriate to free space.

As was the case for electron scattering, we have ignored other effects which might substantially alter those noted above. Nevertheless, there seems to be ample motivation for doing specific nucleon-induced inelastic scattering calculations based on the simple relativistic impulse approximation discussed above.

To summarize, we have shown that the large vector and scalar potentials required in Dirac phenomenology to reproduce simultaneously the phenomenological central and spin orbit strengths of the nucleon-nucleus interaction also enhance the small components of nucleon single-particle wavefunctions. This enhancement can appreciably alter reaction amplitudes in electron-or nucleon-induced inelastic scattering processes. While the present work has ignored other relevant effects, it nevertheless seems likely that relativistic effects will be appreciable and that specific, consistent calculations are warranted. Such calculations will contitute a stringent test of Dirac phenomenology.

REFERENCES

1. e.g., L. G. Arnold et al., Phys. Rev. C 25 (1982) 936.
2. e.g., A. E. S. Green and T. Sawada, Rev. Mod. Phys. 39 (1967) 594.
3. J. D. Walecka, Ann. Phys. 83 (1974) 491.
4. e.g., L. D. Miller and A. E. S. Green, Phys. Rev. C 5 (1972) 241 and C. J. Horowitz and Brian D. Serot, Nucl. Phys. A368 (1981) 503.
5. J. R. Shepard, E. Rost, and D. Murdock, Phys. Rev. Lett. 49 (1982) 14.
6. e.g., E. D. Cooper and H. Sherif, Phys. Rev. C 25 (1982) 3024.
7. E. Rost, J. R. Shepard, and D. Murdock, Phys. Rev. Lett. 49 (1982) 443.
8. C. G. Darwin, Proc. Roy. Soc. (London) A118 (1928) 654.
9. T. W. Donnelly and J. D. Walecka, Ann. Rev. Nucl. Sci. 25 (1975) 329.

296

10. J. V. Noble, Phys. Rev. Lett. 46 (1981) 412.
11. J. A. McNeil, Ph.D. Thesis, U. of Maryland (1979), unpublished;
    and S. J. Wallace, Adv. in Nucl. Phys. 12 (1982) 135.
12. W. G. Love and M. A. Franey, Phys. Rev. C 24 (1981) 1073.
13. A. Gersten, R. H. Thompson, and A. E. S. Green, Phys. Rev. D 3
    (1971) 2076.
14. L. G. Arnold et al., Phys. Rev. C 23 (1981) 1949.
15. G. W. Hoffmann et al., Phys. Rev. Lett. 47 (1981) 1436.

DISCUSSION

Noble: I was just wondering if you could say, in a word, what you
did for the deuteron?

Shepard: What we did was to assume that we can write down the free
Hamiltonians for 2 particles, add these Hamiltonians together and
then assume that there was some interaction between each particle and
the nucleus which we took to be exactly the same as the relativistic
interaction of a free nucleon with a nucleus. We also assume that
there is no internal structure to the deuteron. In the end we get 4
coupled equations which have to be solved.

Brown: I find something is fishy with changing the small component
that much. Let me explain. You have to properly compare with pair
theory in order to see this . You just can't do it in the way you
have. Suppose I start with a wave packet in a plane wave
representation with all the particles in positive energy states.
Then I use the projection operator on the positive energy components
for a free particle . If you want to change the ratio of small to
large components in this theory you can only do it by introducing
negative energy states, because if I have only positive energy states
that ratio is set. The only way of introducing negative energy
states, if I handle pair theory properly, is through the $V_\pm$, with the
Λ^+ on one side and Λ^- on the other. But that involves energy
denominators of two nucleon masses. You haven't got any $V_\pm$ in your
equations, so you can't do that, and if you did put a $V_\pm$ in, the
change in the small components would be damped a lot by these large
energy denominators and so I don't think that you can change the
small components much. I think the large lower components are an
artifact of making a reduction to a one body theory and not handling
virtual pairs correctly.

Shepard: Perhaps a service has been performed by bringing this matter to the attention of the meeting. It's obviously a relevant comment.

Walecka: Maybe Carl will comment on this.

Shakin: Let me say something now. It is very bad to think in that perturbative language. It's much better to discuss and think of this as a theory in which the nucleon mass, which appears as a parameter in the spinor is varied. Now, when you vary the nucleon mass, you can take the new spinor (with the new mass parameter) and if you wish expand it in terms of the spinor solutions of the free Dirac equation parametrized by the vacuum mass. The latter is useful for calculations but it's a very poor way to talk about what's going on.

Friar: Is the Darwin term you talked about, the same Darwin-Foldy interaction that plays such a big role in the hydrogen atom?

Shepard: Yes, it is the same Darwin term that appears in the calculation of the Lamb shift..

Friar: ... not the Lamb shift, the fine structure.

Shepard: Yes, it's exactly the same one. And it is small there because there is only the fourth component of the vector potential present. Here, by introducing vector and scalar potentials of opposite signs you allow that thing to become much larger.

Hwang: I would like to make two comments. The first comment is: I agree with you in that the electro-weak interaction may be the ideal place to test the Dirac phenomenology. However, I think that this game has already been played by several authors, including myself, and for example all the stuff about the convection current is in fact already contained in a published paper. The other comment is that people didn't make clear that the upper component of the Dirac solution can be identified directly with the Schrodinger solution, to obtain a conventional convection current plus a spin current, and this in fact is not as naive as you tried to describe. The second point is: I think that if one tries to become serious about Dirac phenomenology one has to recognize that you also have to try to use consistently the transformation properties of your solutions. In other words you also must introduce a number of recoil corrections. In this language this means doing a number of Lorentz transformations on each constituent particle. It turns out that those terms are not very big, so I conclude that if Dirac phenomenology can ever be tested up to the final form, all these questions have to be adressed in a very careful manner.

Shepard: Well, I never intended this to be a definitive comment on Dirac phenomenology. As applied to nuclear reactions I think that is obvious. This was intended to be an introduction to that particular subject. It certainly was for me.

Feshbach: A lot of us were very impressed with Dirac Hartree. I have
one word of caution, namely don't throw out the energy dependence of a
lot of these parameters because you're throwing out the baby with the
bath water. That energy dependence comes from real physical effects
(it is not just something that comes from eliminating the lower
components). These effects that have to do with the structure of the
nucleus. You just have to be careful and realise these effects are
there. In other words, if you're going to tell me that you've
explained all the energy dependence then I'm going to tell you you're
wrong.

Shepard: Sure the antisymmetrization was never discussed in anything
I said. Whether we use the Dirac equation or the Schrodinger equation
antisymmeterization has to be built in. I guess I'm aware that this
isn't everything.

RELATIVISTIC BRUECKNER-HARTREE-FOCK THEORY:
THEORETICAL FOUNDATIONS AND EMPIRICAL EVIDENCE*

C. M. Shakin
Department of Physics and Institute for Nuclear Theory
Brooklyn College of the City University of New York
Brooklyn, New York 11210

ABSTRACT

It is suggested that the nucleon mass may be quite different
in the interior of a large nucleus than in free space. The
consequences of this assumption are explored using a model based on
the one-boson-exchange model of the nuclear force. We present the
results of parameter-free calculations of nuclear saturation curves,
of the density and momentum-transfer dependence of the effective
force in nuclei, and of the energy and density dependence of the
nuclear optical potential. We also put forth some speculations
concerning the relation of boson-exchange models and quark models of
the nucleon-nucleon interaction.

This talk describes research carried out at Brooklyn College
by Professor L. S. Celenza, Dr. M. Anastasio, Dr. W. S. Pong and
myself. References to those relativistic models which make use of
the Hartree or mean-field approximations may be found in our
published papers. It is anticipated that the work of other authors
in this field will be adequately summarized by the other speakers
at this conference. Therefore, no attempt has been made to review
the large number of works that have been published dealing with
relativistic models of nuclear structure.

There are essentially two ways to introduce the material of
this talk. The first follows the more traditional approach in
which one extends the standard self-consistent field theories
(Hartree or Hartree-Fock) to include correlation effects in order
to describe the physics of nuclei and nuclear matter. We can
assume that most physicists are familiar with the Hartree-Fock
approximation for many-body systems and most nuclear physicists are
familiar with Brueckner's contribution. A new element is introduced when one considers relativistic effects. In the non-
relativistic theories there is no question as to the wave function
of a nucleon in nuclear matter. It is a plane wave multiplied by
the appropriate Pauli spinors describing the spin and isospin state
of the nucleon. If one desires a relativistic formulation one can
replace the non-relativistic wave function by a spinor solution of
the Dirac equation and, indeed, this has been done by several
authors.[1] The essential issue we raise here is the nature of this

*Presented at the Indiana University Cyclotron Facility Workshop:
The Interaction Between Medium Energy Nucleons in Nuclei;
McCormick's Creek State Park, October 28-30, 1982.

relativistic spinor. There are good reasons to believe that one should not use the positive-energy solution of the <u>free</u> Dirac equation, $u(\vec{p},s)$, but one should determine the appropriate spinor by solving a Dirac equation containing a self-energy operator:[2]

$$\left[\, \vec{\gamma}\cdot\vec{p} + m_N + \Sigma(k_F,\vec{p},\{f(\vec{p},s)\})\,\right] f(\vec{p},s) = \gamma^o E(\vec{p}) f(\vec{p},s) \qquad (1)$$

In Eq. (1) we have indicated that the self-energy, Σ, which depends on the density of the medium, is also a functional of the $f(\vec{p},s)$. Thus we have a typical problem of self-consistency <u>even for nuclear matter</u>.

Reasons for considering the solutions of Eq. (1) to be important for the study of nuclear matter have been given elsewhere. Essentially, the solution of Eq. (1) takes one to a self-consistent representation in which the system is stable against the admixture of particle-hole states into the wave function.[2,3] Here we are extending our concept of "hole" to include holes in the negative-energy Dirac sea. From this point of view, conventional calculations[1] which use the $u(\vec{p},s)$ instead of the $f(\vec{p},s)$ are not self-consistent.. These ideas have been discussed at some length in various publications.[2,3]

We would like to take this opportunity to present the material of this talk from the second point of view alluded to above. As we will see, the mass of the nucleon in nuclear matter plays an essential role in this discussion and we will also see that the nucleon mass in nuclear matter (or in finite nuclei) is quite different from that in free space. How are we to understand this phenomenon?

The origin of the nucleon mass is not well understood; however there are several models such as the σ-model or the MIT bag model that give us some insight. Probably the most sophisticated approach exists at the quark level and involves mass generation for the quarks through some mechanism of dynamical symmetry breaking.[4] Here one attempts to understand the origin of what may be termed the "constituent" quark mass. This number is about 340 MeV for the non-strange quarks. In the weak-binding limit the nucleon mass is then $m_N \sim 3m_q^{con}$. (We will not attempt to distinguish between the "constituent" and the "dynamical" quark mass).[4]

Let us, for the moment, assume that one understands mass generation for the quarks (or the nucleon) when these objects are in vacuum. The essential question raised by the relativistic theories is the question of dynamical symmetry breaking and mass generation in the presence of hadronic matter. Is is quite possible that the quark (constituent) mass or the nucleon mass is different in a nuclear environment than in vacuum. In the relativistic model considered here a specific modification for the nucleon mass in hadronic matter is suggested. The equation governing the modification of the mass from the vacuum mass is:[4]

$$\tilde{m}(p^2) = m_N + \tfrac{1}{4} \, \mathrm{Tr} \, \Sigma(p^2,\tilde{m}(p^2),k_F) \qquad (2)$$

where Σ is again the self-energy of the nucleon due to the presence of matter. If we determine the relation between the energy and the momentum, $p^0 \equiv E(\vec{p})$, we can write Eq. (2) as

$$\tilde{m}(\vec{p}) = m_N + \tfrac{1}{4} \, \mathrm{Tr} \, \Sigma(\vec{p}, \tilde{m}(\vec{p}), k_F). \tag{2'}$$

For example, if $\Sigma(\vec{p}) = A(\vec{p}) + \gamma^0 B(\vec{p})$ we would have, for $p^0 > 0$,

$$p^0 = E(\vec{p}) = B(\vec{p}) + \{\vec{p}^2 + [\, m_N + A(\vec{p})\,]^2\}^{\frac{1}{2}} \tag{3}$$

as the dispersion relation relating the quasi-particle energy and momentum in the nuclear matter. Further we see that

$$\tilde{m}(\vec{p}) = m_N + A(\vec{p}). \tag{4}$$

The models to be discussed here are characterized as having large negative values for $A(\vec{p})$, of the order of -400 MeV, and large positive values for $B(\vec{p})$ of about 300 MeV.[5,6]
For the above form of $\Sigma(\vec{p})$ we obtain,

$$f(\vec{p},s) = \sqrt{\frac{E_N(\vec{p})}{m_N} \frac{\tilde{m}(\vec{p})}{\tilde{E}(\vec{p})} \left(\frac{\tilde{E}(\vec{p}) + \tilde{m}(\vec{p})}{2\tilde{m}(\vec{p})}\right)} \left(\begin{array}{c} \chi_s \\ \dfrac{\vec{\sigma}\cdot\vec{p}}{\tilde{E}(\vec{p}) + \tilde{m}(\vec{p})} \chi_s \end{array}\right) \tag{5}$$

$$= \sqrt{\frac{E_N(\vec{p})}{2m_N} \frac{\tilde{\epsilon}(\vec{p})}{\tilde{E}(\vec{p})}} \left(\begin{array}{c} \chi_s \\ \dfrac{\vec{\sigma}\cdot\vec{p}}{\tilde{\epsilon}(\vec{p})} \chi_s \end{array}\right) \tag{6}$$

where $\tilde{\epsilon}(\vec{p}) = [\,\vec{p}^2 + \tilde{m}^2(\vec{p})\,]^{\frac{1}{2}} + \tilde{m}(\vec{p})$, $E_N(\vec{p}) = [\,\vec{p}^2 + m_N^2\,]^{\frac{1}{2}}$
and $\tilde{E}(\vec{p}) = [\,\vec{p}^2 + \tilde{m}^2(\vec{p})\,]^{\frac{1}{2}}$. We have chosen the normalization,

$$f^\dagger(\vec{p},s')f(\vec{p},s) = (E_N(\vec{p})/m_N)\delta_{ss'} \tag{7}$$

for the spinors $f(\vec{p},s)$. Note that if $\tilde{m} \to m_N$ we have $f(\vec{p},s) \to u(\vec{p},s)$, where

$$u(\vec{p},s) = \sqrt{\frac{E_N(\vec{p}) + m_N}{2m_N}} \left(\begin{array}{c} \chi_s \\ \dfrac{\vec{\sigma}\cdot\vec{p}}{E_N(\vec{p}) + m_N} \chi_s \end{array}\right) \tag{8}$$

is the standard, positive-energy solution of the free Dirac equation normalized such that $\bar{u}(\vec{p},s)\, u(\vec{p},s') = \delta_{ss'}$.
The details of the solution of Eqs. (1) and (2) have been discussed extensively elsewhere.[5,6] In this work we want to discuss the consequences which follow when one uses the (self-consistent) spinors $f(\vec{p},s)$ instead of the solutions parameterized by the vacuum mass, the $u(\vec{p},s)$ of Eq. (8). The appearance of the vacuum parameter, m_N, in the definition of $f(\vec{p},s)$ is the result of the normalization choice made in Eq. (7). When calculating physical quantities in nuclear matter the factor $[\, E_N(\vec{p})/m_N \,]^{-\frac{1}{2}}$

appearing in $f(\vec{p},s)$ is cancelled by corresponding factors in the relativistic density of states – see the Appendix.

The calculations we report will be based upon the one-boson exchange model of nuclear forces.[1] The use of the Lagrangian of this model provides a specific scheme in which one can construct solutions to Eqs. (1) and (2). Some speculations concerning the relation of the OBE model to the quark model of nucleon structure will be put forth at the end of this presentation.

We now turn to a comparison of two models for the calculation of various nuclear properties. These models differ in that the spinors of Model 1 are the $u(\vec{p},s)$ while the spinors used in Model 2 are the $f(\vec{p},s)$ obtained from the solution of Eq. (1). For example, the expression for the energy of nuclear matter is:[2,11]

$$E_1 = \sum_s \int \frac{d\vec{q}}{(2\pi)^3} \frac{m_N}{E_N(\vec{q})} \; \bar{u}(\vec{q},s)(\vec{\gamma}\cdot\vec{q} + m_N)u(\vec{q},s)$$

$$+ \tfrac{1}{2} \sum_{ss'} \iint \frac{d\vec{p}}{(2\pi)^3} \frac{d\vec{q}}{(2\pi)^3} \frac{m_N}{E_N(\vec{p})} \frac{m_N}{E_N(\vec{q})}$$

$$\times \; <\bar{u}(\vec{p},s)\bar{u}(\vec{q},s') \mid \hat{M}(1-P_{12}) \mid u(\vec{p},s)u(\vec{q},s')> \tag{9}$$

in Model 1 and

$$E_2 = \sum_s \int \frac{d\vec{q}}{(2\pi)^3} \frac{m_N}{E_N(\vec{q})} \; \bar{f}(\vec{q},s)(\vec{\gamma}\cdot\vec{q} + m_N)f(\vec{q},s)$$

$$+ \tfrac{1}{2} \sum_{ss'} \iint \frac{d\vec{p}}{(2\pi)^3} \frac{d\vec{q}}{(2\pi)^3} \frac{m_N}{E_N(\vec{p})} \frac{m_N}{E_N(\vec{q})}$$

$$\times \; <\bar{f}(\vec{p},s)\bar{f}(\vec{q},s') \mid \hat{M}(1-P_{12}) \mid f(\vec{p},s)f(\vec{q},s')> \tag{10}$$

in Model 2 (see the Appendix). Here $\hat{M}$ is a reaction matrix which includes Pauli Principle and dispersive effects and P_{12} is an exchange operator. (We suppress reference to isospin for simplicity).

In the following we will discuss the topics:
1. Nuclear saturation[5,6] and nuclear compressibility.[7]
2. The density dependence of the interaction[7,8] between nucleons in nuclear matter and finite nuclei.
3. The momentum transfer dependence of the nuclear interaction.[9]
4. The energy and density dependence of the nuclear optical potential.[10]

In each case we will contrast the results of Model 1 and Model 2.

Nuclear saturation curves have been calculated for various one-boson-exchange models.[1,5,6] In general, calculations using

Model 1 yield saturation points for various potentials which lie on a curve ("the Coester line") which does not pass through the empirical values, B.E./A≈15.6 MeV and ρ_0 = 0.17 fm^{-3}. In Fig. 1 we contrast results of calculations[6] using Model 1 (dashed curves) with calculations using Model 2 (solid lines). It is clear that a line drawn through the saturation points for the three Model 2 calculations (solid lines) will yield a new "Coester line" which does pass through the region of the figure specifying the empirical values (rectangular box) for the binding energy and saturation density.

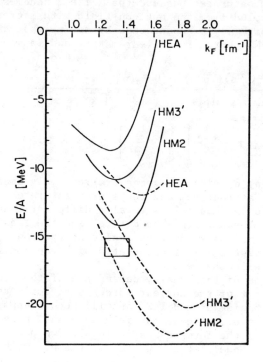

Fig. 1. The dashed lines labeled HEA, HM2 and HM3' denote the saturation curves for Model 1 for the potentials of Refs. 12, 13 and 14, respectively. The solid lines are the saturation curves for these potentials when calculations are performed using Model 2. (This figure appears in Ref. 6).

It is worth noting that at low density (k_F < 1.0 fm^{-1}) the dashed and solid curves for each potential will agree. In a rough approximation the solid curve may be obtained from the dashed curves by adding $\Delta E/A \simeq 3.6 \, (\rho/\rho_0)^{2.4}$ where ρ_0 is the density at k_F = 1.36 fm^{-1}.

The nuclear incompressibility parameter, K_∞, has been calculated in Ref. 7 with the result that K_∞ = 175 MeV for the potential HEA of Ref. 12 and K_∞ = 128 MeV for the potential HM2 of Ref. 13. A value of $K_\infty \sim$ 200 MeV seems to be the generally accepted value for this quantity at this time. It appears that the potential HEA gives an acceptable value for K_∞; however, we should note that the value obtained for this quantity is sensitive to the details of the calculation.

It is worth remarking upon the saturation mechanism in the relativistic theory. In the one-boson-exchange models the mesons σ, ω, ρ and π play important roles.[1] A good deal of the attraction in the nuclear force arises from σ exchange. In the relativistic theory, the system saturates because the σ field becomes increasingly decoupled from the nucleon field with increasing density. This remark may be understood by noting that the ratio of the value of the σNN vertex in nuclear matter and in vacuum is given by

$$R = \frac{\bar{f}(\vec{p},s)f(\vec{p},s)}{\bar{u}(\vec{p},s)u(\vec{p},s)} = \frac{E_N(\vec{p})}{m_N} \frac{\tilde{m}}{\left[\vec{p}^2 + \tilde{m}^2\right]^{\frac{1}{2}}} \tag{11}$$

$$\simeq \left\{ 1 + \tfrac{1}{2}\vec{p}^2 \left[\frac{1}{m_N^2} - \frac{1}{\tilde{m}^2}\right] + \cdots \right\}. \tag{12}$$

Thus with increasing density R becomes progressively smaller, the force becomes less attractive, and the system saturates. The saturation mechanism in Model 2 is totally different than in Model 1. In Model 1 one looks to the "hard core" and the tensor force to provide mechanisms for saturation, while saturation in Model 2 is a purely relativistic effect associated with the modifications of the nucleon mass in the medium.

The explanation of the density dependence of the isoscalar, spin-independent part of the nuclear force is directly related to the saturation mechanism of Model 2, as will become apparent in the following discussion of the quasiparticle interaction in nuclear matter. We consider the <u>forward scattering</u> of two particles at the Fermi surface.[7] In Model 1 the amplitude would be, with $|\vec{p}_1| = |\vec{p}_2| = k_F$,

$$\mathcal{F}_1(\theta) = \left[\frac{m_N}{E_N(k_F)}\right]^2 <\bar{u}(\vec{p}_1 s_1')\bar{u}(\vec{p}_2 s_2')|\hat{M}(1-P_{12})|u(\vec{p}_1 s_1)u(\vec{p}_2 s_2)>, \tag{13}$$

while in Model 2 we would have

$$\mathcal{F}_2(\theta) = \left[\frac{m_N}{E_N(k_F)}\right]^2 <\bar{f}(\vec{p}_1 s_1')\bar{f}(\vec{p}_2 s_2')|\hat{M}(1-P_{12})|f(\vec{p}_1 s_1)f(\vec{p}_2 s_2)>. \tag{14}$$

Note that θ is the angle between $\vec{p}_1$ and $\vec{p}_2$ and is <u>not</u> the scattering angle. It is useful to write

$$\mathcal{F}(\theta) = f(\theta) + f'(\theta)\vec{\tau}_1 \cdot \vec{\tau}_2 + \left[g(\theta) + g'(\theta)\vec{\tau}_1 \cdot \vec{\tau}_2 \right]\vec{\sigma}_1 \cdot \vec{\sigma}_2$$

$$+ \frac{1}{k_F^2} \left[h(\theta) + h'(\theta)\vec{\tau}_1 \cdot \vec{\tau}_2 \right] S_{12}(\vec{p}) + \cdots \tag{15}$$

where S_{12} is the tensor operator and $\vec{p} = (\vec{p}_1 - \vec{p}_2)/2$. Following Migdal,[15] one expands the amplitudes f, f', g, g', etc. in terms of Legendre polynominals:

$$f(\theta) = \sum_{\ell} f_\ell P_\ell (\cos\theta), \cdots . \tag{16}$$

It is also convenient to introduce the quantities[7]

$$F_\ell^{NR} = N_o f_\ell^{NR}, \quad G_\ell^{NR} = N_o g_\ell^{NR}, \text{ etc.,} \tag{17}$$

where

$$N_o = 2k_F m_N / (\pi^2 \hbar^2). \tag{18}$$

Here the superscript NR refers to the amplitudes of Model 1. For Model 2 we define

$$F_\ell = N_o f_\ell, \quad G_\ell = N_o g_\ell, \text{ etc.} \tag{19}$$

In many calculations, the quasiparticle interaction $\mathcal{F}(\theta)$ has been approximated by a zero-range, density-dependent interaction in order to study collective excitations and electromagnetic transitions in the lead region.[16] (That is, the terms with $\ell > o$ and the tensor interactions have been dropped). In these studies the interaction is taken to be

$$V(1,2) = N_o^{-1}\delta(\vec{r}_1 - \vec{r}_2)\left[F_o + F_o'\vec{\tau}_1 \cdot \vec{\tau}_2 + G_o\vec{\sigma}_1 \cdot \vec{\sigma}_2 + G_o'\vec{\sigma}_1 \cdot \vec{\sigma}_2\vec{\tau}_1 \cdot \vec{\tau}_2 \right]. \tag{20}$$

(At this point, we do not distinguish F_o from F_o^{NR}, etc.). Usually G_o, G_o' and F_o' are assumed to be independent of density. Following Migdal's suggestion F_o is taken to be linearly dependent on the density,[16]

$$F_o(r) = F_o^{in} \frac{\rho(r)}{\rho(o)} + F_o^{ex} \left[1 - \frac{\rho(r)}{\rho(o)} \right] \tag{21}$$

Here

$$\frac{\rho(r)}{\rho(o)} = \left[1 + \exp\{(r-R)/a\} \right]^{-1} \tag{22}$$

Recently, empirical studies have found $F_o^{ex} = -3.00$ and $F_o^{in} = -0.30$.[17] There is some uncertainty in these numbers, but it is clear that the isoscalar, spin-independent part of the force is much stronger in the nuclear surface than in the interior of a large nucleus.

In Fig. 2 we show F_0^{NR} (Model 1) and F_0 (Model 2) as a function of k_F. The strong density dependence of F_0 (Model 2) is apparent.

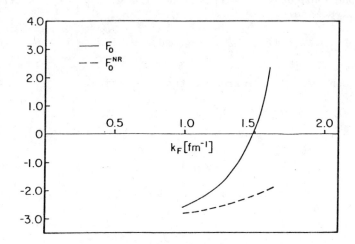

Fig. 2. The parameter F_0 (Model 2) and the parameter F_0^{NR} (Model 1) as a function of k_F. The potential is HM2 of Ref. 13.

A more detailed comparison of the empirical interaction with the calculated values of F_0 and F_0^{NR} is given in Ref. 7.

We have discussed this matter at some length since it is clear that Model 2 can account for most of the strong density dependence found empirically, while the density dependence of Model 1 is rather weak (see Fig. 2). We again refer the reader to Ref. 7 for a very detailed discussion of the Landau-Migdal parameters for the potentials HEA (Ref. 12) and HM2 (Ref. 13).

If the RBHF model discussed here is to be a complete, parameter-free, model of nuclear structure, we should be able to calculate the effective nuclear force as a function of energy, momentum transfer, and density. In Ref. 7 we have presented a complete discussion of the density dependence of the effective force at zero momentum transfer. We now turn to a discussion of the momentum-transfer dependence of the interaction at nuclear matter density and (essentially) at zero energy. (This is the interaction between bound particles and is appropriate for shell-model studies. Calculations of the interaction between inter-mediate-energy projectiles and bound nucleons will be reported elsewhere).

We now extend our consideration to nucleons with momenta $\vec{p}_1 + \vec{q}$ and $\vec{p}_2$. The scattering amplitude in the medium is denoted as $\mathcal{F}(\vec{p}_1, \vec{p}_2 + \vec{q}; \vec{p}_1 + \vec{q}, \vec{p}_2) = \mathcal{F}(\vec{p}_1, \vec{p}_2; \vec{q})$ where $\vec{q}$ is the momentum transfer. Thus in Model 1 we study the amplitude

$$\mathcal{F}(\vec{p}_1,\vec{p}_2;\vec{q}) = \sqrt{\frac{m_N}{E_N(\vec{p}_1)} \cdot \frac{m_N}{E_N(\vec{p}_1+\vec{q})} \cdot \frac{m_N}{E_N(\vec{p}_2)} \cdot \frac{m_N}{E_N(\vec{p}_2+\vec{q})}}$$

$$\times \; <\bar{u}(\vec{p}_1,s'_1)\bar{u}(\vec{p}_2+\vec{q},s'_2)|\hat{M}(1-P_{12})|\, u(\vec{p}_1+\vec{q},s_1)u(\vec{p}_2,s_2)> \quad (23)$$

while in Model 2 we consider

$$\mathcal{F}_2(\vec{p}_1,\vec{p}_2;\vec{q}) = \sqrt{\frac{m_N}{E_N(\vec{p}_1)} \cdot \frac{m_N}{E_N(\vec{p}_1+\vec{q})} \cdot \frac{m_N}{E_N(\vec{p}_2)} \cdot \frac{m_N}{E_N(\vec{p}_2+\vec{q})}}$$

$$\times \; <\bar{f}(\vec{p}_1,s'_1)\bar{f}(\vec{p}_2+\vec{q},s'_2)|\hat{M}(1-P_{12})|\, f(\vec{p}_1+\vec{q},s_1) f(\vec{p}_2,s_2)> \quad (24)$$

We again restrict $\vec{p}_1$ and $\vec{p}_2$ to be at the Fermi surface. (This is the natural extension of the Landau-Migdal analysis of the force which is carried out at $\vec{q} = 0$). Aside from its spin and isospin dependence, the amplitude $\mathcal{F}$ depends on $|\vec{q}|$ and the angles which specify the vectors $\vec{p}_1$ and $\vec{p}_2$. We proceed by fixing the direction of $\vec{q}$ and averaging over the directions of $\vec{p}_1$ and $\vec{p}_2$ with the restriction that $|\vec{p}_1 + \vec{q}|$ and $|\vec{p}_2 + \vec{q}|$ be greater than k_F. This is appropriate since $\mathcal{F}(\vec{p}_1, \vec{p}_2; \vec{q})$ is ultimately used to construct a particle-hole interaction for the study of the nuclear response to electromagnetic or hadronic probes. We denote the amplitude obtained via this averaging procedure as $\bar{\mathcal{F}}(q)$ and write[9]

$$\bar{\mathcal{F}}(q) = \frac{(\hbar c)^3 \pi^2}{2k_F m_N} \left\{ F(q) + F'(q)\vec{\tau}_1\cdot\vec{\tau}_2 + G(q)\vec{\sigma}_1\cdot\vec{\sigma}_2 + G'(q)\vec{\sigma}_1\cdot\vec{\sigma}_2\vec{\tau}_1\cdot\vec{\tau}_2 \right.$$

$$\left. + \frac{q^2}{k_F^2}\left[H(q) + H'(q)\vec{\tau}_1\cdot\vec{\tau}_2\right] S_{12}(\hat{q}) + \cdots \right\}. \quad (25)$$

It is useful to rewrite Eq. (25) using the relation

$$\vec{\sigma}_1\cdot\vec{\sigma}_2 = \vec{\sigma}_1\cdot\hat{q}\,\vec{\sigma}_2\cdot\hat{q} + (\vec{\sigma}_1 \times \hat{q})\cdot(\vec{\sigma}_2 \times \hat{q}) \quad (26)$$

in that part of $\bar{\mathcal{F}}(q)$ proportional to $\tau_1\cdot\tau_2$. We have

$$\bar{\mathcal{F}}(q) = \frac{(\hbar c)^3\pi^2}{2k_F m_N}\left\{\left[F(q) + G(q)\vec{\sigma}_1\cdot\vec{\sigma}_2 + \left(\frac{q}{k_F}\right)^2 H(q)S_{12}(\hat{q})\right]\right.$$

$$\left. + \left[F'(q) + v_\pi(q)\vec{\sigma}_1\cdot\hat{q}\,\vec{\sigma}_2\cdot\hat{q} + v_\rho(q)\,(\vec{\sigma}_1\times\hat{q})\cdot(\vec{\sigma}_2\times\hat{q})\right]\vec{\tau}_1\cdot\vec{\tau}_2\right\} \quad (27)$$

where

$$v_\pi(q) = \left[G'(q) + \frac{2q^2}{k_F^2} H'(q) \right], \qquad (28)$$

and

$$v_\rho(q) = \left[G'(q) - \frac{q^2}{k_F^2} H'(q) \right]. \qquad (29)$$

In Figs. 3–8 we contrast the results for Model 1 (solid line) and Model 2 (dashed line) for the quantities $F(q)$, $G(q)$, $F'(q)$, $G'(q)$, $v_\pi(q)$ and $v_\rho(q)$.[9] (In the last two figures we also give the results for pion and rho exchange in the Born approximation). It is clear that there are major modifications for the $T = 0$ parts of the force. The major reduction in $F(q)$ near $q = 0$ has been noted before and is essential to obtain a positive nuclear compressibility. It is worth remarking on the values of $v_\pi(q)$ and $v_\rho(q)$. In the region $1 < q/k_F < 2$, Model 2 gives values of $v_\pi(q)$ that are consistent with empirical determination of the strength of this part of the force and the force is just weak enough to avoid significant pion-condensation precursor phenomena. (For Model 2 in this momentum range, $g' \simeq 0.6$, if we use the notation of Weise et al.[18]). The force $v_\pi(q)$ for Model 1 appears somewhat too attractive in the light of phenomenological studies of this force in finite nuclei.

Turning to $v_\rho(q)$ of Fig. 8, we see that Model 2 yields a weaker force which actually goes to zero at about $q \sim 430$ MeV/c. The weaker force of Model 2 is consistent with the transverse response function, $S_T(\vec{q},\omega)$, measured in quasi-elastic electron scattering. The fact that the shape of the quasi-elastic peak is fit quite well in the impulse approximation leads one to conclude that the effective force is quite weak in the "rho channel". (See Ref. 19 for a further discussion of this matter).

All in all we can conclude that Model 2 is able to give a good account of the density and momentum-transfer dependence of the effective interaction. The results for Model 2 are consistent with the little that is known concerning the effective interaction. (It is also significant that Model 2 does not introduce any new features of the interaction which might be inconsistent with present knowledge).

It is clearly important to study the interaction for positive energy. Thus far we have directed our attention to a study of the optical potential for projectile kinetic energies less than about 200 MeV. (We have again limited our consideration to symmetric nuclear matter, for simplicity).

In an infinite medium the calculation of the optical potential may be approached via a study of the nucleon self-energy, $\Sigma(k_F,p)$.

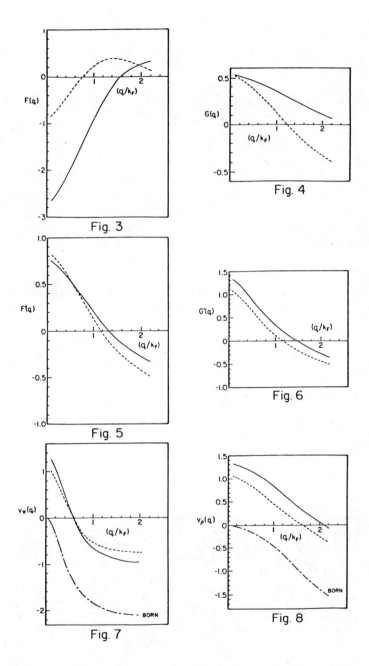

Figs. 3-8. The momentum dependence of the effective quasi-particle interaction in nuclear matter. See Eqs. (25) - (29).

In general one has

$$\Sigma(\vec{p}) \;=\; A(\vec{p}) \;+\; \gamma^{0}B(\vec{p}) \;+\; \frac{\vec{\gamma}\cdot\vec{p}}{m_N}\, C(\vec{p}) \tag{30}$$

where we have suppressed explicit reference to the density dependence of the various terms in Eq. (30).

Again we may contrast the expressions for the self-energy in Models 1 and 2. In Model 1 we have

$$\Sigma_1(p) \;=\; \sum_s \int \frac{d\vec{q}}{(2\pi)^3}\; \frac{m_N}{E_N(\vec{q})}\; <p,\bar{u}(\vec{q},s)|\hat{M}(1-P_{12})|p,u(\vec{q},s)> \tag{31}$$

where $|\vec{q}| \leq k_F$. (In Eq. (30) we have used a mixed notation; however it should be clear that $\Sigma(p)$ so constructed is a 4 x 4 Dirac matrix.) The corresponding self-energy in Model 2 is

$$\Sigma_2(p) \;=\; \sum_s \int \frac{d\vec{q}}{(2\pi)^3}\; \frac{m_N}{E_N(\vec{q})}\; <p,\bar{f}(\vec{q},s)|\hat{M}(1-P_{12})|p,f(\vec{q},s)>. \tag{32}$$

The details of the calculation of the quantities $A(\vec{p})$, $B(\vec{p})$ and $C(\vec{p})$ need not concern us here. The methods of calculation have been discussed extensively in the literature.[11] We have also shown how to construct an optical potential for use in the Schroedinger equation starting from a knowledge of the self-energy.[20]

In the Hartree or mean-field approximation, the central part of the optical potential is given as[21]

$$U_{opt}(r) \;=\; A(r) \;+\; \frac{E(\vec{p})}{m_N}\, B(r) \;+\; \frac{A^2(r) - B^2(r)}{2m_N}\,. \tag{33}$$

We do not use this form since in a Hartree-Fock or RBHF analysis the quantities A, B and C depend on the momentum $\vec{p}$. This implies that the coordinate-space potential for a finite system will be non-local. (The construction of an equivalent local, energy-dependent potential is discussed in Ref. 20). Using the techniques of Ref. 20 we have calculated $A(\vec{p})$, $B(\vec{p})$ and $C(\vec{p})$. The real and imaginary parts of these potentials are given in Fig. 9. We note that $C(\vec{p})$ appears multiplied by $\vec{\gamma}\cdot\vec{p}/m_N$ in Eq. (30) so that this quantity makes only a small contribution to the optical potential if $|\vec{p}|/m_N$ is small.

Using the techniques described in Ref. 20 we have also calculated the optical potential to be used in the Schroedinger equation. We have made calculations using both Model 1 and Model 2. In Fig. 10 we present some of our results. The data

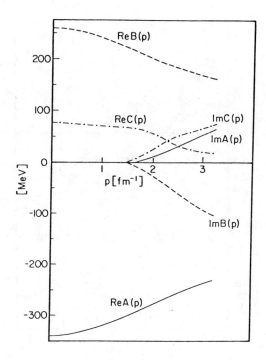

Fig. 9. The real and imaginary parts of $A(\vec{p})$, $B(\vec{p})$ and $C(\vec{p})$ for the potential HEA (Ref. 12) as a function of the quasi-particle momentum.

points shown in the upper part of the figures represent the strength of the central part of an optical potential of the Woods-Saxon form with fixed geometry. The lower part of the figure shows the strength of the imaginary potential determined in the same studies.[22] The dashed line is the result for Model 1 for a calculation, using the potential HEA of Ref. 12, made at nuclear matter density. The solid line is the result for Model 2. The dotted curve in the lower part of the figure represents the imaginary potential for Model 2 multiplied by (m^*/m_N), where m^* is the nucleon effective mass.[23] (The value of m^* is related to the non-locality of the nucleon optical potential and should not be confused with the Dirac mass, $\tilde{m} = m_N + A$.)

It can be seen that Model 2 provides a satisfactory representation of the real part of the potential. The deviation of the theory from the experiment for $\varepsilon > 100$ MeV can probably be understood on the basis of the complicated geometry that is exhibited by the optical potential at the higher energies. For example, as the energy increases the potential begins to develop a "wine-bottle bottom" shape and, in addition, there is an increase in the effective radius parameter. If one fits the data with fixed Woods-Saxon geometry one would have to increase the depth of

312

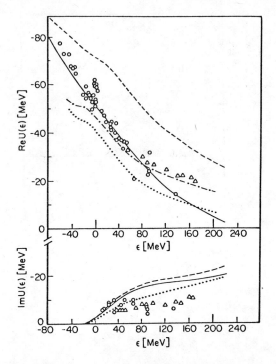

Fig. 10. The real and imaginary parts of the nuclear
optical potential for the interaction HEA of Ref. 12.
The dashed and solid lines are the results for
Models 1 and 2, respectively, at nuclear matter
density, ρ_o. The dash-dotted curve is the result
for ReU in Model 1 at $\rho_o/2$. The corresponding
result for Model 2 is given by the dotted curve. In
the lower half of the figure, the dotted curve
represents the imaginary potential at $\rho = \rho_o$ for
Model 2, multiplied by m^*/m_N (see Ref. 26). The
imaginary potentials for $\hat{\rho} = 1$ for Model 1 (solid
curve) and Model 2 (dashed curve) are also shown.

the potential to compensate for the changing geometry of the
physical potential. This matter requires further analysis: One
could calculate an _effective_ potential depth for data fits using
a Woods-Saxon potential of fixed geometry or one can re-analyze
the data using _theoretical_ models to specify the potential shape
as the projectile energy ϵ is varied. The fit to the imaginary
potential is also complicated by the interplay of surface and
volume absorption. Probably fits to volume integrals of the
absorptive potential would be more meaningful than a comparison of
the empirical strengths of the volume-absorption parameters for
finite nuclei and our nuclear matter results. We hope to complete
such studies in the future.

The origin of "wine-bottle" shape in Model 2 can be understood from the curves shown in Fig. 10. In the upper part of the figure the dash-dotted curve gives the result for ReU_{opt}, calculated at one-half nuclear matter density, with Model 1. The corresponding result for Model 2 is given by the dotted curve. It is clear that the density dependence of ReU_{opt} (ϵ,ρ) is quite different in the two models. For example, at $\epsilon = 160$ MeV, ReU_{opt} $(\frac{1}{2}\rho_o) \approx \frac{1}{2}\text{ReU}_{opt}$ (ρ_o) for Model 1, while for Model 2, ReU_{opt} $(\frac{1}{2}\rho_o) \approx \text{ReU}_{opt}$ (ρ_o).

In a very dilute system one would expect that the optical potential would be proportional to the density. Deviations from this proportionality could be thought of as "medium corrections". In Model 1 these corrections include Pauli Principle and dispersive effects which change the reaction matrix from its free-space value. It is clear that in Model 2 the medium corrections, which include those of Model 1 and those arising from the replacement of the $u(\vec{p},s)$ by the $f(\vec{p},s)$, are very much larger. From our calculations it would appear that a calculation of the optical potential using a multiple-scattering expansion based upon the free-space scattering amplitudes is not a satisfactory procedure for projectile energies less than 300 MeV. It is not clear, from the calculations made thus far, at what energy one can reliably use the free-space scattering amplitudes for the study of nucleon-nucleus scattering. This matter requires further study.

In Table 1 we present the values of the real part of the optical potential, calculated in Model 2, at nuclear matter density, ρ_o, and at one-half nuclear matter density, $\rho_o/2$. (These values have already been shown graphically in Fig. 10). From the knowledge of ReU (ϵ,ρ_o) and ReU $(\epsilon,\rho_o/2)$, we can form a simple expression which allows us to extrapolate these values to other values of the density. We can write

$$\text{Re } U(\epsilon,\hat{\rho}) = a(\epsilon)\hat{\rho} + b(\epsilon)\hat{\rho}^2 \tag{34}$$

where $\hat{\rho} = \rho/\rho_o$ is the ratio of the density to that of nuclear matter. (This formula gives the exact result for ReU $(\epsilon,\hat{\rho})$ for $\hat{\rho} = 1$ and $\hat{\rho} = \frac{1}{2}$). The values of $a(\epsilon)$ and $b(\epsilon)$ are given in Table 1. One may also make use of the following formulas which reproduce the calculated values of $a(\epsilon)$ and $b(\epsilon)$ quite well:

$$a(\epsilon) = 3\epsilon - 4(786.59) \left\{ 1 - \left[1 - \left(\frac{\epsilon + 52.508}{552.66} \right) \right]^{\frac{1}{2}} \right\}$$

$$+ 657.82 \left\{ 1 - \left[1 - \left(\frac{\epsilon + 78.38}{530.33} \right) \right]^{\frac{1}{2}} \right\} \tag{35}$$

$$b(\epsilon) \;=\; -2\epsilon + 4(786.59) \left\{ 1 - \left[1 - \left(\frac{\epsilon + 52.508}{552.66} \right) \right]^{\frac{1}{2}} \right\}$$

$$-2(657.82) \left\{ 1 - \left[1 - \left(\frac{\epsilon + 78.38}{530.33} \right) \right]^{\frac{1}{2}} \right\} \tag{36}$$

It is clear from the values of $a(\epsilon)$ and $b(\epsilon)$ given in Table 1 that the medium corrections, here defined as the terms proportional to ρ^2 in the optical potential, are extremely large for Model 2. Indeed they are not less than 50 percent at nuclear matter densities, $\hat{\rho} = 1$. We also see that when $\hat{\rho} = \frac{1}{2}$ the medium corrections are at least 25 percent. It is also clear from Table 1 that the medium correction becomes relatively more important as ϵ increases from 100 to 200 MeV. On the other hand, inspection of Fig. 10 indicates that for Model 1 medium corrections are significantly less important; these corrections tend to <u>decrease</u> in importance with increasing ϵ for Model 1. (For example, at $\epsilon =$ 170 MeV Model 2 gives $\mathrm{ReU}\,(\epsilon,\hat{\rho}) = -26\hat{\rho} + 18\hat{\rho}^2$ while Model 1 gives $\mathrm{ReU}\,(\epsilon,\hat{\rho}) = -38\hat{\rho} + 8\hat{\rho}^2$. The corresponding values at $\epsilon = 0$ MeV are are $\mathrm{ReU}\,(\epsilon,\hat{\rho}) = -113\hat{\rho} + 42\hat{\rho}^2$ for Model 1 and $\mathrm{ReU}\,(\epsilon,\hat{\rho}) = -102.64\hat{\rho} + 52.09\hat{\rho}^2$ for Model 2).

Our calculations of the optical potential should be compared to those of Brieva and Rook,[24] Jeukeune, Lejeune and Mahaux,[25] Friedman and Pandharipande[26] and Boguta.[27] Friedman and Pandharipande include a phenomenological three-body force in their calculations and obtain results somewhat similar to ours. Our values of $\mathrm{ReU}\,(\epsilon)$ are also in excellent agreement with the parameter-dependent, relativistic calculation of Boguta.

Our study of the optical potential is in early stage and we hope to have more to report in the near future. We are particularly interested in studying the spin-orbit potential since we have reason to expect that Models 1 and 2 will again yield quite different results for this quantity. It appears that the magnitude and energy dependence of the spin-orbit potential is not well understood at this time.[28]

We conclude this presentation with some speculations concerning the relation of the one-boson-exchange model of the nucleon-nucleon force and the quark model of the nucleon. It has been suggested that since it is now generally agreed that nucleons are fairly large objects, one should reject the boson-exchange model. Part of the reasoning behind this conclusion is the assumption that the OBE model requires that the nucleon and meson be very small, possibly point-like objects. We have discussed this matter in Refs. 29 and 30 and have shown that the OBE model is <u>not</u> inconsistent with a picture of the exchange of extended objects (mesons) between sizeable nucleons.

It is possible that the various successes of the OBE model are not accidental, but are actually indicative of how nucleons interact. To be more precise, one may note a complementarity

Table 1. Values of the real part of the optical potential
for nuclear matter as a function of the quasiparticle
energy ε for Model 2. The quantities $a(\varepsilon)$ and $b(\varepsilon)$ are
discussed in the text. All values are in MeV units. The
potential is HEA of Ref. 12.

ε [MeV]	Re $U(\rho_o)$	Re $U(\rho_o/2)$	$a(\varepsilon)$	$b(\varepsilon)$
0.00	−50.55	−38.30	−102.64	+52.09
10.00	−47.31	−35.82	−95.96	+48.65
20.00	−44.14	−33.41	−89.51	+45.37
30.00	−41.05	−31.09	−83.30	+42.24
40.00	−38.05	−28.85	−77.33	+39.28
50.00	−35.13	−26.69	−71.62	+36.49
60.00	−32.30	−24.62	−66.17	+33.87
70.00	−29.56	−22.64	−60.99	+31.43
80.00	−26.92	−20.75	−56.09	+29.17
90.00	−24.37	−18.96	−51.47	+27.10
100.00	−21.93	−17.27	−47.16	+25.23
110.00	−19.60	−15.69	−43.16	+23.56
120.00	−17.38	−14.21	−39.48	+22.09
130.00	−15.28	−12.85	−36.13	+20.85
140.00	−13.30	−11.61	−33.12	+19.82
150.00	−11.45	−10.48	−30.48	+19.03
160.00	−9.74	−9.49	−28.21	+18.47
170.00	−8.17	−8.63	−26.34	+18.16
180.00	−6.76	−7.91	−24.87	+18.11
190.00	−5.50	−7.33	−23.82	+18.32
200.00	−4.41	−6.91	−23.22	+18.81

316

between a nucleon-meson model and a quark-gluon picture. In Fig. 11(a) we show the nucleon potential as generated by meson (σ, π, ρ, ω.....) exchange and in Fig. 11(b) we show a complementary picture of how this force could be generated on the quark level. (The wavy lines denote gluons).

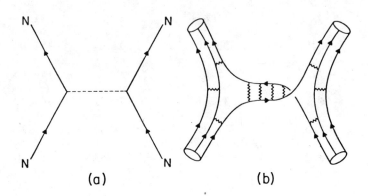

(a)   (b)

Fig. 11. (a) The nucleon-nucleon force in the OBE model. (b) Nucleon-nucleon interaction mediated by the exchange of a strongly interacting q$\bar{\text{q}}$ system.

Essentially we assume that the quark-quark interaction is dominated by the t-channel singularities which are bound states of q$\bar{\text{q}}$ pairs.[29] When we first discussed this idea in Ref. 29, the σ meson did not seem to fit naturally into this picture as it was thought to be a qq q$\bar{\text{q}}$ system. However, recent work by Scadron[31] on the scalar nonet indicates that the σ is indeed a q$\bar{\text{q}}$ state. He predicts an energy of about 750 MeV for the (quite broad) σ meson and this value is not inconsistent with OBE phenomenology.

Another feature of this speculation concerning the nature of the nucleon-nucleon force is the absence of an explicit bag model for the nucleon. It is clearly extremely difficult to calculate the N-N force starting with the bag picture. The chiral bag models which have mesons coupled to the bag surface (in order to provide a continuous axial current) also seem more suited to the calculation of static properties of the nucleon than for the calculation of the nucleon-nucleon interaction.

It is also worth remarking that the model in which mesons are effectively coupled to the quarks[29] has another interesting aspect. If the σ field is effectively coupled to the nucleon with a coupling constant $g_{\sigma NN}$, the corresponding coupling constant at the quark level is $g_{\sigma qq} = g_{\sigma NN}/3$. Thus we can envision an equation analogous to Eq. (2) operative on the quark level.

$$\tilde{m}_q^{\text{con}}(p^2) = m_q^{\text{con}} + \tfrac{1}{4}\text{Tr}\Sigma_q \, (p^2, \tilde{m}_q^{\text{con}}(p^2), k_F) \qquad (37)$$

In a first approximation, this equation is obtained from Eq. (2) by dividing by 3. For example, from Fig. 10 we see the $\text{ReA}(\vec{p} = 0) = -330$ MeV. Therefore we expect that the last term in Eq. (37) would be about -110 MeV. Thus with $m_q^{con} = 340$ MeV[4], we would have $\tilde{m}_q^{con} \simeq 230$ MeV, giving rise to $\tilde{m}_N \simeq 690$ MeV. In this discussion we have not included the energy splitting associated with gluon exchange which would reduce $3m_q^{con} = 1020$ MeV to the empirical value for the nucleon mass, $m_N \cong 940$ MeV and might, at the same time, reduce $\tilde{m}_N$ to about 600 MeV, the value obtained from the solution of Eq. (2) - see Fig. 10.

Some aspects of the model in which the meson fields are coupled to the quarks are discussed in Refs. 29 and 30. As noted above, these ideas are speculative; however in the light of our studies we feel that the adoption of a modern viewpoint concerning the size and underlying quark structure of the nucleon does not mean that we must discard the OBE model of the nuclear force. It is possible that much will be learned if we are able to understand why the OBE model works as well as it does.

Acknowledgements

The work described here was carried out in collaboration with Professor L. S. Celenza, Dr. Michael Anastasio and Dr. Won-Sing Pong. This work was supported in part by the National Science Foundation and the City University of New York Faculty Research Award Program.

REFERENCES

1. See for example, K. Erkelenz, Phys. Rep. C13, 191 (1974).
2. L. S. Celenza and C. M. Shakin, Phys. Rev. C24, 2704 (1981).
3. L. S. Celenza, B. Goulard and C. M. Shakin, Phys. Rev. D24, 912 (1981).
4. For a review, see M. Scadron, Rep. Prog. Phys. 44, 213 (1981); R. Delburgo and M. D. Scadron, Phys. Rev. Lett. 48, 379 (1982). See also talk given by M. Scadron at the Fifth Warsaw Symposium on Elementary Particle Physics, Kazimierz, Poland, May 25-27, 1982 (unpublished).
5. M. R. Anastasio, L. S. Celenza, and C. M. Shakin, Phys. Rev. Lett. 45, 2096 (1980).
6. M. R. Anastasio, L. S. Celenza and C. M. Shakin, Phys. Rev. C23, 2273 (1981).
7. L. S. Celenza, W. S. Pong, and C. M. Shakin, Phys. Rev. C25, 3115 (1982).
8. L. S. Celenza, W. S. Pong, and C. M. Shakin, Phys. Rev. Lett. 47, 156 (1982).
9. L. S. Celenza, W. S. Pong, and C. M. Shakin, Brooklyn College Report, B.C.I.N.T. 82/071/113, submitted to Physical Review C.
10. L. S. Celenza, W. S. Pong, and C. M. Shakin, in preparation.
11. M. R. Anastasio, L. S. Celenza, and C. M. Shakin, Phys. Rev. C23, 2258 (1981).

318

12. K. Holinde, K. Erkelenz and R. Alzetta, Nucl. Phys. A198, 598 (1972).
13. K. Holinde and R. Machleidt, Nucl. Phys. A256, 479 (1976).
14. The potential HM3' refers to the potential discussed in Sec. 2.3. of K. Holinde, Phys. Rep. C68, 122 (1981), and whose parameters are given in Table 1 of that reference under the column labeled Λ_π = 1265.
15. A. B. Migdal, Theory of Finite Fermi Systems and Applications to Atomic Nuclei (Wiley, New York, 1967); Nuclear Theory: The Quastiparticle Method (Benjamin, New York, 1968).
16. See for example, G. A. Rinker and J. Speth, Nucl. Phys. A306, 360 (1978); J. Speth, E. Werner and W. Wild, Phys. Rep. C33, 127 (1977).
17. S. O. Backman, G. E. Brown, V. Klemt, and J. Speth, Nucl. Phys. A345, 202 (1980).
18. H. Toki and W. Weise, Z. Physik A292, 389 (1979); A295, 187 (1980); Phys. Lett. 92B, 265 (1980); Phys. Rev. Lett. 42, 1034 (1979). For a review of pionic modes of excitation see E. Oset, H. Toki and W. Weise, Univ. of Regensburg preprint; submitted to Physics Resports.
19. L. S. Celenza, W. S. Pong, and C. M. Shakin, "Longitudinal and Transverse Response Function for Quasi-Elastic Electron Scattering", Brooklyn College report: B.C.I.N.T. 82/072/114; submitted to Physical Review C.
20. M. R. Anastasio, L. S. Celenza, and C. M. Shakin, Phys. Rev. C23, 2606 (1981).
21. R. L. Mercer, L. G. Arnold, and B. C. Clark, Phys. Lett. 73B, 9 (1978); L. G. Arnold, B. C. Clark and R. L. Mercer, Phys. Rev. C23, 1949 (1981);ibid. C23, 15 (1981); ibid. C21, 1899 (1980); ibid. C19, 917 (1979); Lett. Nuovo Cimento 18, 151 (1977). J. M. Noble, Nucl. Phys. A329, 354 (1979).
22. M. M. Giannini, G. Ricco and A. Zucchiatti, Ann. Phys. (N.Y.) 124, 208 (1980); ibid, 102, 458 (1976).
    A. Nadasen, P. Schwandt, P. P. Singh, W. W. Jacobs, A. D. Bacher, P. T. Debevec, M. D. Kaitchuck, and J. T. Meek, Phys. Rev. C23, 1023 (1981).
23. See for example, S. Fantoni, B. L. Friman and V. R. Pandharipande, Phys. Lett. 104B, 89 (1981).
24. F. A. Brieva and J. R. Rook, Nucl. Phys. A291, 299 (1976).
25. J. P. Jeukeune, A. Lejeune and C. Mahaux, Phys. Rep. C25, 83 (1976).
26. B. Friedman and V. R. Pandharipande, Phys. Lett. 100B, 205 (1981).
27. J. Boguta, Phys. Lett. 106B, 250 (1981).
28. P. Schwandt, H. O. Meyer, W. W. Jacobs, A. D. Bacher, S. E. Vigdor, M. D. Kaitchuck, and T. R. Donoghue, Phys. Rev. C26, 55 (1982).
29. L. S. Celenza, W. S. Pong, and C. M. Shakin, "The Quark Model and the Nucleon-Nucleon Interaction," Brooklyn College

Report: B.C.I.N.T. 81/061/109; submitted to Zeit. für Phys.

30. L. S. Celenza, W. S. Pong, and C. M. Shakin, "Electromagnetic and Hadronic Form Factors of the Nucleon," Brooklyn College Report: B.C.I.N.T. 81/101/110; submitted to Zeit. für Phys.

31. M. Scadron, Phys. Rev. D26, 239 (1982).

APPENDIX

We have used a somewhat unnatural normalization for the spinors $f(\vec{p},s)$ in the text. We note that the $f(\vec{p},s)$ defined earlier could be written as

$$f(\vec{p},s) = \sqrt{\frac{E_N(\vec{p})}{m_N} \frac{\tilde{m}(\vec{p})}{\tilde{E}(\vec{p})}} \; u(\vec{p},s,\tilde{m}(\vec{p})) \qquad (A1)$$

where $u(\vec{p},s,\tilde{m}(\vec{p}))$ is the positive-energy solution of the free Dirac equation with m_N replaced everywhere by $\tilde{m}(\vec{p})$. If we use Eq. (A.1) in Eq. (10) we have

$$E_2 = \sum_s \int \frac{d\vec{q}}{(2\pi)^3} \left[\frac{\tilde{m}(\vec{q})}{\tilde{E}(\vec{q})}\right] u(\vec{q},s,\tilde{m}(\vec{q})) \left[\vec{\gamma}\cdot\vec{q} + m_N\right] u(\vec{q},s,\tilde{m}(\vec{q}))$$

$$+ \frac{1}{2} \sum_{ss'} \iint \frac{d\vec{q}}{(2\pi)^3} \frac{d\vec{p}}{(2\pi)^3} \left[\frac{\tilde{m}(\vec{q})}{\tilde{E}(\vec{q})}\right] \left[\frac{\tilde{m}(\vec{p})}{\tilde{E}(\vec{p})}\right] \qquad (A2)$$

$$\times \; \langle \bar{u}(\vec{p},s,\tilde{m}(\vec{p})), \; \bar{u}(\vec{q},s',\tilde{m}(\vec{q})) | M(1-P_{12}) | u(\vec{p},s,\tilde{m}(\vec{p})) \; u(\vec{q},s',\tilde{m}(\vec{q})) \rangle.$$

We remark that if $\tilde{m}$ is replaced by m_N in Eq. (A.2), $E_2 \rightarrow E_1$.

In the text we have used the same notation as in the published literature for ease of comparison; however, the physical interpretation of Eq. (A.2) is more easily made than the same interpretation for the corresponding equation in the text.

DISCUSSION

Miller: I'm a little bit puzzled about the absence of explicit delta's in your theory. Presumably you have them through σ and ρ effects, at least as far as the nucleon-nucleon scattering problem is concerned. But how do you know when you take your matrix elements in nuclear matter where things are off-shell, self-consistent and all that, that the σ exchange forces will behave like (what I believe is more correct) the two pion exchange potential.

Shakin: There are several assumptions here. The underlying
assumption is to extrapole in a rather specific way the forces due to
the σ mesons to where the nucleon goes off its mass-shell. So, the
model is not free of assumptions. The force is fit for the case where
the nucleon spinors are parameterized by the vacuum mass. To just go
ahead and say this field theory works when you change the mass is an
assumption which you needn't accept, but I think one nice thing about
it is that it is a simple field theoretic model which has some hope of
making a connection to a microscopic theory as opposed to a potential
model such as Hamada-Johnston which is so removed from any field
theoretic description that there is no hope of making that contact.

Sauer: What is actually your relativistic many-body theory which you
could simplify to the two nucleon problem, and to which you then make
certain approximations so as to get out your equations?

Shakin: We've written three papers on how to derive this theory. One
is a formal Green function technique where the self energy appears
very naturally in the equations for the Green functions. When you do
this it's obvious that you are neglecting something. What you are
neglecting most clearly is the change in the vacuum energy when you
have several nucleons interacting with each other. Presumably, if you
have one nucleon in vacuum the use of the nucleon mass parameter
already has subsumed the effects of the local change of the vacuum.
Now, when you put many nucleons together, you are still using the
nucleon mass parameter and introducing the interaction between them.
You could argue that, while the nucleons are interacting in positive
energy states there is some further modification of the vacuum energy
because they are interacting and are near each other, that's the thing
that's left out. Otherwise, this follows from standard manipulations.
The other way of doing this, is to start with the Hamiltonian and, as
in the textbooks, rearrange the Hamiltonian, isolate the residual
interaction and then you can show that if you choose the right
representation you diagonalize the rest of the Hamiltonian, so that
the residual interaction just excites the two-particle two-hole states
while the rest of the interaction is in diagonal form. The equation
that takes you to that diagonal form is just the magic equation that I
pointed out at the first step. Now in that case, to eliminate the
calculation of the vacuum energy, you have to define a normal ordering
with respect to the new vacuum. There is still a third way to this,
which is most amusing, which is a, field theory model, which is an
ancient model (more than 20 years old). You can do it in terms of the
Gell-Mann, Levy σ-model, and generate the nucleon mass in vacuum
through symmetry breaking and then when you change the scalar density
by adding nucleons you have to redo the symmetry breaking and because
of the presence of the σ field, the whole thing is modified again now
leading to the same mass modification equation. So, there are three
ways to derive this, and the real problem you might argue is not a
renomalizable field theory, and I think that is a hopeless enterprise
at the level of nucleons and mesons, although, that's a point of
controversy.

Brown: A couple of comments. The first has to do with G. Miller because we showed in covariant calculations that virtual isobars could be reexpressed in terms of the exchange of σ degrees of freedom, distributed mass, as long as you are working in the purely nucleon sector. And contributions that came from very off-shell intermediate states, as long as you are working in the low-energy domain should be good approximations to handle those ...

Miller:...but the initial and final states are also very far off shell...

Brown: ... no, but not like these. These are off-shell like GeV intermediate energy states. This is essentially why the isobars give you mainly only σ degrees of freedom and do not contribute to the ρ. This is a good calculation done by Durso and other people and is done in a covariant dispersion theoretical form and anybody is willing to check it if they want to take a few years off. Second, I think that you've answered my question most of the way, that is that we should put the scalar potential into the denominator but not the vector, because when you take the trace you throw away the vector part.

Shakin: Whenever the mass appears, it goes along with the trace of the self energy.

Brown: So that's probably more straightforward than I thought. I had one worry there at the time, a small one, I didn't see anywhere a pseudoscalar potential but we know that the difference between triplet and singlet binding energies are due to the π meson.

Shakin: There is a π meson in the theory.

Brown: Yes but the point is that your effective scalar interaction at low energies involves quite a large contribution from the tensor coupling of the π meson in second order.

Shakin: That's comparing in the iteration of the one boson exchange model. That is already in what the German (Bonn) group has done.

Brown: But when you go to an effective scalar potential, as people seem to do, you ought to be careful and realize that part of that effective scalar potential has come from the iteration of the pseudoscalar.

Shakin: That's a common error and we didn't make it

Brown: When you take that empirical scalar potential, a good chunk of that hasn't come from scalar interactions, and therefore should not go into this reduction of small to large components.

Shakin: We took the force that fits the free scattering data so all the iterations of the pion field are contained within that force and they are not double counted in the coupling constants of the σ meson.

Brown: How are they treated? As scalars?

Shakin: You have a σ meson and a pion. The people that fit the
nucleon-nucleon data only work within the space of positive energy
spinors where pseudovector coupling is the same as pseudoscalar
coupling. Now, to do this type of problem you must use pseudovector
coupling and if you use pseudoscalar coupling all hell breaks loose.

Friar: Ordinary Hartree Fock calculations have a variational basis.
Do your calculations have a variational basis?

Shakin: The Brueckner-Dirac-Hartree-Fock?

Friar: Yes.

Shakin: Well. This variational principle is the one I stated, that
essentially the calculation is stabilized against variations of this
effective mass parameter. If you neglect rearrangement the variations
of the energy with respect to the mass parameter will give you the
equation that I wrote down, that determines the self-consistent
spinor, but a more precise way to say it, is in the representation in
which the corrections to this calculation involve the excitation of
two-particle two-hole states so we've eliminated the tadpole diagrams;
those that involve all the particles in the system giving a one body
field that excites particle-hole states, those currents are gigantic
in any theory. Well, in any theory with negative energy states you are
not going to have a lower bound.

Lomon: In dealing with the two-nucleon problem relativistically, one
starts with the Bethe-Salpeter equation and then does the
Blankenbeckler-Sugar reduction and in that way one is dealing only
with positive energy correspondence, and in that way the effect of the
negative energy states is stuffed into the effective potential. It
turns out that the two-meson exchange and it's repulsive energy
dependence, give a lot of the effects one wants. My question is
simple. Is this Hartree the same self-consistent interaction, or is
there's some additional repulsive ...

Shakin: Well, the way I view this (may not like this) is that the
mesons that are being exchanged are objects that are defined by the
reduction procedure itself. So, that if you use a different reduction
procedure starting from the Bethe-Salpeter equation, you will get some
different object with different coupling constants. So, we have the
parameterization at the level of the scattering amplitudes. I'm
saying that its probably impossible define what you really mean by a
Bethe-Salpeter equation and sum all those ladder diagrams or crossed
ladder diagrams...

Lomon: These two things overlap at least partially. I think it's an
interesting comment that this is another way of getting the saturation
properties and right at the beginning the two-nucleon forces are non-
realistic two-meson exchange.....

Shakin: Sure, this takes us deeply into the technicalities of how you're going to do the two-body problem.

Noble: I'd like to show just one slide. I worried a lot when I first began to look at this theory some years ago about this business of the enhancement of the small components and whether they're observable or not. One of the places where you can observe them is in the dipole moment of the axial vector density, (the timelike component of the axial vector current) that occurs in the asymmetry parameter in weak decays, and also in magnetic moments. The problem is that one never looks at magnetic moment effects, except for valence nucleons, at least with the data that exist so far, so you don't have to worry about this tremdous enhancement in the density, even if maybe the calculations that I've done about renormalisation of magnetic moments are wrong. But also less people worry about whether this reduction of the mass in the nucleus is a physical effect or not, let me show you what you see when you do quasi elastic electron scattering. Basically, you see data that look like this for, let's say, the longditudinal structure function (and, for the moment forget about whether or not they've found all the strength and whether you have to renormaliz se the form factor) the main thing I want you to look at is the shape of the structure function. Essentially, when you do the relativistic Fermi gas with a mass which is reduced from the free nucleon mass, it reproduces the shape of this structure function admirably. It fits the quasi elastic scattering peak at essentially $q^2/(2m)$ and if you do it with the free nucleon mass the structure function has a peak at much too low an energy loss, and doesn't look at all like the data. People have traditionally put in something called an effective binding correction of 35 MeV to essentially just shift this curve up so that the peaks fit, but I think this is a very unphysical thing to do.

324

THE SCHRÖDINGER EQUIVALENT COMPLEX POTENTIAL AND THE DIRAC PHENOMENOLOGY

M. Jaminon,
Institut de Physique B5, Université de Liège, B4000 Liège 1, Belgium

ABSTRACT

We show that the phenomenological analyses of nucleon-nucleus elastic scattering data using a single-particle Dirac equation are not affected by the inclusion of a complex nuclear field that behaves like the space-like components of a vector.

Recently, phenomenological analyses[1] have shown that excellent fits to nucleon-nucleus scattering data can be obtained by using a single-particle Dirac equation in which the local relativistic average potential is the sum of a complex scalar field U_s^{PH} and of the time-like component U_o^{PH} of a complex vector field. However, microscopic calculations[2] have exhibited that the most general Dirac average potential felt by the scattered nucleon can also contain a complex space-like field U_v and a complex tensor field U_T. Therefore, these two new potentials should be included in the phenomenological Dirac equation which then reads :

$$\{\vec{\alpha}.\vec{p} + \gamma^0[m + U_s^{PH}(r) + \gamma^0 U_o^{PH}(r) + \gamma^r U_v^{PH}(r) + i\, \gamma^0\gamma^r\, U_T^{PH}(r)]\} \times$$
$$\phi^D(\vec{r}) = E\, \phi^D(\vec{r}) \ . \tag{1}$$

In this equation, the index PH refers to "phenomenology" and the i before the last term in the brackets provides real eigenvalues when all the quantities U_s^{PH}, U_o^{PH}, U_v^{PH} and U_T^{PH} are real.

Our purpose is to show that the calculations of the phase shifts and thus the fits of the various observables involved in elastic scattering are not affected by the complex space-like potential U_v^{PH} of a finite range. A similar result is already known in the case of the bound single-particle states whose energies have been shown to be independent of the space-like potential.[3]

The Dirac equation (1) is equivalent to two coupled equations for the large (ϕ_u^D) and the small (ϕ_ℓ^D) components of the wave function. The elimination of ϕ_ℓ^D between these two equations yields a Schrödinger-type equation

$$[\vec{p}^2 + 2m\, \tilde{U}_e(r,\varepsilon,\vec{p}) + \frac{2m}{r}\, U_{so}(r,\varepsilon)\, \vec{\sigma}.\vec{L}]\, \phi_u^D(\vec{r}) = (E^2-m^2)^{\frac{1}{2}}\, \phi_u^D(\vec{r}) \ , \tag{2}$$

where the central part U_e of the potential has the drawback of being a nonlocal operator. It can therefore not be identified with the central part of the phenomenological optical-model potential which is most of the time assumed to be local. In order to get a local 2×2 average nucleon-nucleus potential that can be useful for the comparison with empirical potentials, we define a new wave function

$$\phi^S(\vec{r}) = A^{-1}(\infty,\varepsilon)\, A(r,\varepsilon)\, \phi_u^D(\vec{r}) \ , \tag{3}$$

$$A(r,\varepsilon) = [m + E + U_s(r) - U_o(r)]^{-\frac{1}{2}} \exp[i \int_o^r U_v(s) \, ds]$$

$$= [D(r,\varepsilon)]^{-\frac{1}{2}} \exp[i \int_o^r U_v(s) \, ds] \quad , \qquad (4)$$

which satisfies the Schrödinger-type equation :

$$[\vec{p}^2 + 2m \, U_e(r,\varepsilon) + \frac{2m}{r} U_{so}(r,\varepsilon) \, \vec{\sigma}.\vec{L}] \, \phi^S(\vec{r}) = (E^2-m^2)^{\frac{1}{2}} \phi^S(\vec{r}) \quad . \quad (5)$$

In Eq. (4), we have omitted the upper indices PH for simplicity. The average potential is now local. Its analytical expression in terms of the real and imaginary parts of the relativistic fields can be found in Ref.[4]. A remarkable characteristic is that the local potential is independent of the complex field $U_v(r)$. Since the relativistic fields $U_i(r)$ (i = s,o,v) have finite range, the wave functions ϕ_u^D and ϕ^S have the same asymptotic behaviour and the elastic scattering phase shifts determined from the original Dirac equation (1) are identical to the ones calculated from the Schrödinger equation (5). Since the average potential entering in the latter does not involve the complex space-like potential $U_v(r)$, we immediately deduce that the phase shifts are also independent of this quantity $U_v(r)$. Therefore, a fit of the observables involved in the elastic nucleon-nucleus scattering should not provide any information on the magnitude of the nuclear space-like field U_v .

The only effect of the potential $U_v(r)$ and more precisely of its imaginary part $W_v(r)$ is to modify the amplitude of the wave function $\phi_u^D(\vec{r})$ inside the nucleus. In order to exhibit this property in a simple way, let us suppose that $U_s(r) = U_o(r) = \text{Re} \, U_v(r) = 0$. In that case, $\phi^S(\vec{r})$ is a plane wave of momentum $(E^2-m^2)^{1/2}$ [see Eq. (5)]. Equations (3), (4) then give :

$$\phi_u^D(\vec{r}) = C[\exp \int_o^r W_v(s) \, ds] \, \phi^S(\vec{r}) \quad , \qquad (6)$$

where

$$C = [\exp \int_o^\infty W_v(s) \, ds]^{-1} \quad . \qquad (7)$$

These equations show that at the surface of the nucleus the wave function $\phi_u^D(\vec{r})$ has nearly the same amplitude as $\phi^S(\vec{r})$ while the amplitude of the former is damped or amplified inside the nucleus if $W_v(r) > 0$ or $W_v(r) < 0$ respectively. If we assume $W_v > 0$ and constant on a short distance Δr , Eq. (6) gives :

$$\frac{\phi_u^D(r-\Delta r)}{\phi_u^D(r)} = e^{-W_v\Delta r} \frac{\phi^S(r-\Delta r)}{\phi^S(r)} \qquad (8)$$

for the radial parts of the various wave functions. Hence, the amplitude of the incoming Dirac wave function ϕ_u^D is damped by a factor

$e^{-\Delta r W_v^D}$ on the distance Δr . Inversely, the outgoing wave function ϕ_u^D increases by a factor $e^{W_v \Delta r}$. This behaviour of the wave function is similar to what happens in the Perey effect[5] in a nonrelativistic theory.

Let us try to make a connection with nuclear matter. In that case, the Dirac equation reads :

$$\{\vec{\alpha}.\vec{k} + \gamma^0[m + \overline{U}_s(k) + \gamma^0 \overline{U}_o(k) + \vec{\gamma}.\frac{\vec{k}}{k} \overline{U}_v(k)]\} u^D(\vec{k}) = E_k u^D(\vec{k}) , \quad (9)$$

where $u^D(\vec{k})$ is the self-consistent positive energy Dirac spinor. Let us still assume that $\overline{U}_s(k) = \overline{U}_o(k) = \mathrm{Re}\, \overline{U}_v(k) = 0$ while $\mathrm{Im}\, \overline{U}_v(k) \equiv \overline{W}_v(k) \neq 0$. In that simple case, Eq. (9) is identical to the dispersion relation

$$E^2 - m^2 = k^2(1 + i \frac{\overline{W}_v}{k})^2 \qquad (10)$$

which implies :

$$k = k_R + i k_I = (E^2 - m^2)^{\frac{1}{2}} - i \overline{W}_v \qquad (11)$$

in the limit $\overline{W}_v \ll k$. In nuclear matter, the wave function is therefore a plane wave whose momentum is modified by the inclusion of $\overline{W}_v$:

$$\phi_u^D(z) = u^D(\vec{k}) e^{i[(E^2 - m^2)^{\frac{1}{2}} - i \overline{W}_v] z} \qquad (12)$$

where Oz has been taken along the propagation axis $\vec{k}$. If $\overline{W}_v > 0$ [6,7] Eq. (12) corresponds to a plane wave whose amplitude increases with increasing z , whence to a steadily increasing flux. This suggests that the quantity $\overline{U}_v$ should not be identified with U_v introduced in Eq. (1). Work on this problem is in progress.[4]

We acknowledge stimulating discussions with Dr. C.M. Shakin, Dr. C.J. Horowitz, Dr. G. Garvey, B.C. Clark and C. Mahaux.

REFERENCES

1. B.C. Clark, S. Hama and R.L. Mercer, IUCF Workshop, Bloomington, Indiana, October 28-30, 1982 (to be published by the American Institute of Physics).
2. M. Jaminon, C. Mahaux and P. Rochus, Nucl.Phys. A365, 371 (1981); L.D. Miller and A.E.S. Green, Phys.Rev. C5, 241 (1972); M.R. Anastasio, L.S. Celenza and C.M. Shakin, Phys.Rev. C23, 2273 (1981).
3. L.D. Miller, Ann.Phys. 91, 40 (1975).
4. M. Jaminon, to be published.
5. F.G. Perey, "Direct Interactions and Nuclear Reaction Mechanisms", edited by E. Clementel and C. Villi (Gordon and Breach, New York, 1963), p. 125.
6. L.S. Celenza, W.S. Pong and C.M. Shakin, Phys.Rev.Lett. (submitted).
7. C.J. Horowitz, Phys.Lett. 117B, 153 (1982).

DISCUSSION

Horowitz: I don't understand the physics of eliminating the imaginary Three-vector potential as opposed to eliminating the real three-vector potential. If the imaginary parts of the scalar and time-like vector potentials were zero, what would you say the effective non-relativistic absorption would be?

Jaminon: Nothing.

Horowitz: That's the thing I don't understand. The eigenvalue of the Dirac equation is clearly complex if there is a non-zero imaginary part to the three-vector potential, so that if you have a finite relaxation time or mean free path, just looking at single nucleon eigenvalues. Then you say you are going to compare this to a Schrodinger equivalent equation with no absorption? That seems confusing.

Jaminon: This seems to be a technical point and we should discuss this later.

Mahaux: I think that indeed because of W_v you do get a mean free path and your wave function is effective, but then your scattering matrix would be unitary, so you don't have real absorption because of W_v. You have a damping of the wavefunction which is not the same thing, the phase-shift remains real. It's very striking. If you take the Dirac equation with a vector potential, you can multiply that vector potential by 100 and the binding energies all remain the same. Here the phase-shifts will remain independent of the vector potential. But it's true that the value of the wavefunction inside the nucleus is affected by W_v.

Horowitz: If the eigenvalue of the Dirac Equation is complex, why do you say you have no absorption?

Mahaux: The wavefunction is damped, in a finite system you get a unitary scattering matrix.

Horowitz: I disagree.

Noble: Write down the flux consevation equation which will come out to be

$$\frac{d}{dt}\ \Psi^\dagger\Psi + \underline{\nabla}\cdot\Psi^\dagger\underline{\alpha}\Psi = -2\ \Psi^\dagger\underline{\alpha}\cdot\underline{W}_v\Psi$$

The right hand side is not a positive definite or negative definite operator and so it doesn't produce any net absorption the way the imaginary parts of the scalar and time like vector potentials do.

Horowitz: Would you say that the S matrix is unitary, the transformation required to eliminate the imaginary part of the

three-vector potential is no longer a unitary transformation? It's
unitary for a real three-vector potential.

Mahaux: Yes, but that doesn't matter because you don't change the
asymptotic values of the wave function at large distances from the
nucleus. The problem lies in going from nuclear matter to the finite
system, because for a finite system the situation is fairly clear. You
damp the wavefunction inside the nucleus because of W_v, but you don't
change the asymptotic behavior except for the normalization which
doesn't matter.

THE IMAGINARY PART OF THE RELATIVISTIC OPTICAL POTENTIAL*

C. J. Horowitz

Center for Theoretical Physics
Laboratory for Nuclear Science and Department of Physics
Massachusetts Institute of Technology
Cambridge, Massachusetts 02139

ABSTRACT

The imaginary past of the relativistic optical potential is calculated in nuclear matter for the Walecka $\sigma+\omega$ model. Large Lorentz scalar and vector contributions are found in addition to an important three-vector potential.

I. INTRODUCTION

Microscopic relativistic Hartree [1,2] and Hartree-Fock [3,4] calculations have verified many features of the <u>real</u> part of the relativistic optical potential. In these calculations large scalar and vector potentials arise from meson exchanges. The Lorentz transformation properties of these strong interactions provide a natural description of the spin-orbit potential and much of the energy dependence of nonrelativistic optical models.

Although there is considerable interest in the imaginary parts of relativistic optical models, there have been no previous microscopic calculations[†]. Thus one purpose of this first calculation [5] is to examine the qualitative mixture of Lorentz scalar and vector imaginary self-energies.

II. IMAGINARY SELF-ENERGY

The nucleon self-energy, Σ, is a complex 4 x 4 Dirac matrix. Rotational invariance of infinite nuclear matter implies Σ has the following form [4].

$$\Sigma(\underline{k},E) = U^S(k,E) + \gamma_4 U^o(k,E) + i\underline{\gamma}\cdot\underline{k}U^V(k,E)$$
$$U^S(k,E) = V^S + iW^S(k,E)$$
$$U^o(k,E) = V^o + iW^o(k,E)$$
$$U^V(k,E) = iW^V(k,E) \tag{1}$$

* This work is supported in part through funds provided by the U.S. DEPARTMENT OF ENERGY (DOE) under contract DE-AC02-76ER03069.

† At this conference I learned of a latter independent calculation by Shakin et al[6] that gives similar results.

Hartree-Fock calculations give a very small real part of U^V which we will neglect.[4] Furthermore, for simplicity, we assume the real parts of U^S and U^O, (V^S and V^O) to be independent of energy.[5]

In order to examine the signs and magnitudes of W^S, W^O and W^V we calculate the lowest order contributions in the Walecka $\sigma+\omega$ model [7] for nuclear matter. The results we present depend more on the Lorentz transformation nature of the model (i.e. that it involves strong vector-scalar cancellations), then on details of the effective interaction (such as the values of the coupling constants). Thus these calculations should provide a qualitative guide to the kinds of imaginary self-energies to be expected.

The lowest order contributions to Σ come from the Feynam diagrams in Fig. (1a-c).

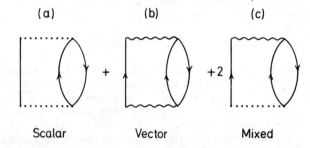

Fig. 1. Feynman diagrams for the nucleon self energy. The dotted lines in Fig. (1a) describe scalar or sigma mesons while the wavy lines of (1b) represent vector omega mesons, and the interference term of Fig. (1c) is due to vector scalar mixing.

Figure (1a) describes the exchange of scalar mesons [6], (1b) vector mesons (a) and Fig. (1c) describes vector-scalar mixing. These contributions can be evaluated with the Feynam rules in [7] The different components of Σ are projected out by taking the appropriate traces,

$$W^S = \tfrac{1}{4} \, \mathrm{tr}(\mathrm{Im}\Sigma),$$
$$W^O = \tfrac{1}{4} \, \mathrm{tr}(\gamma_4 \mathrm{Im}\Sigma),$$
$$W^V = \tfrac{1}{4k^2} \, \mathrm{tr}(i\underline{\gamma}\cdot\underline{k} \, \mathrm{Im}\Sigma). \tag{2}$$

For simplicity we show the results for W^S only in the limit $E-\tilde{E}_f \ll \tilde{E}_f$ (where $\tilde{E}_f$ is the Fermi energy).

$$W^S(k_f,E)\Big|_{E\approx\tilde{E}_f} = \frac{-(E-\tilde{E}_f)^2 M_*}{16\Pi^3 k_f} \int_0^{2k_f} dq$$

$$\left\{ g_s^4\left[M_*^2 + q^2/4\right]\Delta^2(q) + g_v^4\left[M_*^2-q^2/2\right]D^2(q) - 2g_s^2 g_v^2 E_f^2 D(q)\Delta(q) \right\} \tag{3}$$

$$E_f^2 = k_f^2 + M_*^2 \tag{4a}$$

$$D(q) = -\left[q_\mu^2 + M_v^2\right]^{-1} \tag{4b}$$

$$\Delta(q) = -\left[q_\mu^2 + M_s^2\right]^{-1} \tag{4c}$$

$$M_* = M + V_s \tag{4d}$$

Here $g_s(g_v)$ and $M_s,(M_v)$ are the scalar (vector) meson coupling constant and mass. The first term in curly brackets is due to Fig. (1a), the next Fig. (1b) and the last term comes from the mixed exchange in Fig. (1c). Using equation (4a) the large M_*^2 terms can be collected into a perfect square so equation (3) can be rewritten

$$W^S(k_f,E) \underset{E \approx \tilde{E}_f}{\approx} \frac{-(E-\tilde{E}_f)^2 M_*}{16\Pi^3 \, k_f} \int_0^{2k_f} dq \left[M_*^2 (g_s^2 \Delta - g_v^2 D)^2 - 2g_s^2 g_v^2 k_f^2 D\Delta \right] \tag{5}$$

Since the interaction involves large σ,ω cancellations,

$$g_s^2 \Delta(q) - g_v^2 D(q) \approx \frac{1}{10} g_s^2 \Delta(q), \tag{6}$$

the squared term in equation (5) is dominated by the 2nd term even though k_f^2 is only 10 percent of M_*^2. The extra term arises because of the different Lorentz transformation properties of Fig. (1c) compared to Fig. (1a) or (1b). Therefore $\underline{W^S}$ $\underline{will}$ $\underline{be}$ $\underline{positive}$ $\underline{and}$ $\underline{large}$.

In a similar way we find a positive W^V and a large and negative W^O which more than compensates the probability source effects of W^S (and W^V) leading to an effective nonrelativistic absorption similar to conventional optical models [5].

III. THREE-VECTOR POTENTIALS

There is some controversy concerning the effect of a three-vector potential. Jaminon [8] in a contribution to this conference assumes W_V in a finite system points in the $\underline{radial}$ direction.

$$\gamma \cdot \hat{r} |\underset{\sim}{k}| U_r(r) \tag{7}$$

She then finds no contribution of such a potential to the phase shifts. Physically this corresponds to making one side of the nucleus a prob. sink and the other side a prob. source.

However, since $\hat{r}$ does not appear in nuclear matter it is clear that a local density approximation will never give such a potential. Instead, our calculations give a three-vector potential in the direction of the $\underline{momentum}$. In a finite system this should correspond to a potential in the direction of the gradient acting on the ware function,

$$\underset{\sim}{\gamma} \cdot U_r(r) \frac{1}{i} \underset{\sim}{\nabla}, \tag{8}$$

and such a potential will contribute to the scattering phase shifts.

The Dirac equation for scattering from a spherical nucleus reads,

$$\left\{ \left[1 + U_v(r) \right] \underset{\sim}{\alpha} \cdot \frac{1}{i} \underset{\sim}{\nabla} + \gamma_4 \left[M + U_s(r) \right] + U_o(r) \right\} \psi = E\psi. \tag{9}$$

Dividing through by $\left(1 + U_v(r) \right)$ leads to effective scalar $(\tilde{U}_s)$ and vector $(\tilde{U}_o)$ potentials,

$$\left\{ \underset{\sim}{\alpha} \cdot \frac{1}{i} \underset{\sim}{\nabla} + \gamma_4 \left[M + \tilde{U}_s(r) \right] + \tilde{U}_o(r) \right\} \psi = E\psi, \tag{10}$$

$$\tilde{U}_s(r) = \frac{U_s(r) - U_v(r)M}{1 + U_v(r)} \tag{11}$$

$$\tilde{U}_o(r) = \frac{U_o(r) + EU_v(r)}{1 + U_v(r)} \tag{12}$$

Equations (11) and (12) can be directly compared to phenomenological optical models which use only scalar and 4th component of a vector potentials. Simply neglecting $U_v(r)$ will lead to errors of almost 100 percent in the (theoretical) total absorption.

IV. RESULTS AND CONCLUSIONS

In Table I we have collected values for the potentials (see Ref. 5). These are similar to phenomenological fits (at r=o) to p-^{40}Ca elastic scattering [9] and the large values imply an important imaginary spin-orbit potential.

Table I. Effective Imaginary Self-energies (In MeV). The dimensionless vector potential, W_v, has been used to calculate $\tilde{W}_s$ and $\tilde{W}_o$ through Equations (11) and (12).

E	MW_v	W_s	W_o	$\tilde{W}_s$	$\tilde{W}_o$
50	23	28	-40	13	-22
100	43	71	-93	43	-56
181	70	146	-180	100	-114

In conclusion we have calculated the imaginary part of the relativistic self-energy in nuclear matter. We find due to the Lorentz transformation nature of the σ+W interaction:

(i) W_S is positive (a prob. source).

(ii) Both W_S and $-W_O$ are large.

(iii) W_V is important

REFERENCES

1. C. J. Horowitz and B. D. Serot, Nucl. Phys. A368, 503 (1981)
2. M. Jaminon, C. Mahaux and P. Rochus, Phys. Rev. Lett. 43, 1097 (1979)
3. M. Jaminon et al, Phys. Rev. C24, 1353 (1981); Nucl. Phys. A365, 371 (1981).
4. C. J. Horowitz and B. D. Serot, Nucl. Phys. (in press).
5. C. J. Horowitz, Phys. Lett. B (in press).
6. C. Shakin et al, Invited Talk this Conference.
7. J. D. Walecka, Ann. of Phys. 83, 491 (1974).
8. M. Jaminon, Contributed paper this Conference.
9. L. G. Arnold et al, Phys. Rev. C23, 1949 (1981).

DISCUSSION

Mahaux: As you quite rightly say the agreement with experiment is embarrasingly good. So, I have a question to ask here partly to you but perhaps mainly to Carl Shakin. In the non-relativistic Brueckner expansions we know that there exist corrections to the Brueckner-Hartree-Fock approximation. This is certainly also the case in the relativistic Brueckner-Hartree-Fock calculations. You have, what you alluded to as some re-arrangement terms and you have renormalization, although not in the sense of Quantum Field Theory, these things will decrease by something like 50% the imaginary part of the optical potential so I'm puzzled by the very good agreement with empirical values that you obtain and wonder whether you may say something about the convergence of the kind of expansion of which you take the first term.

Horowitz: You know that it's easier in the relativistic formalism to get the real part of the optical potential with simple calculations than it is in the non-relativistic formalism. Already the simple relativistic calculations are doing better so you might wonder if the same thing doesn't happen for the imaginary part also, and if the relativistic calculations aren't easier and a more natural starting point.

Brown: My comment is that I've never heard of that dirty word "re-arrangement" terms. I don't know what it means. There are no re-arrangement terms.

Shakin: Well, I think he's talking about the rearrangement diagrams which are many body diagrams where the particles go forward and back.

Shakin: I think Mahaux is asking a profound question and in Brueckner type analysis you do have such diagrams in the σ-particle expression, where you try to calculate the self energy to the next order. It's a good question and I don't really know the answer.

Walecka: There's a simple physical interpretation of where the imaginary parts come from, namely from those diagrams that produce the quasi-elastic peak.

Shakin: I think he said that there are additional diagrams. That's the second order if you allow the line to go forward in time. But you can also allow them to go backwards in time even in Brueckner theory, and then you have re-arrangement diagrams to that order in the interaction. They don't contribute to the imaginary part, just the real part.

Pandharipande: Rather than having this discussion, can't you just compute the real part of these fourth order diagrams and just tell us what you end up with?

Horowitz: The finite nucleus calculation is difficult. The real part is as difficult as it is non-relativistically, it's a 5 dimensional integral or something. You can do it, it's just numerically a little bit involved exactly as the real part of the same diagram is non-relativistically.

Pandharipande: Oh come on! We calculate that real part on every third day.

Horowitz: I've calculated the binding energy contribution of the diagrams.

Pandharipande: How big is it?

Horowitz: That's going to depend on the coupling constants. Something like 40 MeV for the sum of the scalar and vector and mixed contributions.

Pandharipande: So you're saying that your first order mean field theory gives you a binding energy of -16 MeV, the second order gives you -40 MeV?

Horowitz: There's the exchange operator plus the exchange... But that's not the point. You have very large potentials; to calculate the binding energy from first principles is perhaps not the appropriate thing to do.

Pandharipande: Come on, I do it every time.

Horowitz: I have a potential of 500 MeV. Are you going to tell me you can calculate a binding energy accurate to 1% of that 500 MeV? That's still 5 MeV.

Pandharipande: I want to understand the physics of this and what I'm asking is is there any established convergence critereon?

Horowitz: For the binding energy?

Pandharipande: For any diagrams that you draw. If the real part is -40MeV, what about the 6th order diagrams, what about the 8th order diagrams?

Horowitz: The calculation I'm trying to do is to determine the Lorentz structure. Now, I claim that if I use an interaction which fits the real part of the optical potential to lowest order then I can use that calculation in this simple expression to see what the effective interaction gives me for the ratio of W_O/W_V.

Pandharipande: I completely fail to understand this because, if you cannot treat the particular Hamiltonian in second order perturbation theory, how are you going to trust the result you get from these diagrams?

Horowitz: My claim is, which is based on the Lorentz covariance of the interaction, if I have an interaction, no matter how I got it, that has strong Lorentz vector and Lorentz scalar contributions, then my claim is that that particular interaction will give you large imaginary optical potentials of opposite sign in low order. If you go to higher order it would take a miracle for each of the large numbers to go back and give you a small number.

Shakin: I'd like to repeat my comment that you can't understand this as a Hartree theory, you have to understand these as effective t-matrices that you're fitting to the ground state properties of nuclear matter. Once you do that you have then to define how you're going to go further and find some form of perturbation theory with those potentials which have adjusted parameters. You have to define mathematically what you're doing.

Pandharipande: Just one thing. I love the simplicity of Dirac-Hartree-Fock. I like the results but I don't understand. There is no scalar field that we know of. There is no σ-meson. If you take a two-quark state and want to make a 0^+ out of it there you can satisfy the requirements in three ways. The main attraction in the two-nucleon interaction comes through multiple pion exchanges. You cannot prove that the Lorentz invariance of that attraction corresponds to a scalar field. You have some mass renormalisations here it looks like a whole big castle although really I don't see what it's founded on.

Brown: You can show, I believe, that if you take the two pion-exchange with isobars and crossed isobars, do that covariarantly and relativistically, you can reduce that to an effective scalar exchange which transforms as a scalar.

Pandharipande: You mean to say it couples to the mass? Can you prove it?

Walecka: With zero width?

Brown: Not with zero width. In the chiral theory like Earl Lomon and Partovi and various people since then have shown, it is the σ-meson which gets a distributed mass and vertex correction though the mass correction acquires effectively a long range. These are the remnants of the σ-meson in a chiral theory which you have as an effective σ-meson in these calculations.

Pandharipande: A spin-flip for the isobar?

Brown: You flip into the isobar and back again. We made this tremendous calculation with Durso, Tzara and myself, setting your paper right.

PION DYNAMICS IN THE RELATIVISTIC NUCLEAR MANY-BODY PROBLEM*

Brian D. Serot**
Institute for Theoretical Physics, University of California
Santa Barbara, CA 93106

and

Institute of Theoretical Physics, Department of Physics
Stanford University, Stanford, CA 94305

ABSTRACT

Pion interactions in the nuclear medium are studied using renormalizable relativistic quantum field theories. Previous studies using pseudoscalar πN coupling encountered difficulties due to the large strength of the πNN vertex. These difficulties may be avoided by formulating renormalizable field theories with pseudovector πN coupling using techniques introduced by Weinberg and Schwinger. Calculations are performed for two specific models: the scalar-vector theory of Walecka, extended to include π mesons in a non-chiral fashion, and the linear σ-model with an additional neutral vector meson. The pion propagator is evaluated in the one-nucleon-loop approximation, which corresponds to a relativistic random-phase approximation built on the Hartree ground state. Virtual $N\bar{N}$ loops are included, and suitable renormalization techniques are employed. The local-density approximation is used to compare the threshold pion self-energy to the s-wave pion-nucleus optical potential.

This work was carried out at Stanford University in collaboration with Dr. Tetsuo Matsui.

Stanford ITP-726 and NSF-ITP-82-138   11/82
 *Supported in part by National Science Foundation Grants NSF PHY
  81-07395 and NSF PHY 77-27084
**Alfred P. Sloan Foundation Research Fellow
  Contribution to 1982 IUCF Workshop, Bloomington, Indiana, October
  28-30, 1982

The traditional approach to nuclear structure involves nucleons interacting through static potentials. To avoid some of the approximations inherent in this approach, relativistic quantum field theories have recently been developed to describe the nuclear many-body problem [1-6]. The present work investigates pion propagation in the nuclear medium in the framework of a renormalizable relativistic quantum field theory.

There are several advantages to the relativistic field-theoretic approach. First, mesonic degrees of freedom can be included explicitly as dynamical coordinates. Second, it is important to have a manifestly Lorentz covariant formalism for the examination of nuclear systems under extreme conditions. One may also find new insight into relativistic effects in ordinary nuclei, a subject which is still a matter of controversy.

A renormalizable field theory is important for obtaining finite, well-defined results with a minimum number of parameters in calculations containing divergent vacuum amplitudes. Renormalizability also limits the types of theories and interactions available.

Recently, a model relativistic quantum field theory containing baryons and neutral scalar and vector mesons was proposed by Walecka to study the bulk properties of high-density nuclear matter [1]. This model can be solved exactly in the mean-field approximation, by replacing the meson field operators with their classical expectation values. The resulting mean-field theory (MFT) provides a reasonable equation of state for nuclear matter [1,7], and extensions to finite systems (i.e., the Hartree approximation) agree well with data on the bulk properties of doubly magic nuclei [8]. Since the model is renormalizable, there is a well-defined procedure for calculating corrections to the MFT, and calculations of vacuum fluctuation [4,5], self-consistent exchange [9], and correlation [10] corrections indicate that the MFT equation of state becomes more accurate as the nuclear density increases.

The original model of Walecka has been extended to include isovector π and ρ mesons in a renormalizable fashion [6]. Calculations in the mean-field approximation to the extended model, however, have focused on systems whose ground states are parity eigenstates, in which pionic effects are absent. Whereas this may be a reasonable approximation for spin-saturated nuclear matter, one expects pions to play a significant role in the structure of finite nuclei. Unfortunately, attempts to include pionic effects in the relativistic nuclear many-body problem have encountered profound difficulties when a renormalizable pseudoscalar pion-nucleon (πN) coupling is used.

As an example, relativistic one-pion-exchange corrections to the MFT of nuclear matter were studied in the Dirac-Hartree-Fock approximation [9]. It was found that pseudoscalar πN interactions modify the self-consistent nucleon spectrum drastically at normal densities. In fact, the resulting many-body ground state is a Fermi "shell" rather than the usual Fermi "sphere." This Fermi shell may be ruled out by existing quasielastic electron-scattering data. Similar unrealistic results were obtained in investigations of nucleon-nucleon scattering using the Bethe-Salpeter equation [11] and in studies of (p,π) reactions [12].

The common problem in these calculations is the large strength of the πN interaction in models with pseudoscalar coupling. This large coupling must ·be explicitly cancelled to yield reasonable results. For πN interactions in free space, for example, this may be achieved by introducing a nonlinear coupling between pions and a scalar meson, as in the model of ref. [6] or in the chirally symmetric linear σ-model [13]. Unfortunately, the cancellation mechanism in the nuclear medium (if it indeed exists) is poorly understood and thus difficult to implement. The result is that a reasonable initial approximation is still needed for pseudoscalar πN interactions in the relativistic nuclear many-body problem.

On the other hand, the foregoing problems can be remedied by using a pseudovector coupling of pions to nucleons. Such a coupling leads naturally to a weak πN interaction for low-energy pions. It is well known, however, that a straightforward theory with pseudovector πN coupling is nonrenormalizable. One must therefore restrict calculations to "tree level," since divergent loop integrals require either ad hoc cutoffs or an infinite number of renormalization conditions as one proceeds to higher and higher order.

The present work follows the approach introduced by Weinberg [14] and Schwinger [15] to formulate renormalizable field theories with pseudovector πN coupling. Starting with any renormalizable pseudoscalar model (which contains an additional scalar meson to yield reasonable πN interactions in free space), we redefine the fields using a nonlinear chiral transformation to achieve a representation with pseudovector coupling. Since the latter representation has a naturally weak πN interaction, simple one-pion exchange serves as an adequate starting point for the relativistic treatment of pions in the nuclear medium [9]. Furthermore, since the model is still renormalizable in the pseudovector mode (as long as the scalar meson mass is finite), corrections to this approximate starting point involving loop integrals may be calculated in a well-defined fashion. We explicitly use the renormalization technique in calculating one-nucleon-loop corrections to the pion propagator in nuclear matter, including the effects of virtual nucleon-antinucleon pairs ("vacuum fluctuations").

Explicit calculations are performed for two model lagrangians. (For details, see ref. [16].) The first is the non-chiral model of ref. [6], which contains σ, ω, π, and ρ mesons. This model is of interest because it successfully describes nuclear matter and finite

nuclei in the Hartree approximation. Furthermore, the model can qualitatively reproduce low-energy πN phenomenology in free space. It is
thus worthwhile to investigate the predictions of this model for the
properties of pions in the nuclear medium.

The second model is the linear σ-model [13] with additional vector meson (ω) interactions. The correct πN phenomenology is achieved
in this case through the constraints of chiral symmetry. Furthermore,
by suitably choosing the free parameters, the saturation properties of
nuclear matter can be reproduced in the relativistic Hartree approximation.

The pion self-energy in nuclear matter is evaluated in the one-
nucleon-loop approximation. This includes both the scattering of
pions from nucleons in the filled Fermi sea and the effects of pion
interactions with <u>virtual nucleon-antinucleon pairs</u> in the medium.
The one-nucleon-loop diagrams are a relativistic generalization of the
random-phase approximation that is often used in nonrelativistic many-
body theory. Consistent with this approximation, the nuclear matter
ground state is treated using the relativistic Hartree formalism [4,5],
which includes the effects of the medium on antinucleon wave functions.

We investigate the density dependence of the pion self-energy
$\Pi(q;k_F)$ and compare the predicted threshold self-energy at normal

nuclear density to experimental data on the pion-nucleus optical potential. This is done using an "effective" πN scattering length in the
medium,

$$\tilde{b}_0(k_F) \equiv \frac{-\Pi(\underset{\sim}{q}=0,\ q^0=m_\pi;k_F)}{4\pi\rho_B} \quad ,$$

which is compared to the phenomenologically determined s-wave scattering length in the medium, b_0. (See fig. 1.) Our results may be summarized as follows.

The non-chiral model predicts large corrections when the pion-
nucleon interaction is extrapolated from free space to nuclear matter
density. In the pseudoscalar mode, this occurs because the strong
πN interactions are incompletely cancelled in the medium and because
of large effects involving $N\overline{N}$ pairs. The pseudovector approach is
favored because the πN interactions are weak in the medium (as expected
from the derivative coupling); however, the effects of $N\overline{N}$ pairs and
of nonlinearities induced by the transformation of the lagrangian are
substantial. It appears that fixing the πN interactions to be weak
in free space is not sufficient to render them weak in the nuclear
medium, even in the pseudovector representation. Further sensitive
cancellations are needed, and the predicted pion self-energy in this
model is highly dependent on the renormalization of the nonlinear
meson vertices. With a suitable finite renormalization of the $\sigma^2\pi^2$

vertex, one can reproduce the empirically small s-wave piece of the pion-nucleus optical potential (see fig. 1), but due to the different density dependence of various contributions to the self-energy, pion interactions are much larger at other densities.

The chiral model leads automatically to much smaller effects in the pseudovector scheme, since the self-energy is proportional to the square of the pion four-momentum q and vanishes in the soft-pion $(q^2 \to 0)$ limit. In the pseudoscalar representation, the self-energy depends sensitively on the nucleon effective mass, and vanishes for $q^2 \to 0$ only if the mass is chosen self-consistently. (That is, the mass must be determined by solving the chiral relativistic Hartree equations for the nuclear matter ground state.) In self-consistent calculations in the exact chiral limit ($m_\pi = 0$), the self-energy vanishes at threshold, indicating that the pion remains a (massless) Goldstone boson in the nuclear medium.

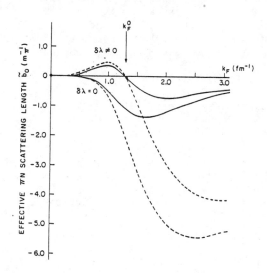

Fig. 1: Effective πN scattering length $\tilde{b}_0$ in units of m_π^{-1} as a function of Fermi wavenumber. Results are indicated for pseudoscalar (dashed) and pseudovector (solid) representations of the non-chiral model [6]. The dots represent experimental scattering lengths in free space ($k_F=0$; $a_0^{(+)} \stackrel{\sim}{=} -0.01\ m_\pi^{-1}$[17]) and at nuclear matter saturation density ($k_F^0=1.30\text{fm}^{-1}$; $b_0=-0.03\ m_\pi^{-1}$[18]).

The curves labeled $\delta\gamma=0$ were renormalized to minimize the nonlinear $\sigma^2\pi^2$ coupling, while those labeled $\delta\gamma \neq 0$ were renormalized to agree with the phenomenological value $b_0= -0.03 m_\pi^{-1}$.

For a given model lagrangian, the pseudovector and pseudoscalar representations yield identical results for on-mass-shell amplitudes but differ in their off-shell predictions. Our results show that pion interactions in the nuclear medium can distinguish between the representations and choose the one more suitable for application to the nuclear many-body problem. In particular, realistic results can be obtained with pseudoscalar coupling only if the nuclear matter and pionic aspects of the problem are treated self-consistently. In addition, the interacting pion propagator can imply important constraints on the model lagrangian used to describe the nuclear ground state. These constraints can be calculated in a reliable, systematic fashion using the techniques discussed above.

References

1. J. D. Walecka, Ann. of Phys. 83 (1974), 491.
2. T. D. Lee and G. C. Wick, Phys. Rev. D9 (1974), 2291.
3. T. D. Lee and M. Margulies, Phys. Rev. D11 (1975), 1591.
4. S. A. Chin and J. D. Walecka, Phys. Lett. 52B (1974), 24.
5. S. A. Chin, Phys. Lett. 62B (1976), 263; Ann. of Phys. 108 (1977), 301.
6. B. D. Serot, Phys. Lett. 86B (1979), 146; Erratum: Phys. Lett. 87B (1979), 403.
7. R. A. Freedman, Phys. Lett. 71B (1977), 369.
8. C. J. Horowitz and B. D. Serot, Nucl. Phys. A368 (1981), 503.
9. C. J. Horowitz and B. D. Serot, Phys. Lett. 108B (1982), 377; Phys. Lett. 109B (1982), 341.
10. M. Brittan, Phys. Lett. 79B (1978), 27; Ph.D. Thesis, Stanford Univ., unpublished.
11. J. Fleischer and J. A. Tjon, Nucl. Phys. B84 (1975), 375; Phys. Rev. D15 (1977), 2537; Phys. Rev. D21 (1980), 87.
12. E. D. Cooper and H. Sherif, Phys. Rev. Lett. 47 (1981), 818.
13. J. Schwinger, Ann. of Phys. 2 (1957), 407; M. Gell-Mann and M. Lévy, Nuovo Cimento 16 (1960), 705.
14. S. Weinberg, Phys. Rev. Lett. 18 (1967), 188; Phys. Rev. 166 (1968), 1568; Physica 96A (1979), 327.
15. J. Schwinger, Phys. Lett. 24B (1967), 473.
16. T. Matsui and B. D. Serot, to be published in Annals of Physics (1982).
17. H. Pilkuhn, W. Schmidt, A. D. Martin, C. Michael, F. Steiner, B. R. Martin, M. M. Nagels, and J. J. deSwart, Nucl. Phys. B65 (1973), 460.
18. M. Krell and T. E. O. Ericson, Nucl. Phys. B11 (1969), 521; K. Stricker, H. McManus, and J. A. Carr, Phys. Rev. C19 (1979), 929; C22 (1980), 2043; R. Seki, K. Masutani, M. Oka, and K. Yasaki, Phys. Lett. 97B (1980), 200.

DISCUSSION

Redish: Renormalizability doesn't guarentee convergence. How do these transformed Feynman graphs converge?

Serot: As far as I know, no Quantum Field Theory converges.

Redish: But it may be an assymptotic series, what is the expansion parameter?

Serot: In some sense, when you start with the original pseudoscalar case, you're dealing with the $g_{\pi NN}$ which is large. When you make the transformation to pseudovector coupling you have a smaller coupling constant.

Lomon: Some of the higher terms get back to pseudoscalar coupling, so you just remove it from the lowest term.

Serot: No, I disagree. The pseudoscalar coupling is done to all orders in the Lagrangian.

Rho: When you do the non-linear transformations, you say that you only do scaling on the counter term. Now, a non-linear transformation usually involves a square root of one plus something. How many counter terms come in and how do you cancel them?

Serot: What saves us is that we're solving linearizsed pion field equations. So I take the pseudoscalar model. I write down the linearized pion field equations . I calculate the counter terms. Then I transfom them and I only get linear terms.

Rho: Yes but that's the way to do it. Those who normalize this whole theory in many cases had to go through a symmetric theory, renormalise the whole symmetric theory and go back to the non-symmetric theory to make sense and to satisfy the Ward identities, there is no other way. When somebody does this in Mean Field Theory and fits the data, and thus gets an effective theory, I don't see the merit of trying to do the renormalization with a theory which is based on an effective Lagrangian. Why do you need a renormalizable theory?

Serot: I can answer that on two levels. The first is a practical one. It limits the theory and it lets me calculate corrections without imposing cut-offs. Secondly, I know of no arguement that has anything to do with the relation between fundamentality, or effective interactions, and renormalizability. Basically, every quantum field theory which has ever been written is meant to work over a particular range. As far as I know the strongest argument for renormalizability is one in which the constituents have sub-constituents. Basically, this is just a model. It is an extended version of good old nuclear physics where I have some extra degrees of freedom which I know how to calculate.

Noble: I think what you say about the Gellmann Levy σ-model plus an ω coupling to the conserved nuclear current, is that it essentially describes all of low energy (up to 1 GeV) dynamics that's relevant to nuclear physics. That is to say it describes π-π scattering, π-N scattering and N-N scattering pretty adequately . The reason why renormalizability is useful is that it makes any results you calculate insensitive to cut-offs. So you don't worry about whether you have a large or small bag in there, you can deffer those questions to those people who want to play with bags. That's really all you can say about it: its a good phenomenological theory which allows you to deal with pions in nuclei.

Miller: I think I might bring up the question which V.J. Pandharipande brought up before. Let's forget for a moment about renormalization so that we only take into account higher order diagrams which are finite. There are a large number of such things in any effective field theory renormalizable or not. The question is, we have had a speaker that very confidently exclaimed that he thinks he has the answer. So, I'd like to ask him, let's go back to the nuclear part when he speaks about the t-matrix, and ask him why he thinks he has the answer.

Shakin: I think that the answer exists at the level on which one calculates the tree diagrams. I think that the enterprise of trying to create a renormalizable field theory is an interesting intellectual exercise, but it's also very difficult. And also the physics, in my opinion, is not right. Because, for example, if you want to describe nuclei you have to introduce the ρ-mesons. These are vector mesons, if you want to give them mass, you have to introduce them as gauge bosons and introduce Higgs fields to get the mass, otherwise you don't have a renormalizable theory of vector mesons.

Brown: That's simply not true, Carl, because the ρ-meson shows up in the π-π scattering as resonances in the σ-model.

Shakin: The question I posed myself was the following. The old Brueckner programme was to take something which fits the scattering data and calculate the properties of nuclear matter and finite nuclei. We have done this in a model in which you define the calculational procedures. It's a model. I never stated that it was a theory. If it was a theory then you could calculate all the corrections. None of us are working with theories. In the context of this specific model with the calculational prescriptions, I've defined, you can calculate in a parameter free manner all of the interesting properties of nuclei. I have the answer to the question I posed myself at the beginning. The next question I would pose myself is: why does this work? What is the relation of this simple model, that does so well to a more fundamental level which would exist at the level of QCD, quarks and gluons? I've answered the question to myself, I may not have answered the questions that other people in the audiance may have.

Pandharipande: You can do rigorous theory. The problem with the old approach was to get nuclear matter properties from two-body forces. However, nobody ever claimed that in nucleons, since they are composite objects, not fundamental entities, that there are no three-body forces. So if you want to continue the old approach in a rigorous way, what you should do is look at Δ and three and four body systems, then write a Hamiltonian for which you can solve the Schrodinger equation. A second approach is within non-relativistic quantum mechanics, a rigorous approach which can be pursued and there are people pursuing it (like myself). The second item; as far as the relativistic corrections are concerned I would completely agree that that's a very good field. To pretend that nuclear matter is completely a non-relativistic subject is certainly an oversimplification. If you want to build a theory which will last for 15 or 20 years, I think you have to answer these questions. It's a composite object you are coupling to a scalar and vector field, what are these elementary fields?

Shakin: An excellent question. I think that's the question one should think about.

Walecka: I'd like to say a little about my philosophy and how I got into this. We have a very narrow window on the nuclear equation of state, the binding energy and the density. We've spent a long time trying to calculate that from a N-N interaction potential. When neutron stars were discovered this meant a tremendous extrapolation in the equation of state for nuclear matter into new conditions, up to an order of magnitude in the density, everthing held together by gravitation. I wanted to try and make a model which starts building with the basic symmetries which you need and also with the right number of degrees of freedom built in. I happen to believe that for nuclear physics the right degrees of freedom are mesons and baryons. If you take potential calculations for neutron stars they violate causality because at the high matter density the speed of sound becomes greater than the speed of light. So if you take potentials as a fundamental theory at some point you get into serious problems. I wanted a theory that I can solve with the basic symmetries built in, what I believe are the basic degrees of freedom. So I said you need baryons, you need mesons, the simplest mesons you have are scalar, vector and the pions. It turns out that for scalars and vectors there is a solution to that problem, the field theory, which happens to be valid at high baryon density, and this mean field s solution, in fact, as far as we know, is the solution to the field theory at very high density, where we have baryons and mesons whose quantum fluctuations are negligible.

Brown: But many times greater than nuclear matter...

Walecka: Many times, yes. But at least we have a starting point at high density, which I believe starts the solution of the iteration. The model also happens to saturate, actually because of these velocity dependent terms. Carl showed us that the strong velocity dependence is essentially because of this term. But it's got Lorentz invariance

built into it, which I think any theory really has to have, potential
breakdown also if you are doing rigorous things at certain densities
(i.e. high densities). It's got retardation effects. We don't have the
right degrees of freedom...

Pandharipande: I completely agree.

Walecka: We have just kept calculating things, and the nicest thing,
I think, is this relativistic Hartree. You have a set of field
equations, you need the sources for the meson field, the local
sources; how do you compute that? Will you merely compute it by
adding up all the baryons contributions to that source? (the scalar
source and the vector source). There are no non-localities in the
theory at that point. You simply calculate the source, solve the meson
field equations and you recompute the source and do it self
consistently; and thus we get beautiful mass densities and charge
densities for nuclei. If I calculate the second order energy, it's big
because the coupling is big, as in the strong coupling theory. But I
have a completely different philosophy. I don't say that I solved the
two body problem and try to reproduce nuclear matter. I give up. I say
I always renormalize the theory to fit nuclear matter and then I ask
the other things to change. So, in that sense it's a choice of
renormalization points. You always come back and fit the properties of
nuclear matter. I calculate the high order effects, change the
coupling constants, and come back and fit nuclear matter and look how
things change. It's a different model and a different choice of the
renormalization point. I claim, I don't know how to do the two-body
problem correctly.

Brown: At this point it's where V.J. and I would say you're making a
model. It's a model calculation because your coupling constants don't
have anything directly to do with the coupling constants for mesons to
baryons you renormalize in nuclear matter, which means that all kinds
of correlations have gone into these coupling constants and may work
over a narrow range, but you really would have to establish if they
work over a wide range since if you put the correlations in... , since
any mean field theory certainly is not really valid at the densities
you're using it. You've to get to densities which are an order of
magnitude higher before you can have any indication of convergence.
So, it's a model theory, and does not work as you do, and I think,
ultimately, people should do it with real couplings to the mesons and
baryons.

Walecka: But it's also a framework in which you can calculate
corrections. It's a strong coupling theory; big coupling constants, so
you can calculate your higher order diagrams and come back to nuclear
matter, and see how things change.

Brown: It's a model in which you can extrapolate and intrapolate if
you don't go very widely, but it does not come down to fundamental
quantities that you would measure in field theory or in particle
theory.

Walecka: Well, I claim the two-nucleon problem is very difficult.
Carl has not justified how to deal with the two-nucleon problem that
solves the Bethe-Salpeter equation with meson exchange. You've got to
do some justifying of the big coupling constants.

Lomon: A lot of those justifications exist to a large extent. The
nucleon problem, the relativistic correction.... The question is, if
one wants to further relate phenomena to fundamental constants and
fundamental interactions and further improve the accuracy and detail,
which I think is the best procedure. Some measurements just calculate
easily by the Hartree perturbation, but there are nevertheless certain
discrepancies, and so, as I said, becomes positive a question where to
find fundamental nuclear descriptions and fundamental
Lagrangians.........

Walecka: I want to make a further comment. One to Gerry and is about
that he's right when he says the coupling constants meaning is, they
are effective coupling constants of the field and are not directly
related to one boson exchange. But they are in the bulk. They are
different within a factor of two, but not an order of magnitude
different.

Miller: Nonetheless, that is serious.

Shakin: I think that is not the issue. I think the issue is, if you
take these coupling constants and we will not even be able of fitting
scattering data, if you take that potential and pretend that it fits
the elastic scattering data and do Hartree or Hartree-Fock
consistently ......We did a calculation with one seven-boson exchange
potential..............

Someone: One little comment I see in renormalizability. If you
believe in quarks... long distance appear to have nothing to do with
mesons and quarks in bags. Why are you so anxious to get a
renormalizable theory?

Walecka: That is a tough question. I think nobody can answer that
question. You don't see quarks You have to infer the existence of
quarks from the experimental result. Now suppose you lay yourself in
the nuclear domain. You have to find a signal, I think, that tells you
that these are quarks, something about the quarks. And how do you
decide to that? Well, chiefly by taking degrees of freedom in which
you believe, and I ask them to be mesons and baryons that is what I
see.... You get a theory where you can estimate things and compare
with the quark model. I got a theory where you can calculate, make
some predictions and then compare and see.

Brown: I insist. You don't have a theory. You have a model, because
these interactions are the interactions of the vector mesons and the
nucleon, or the scalar mesons and the nucleon. Highly renormalizable,
but it's a model, and I think the thing that upsets V.J. is that your

348

approximations are uncontrollable. You can calculate them, you can see what they give, but he can really start with a two body Hamiltonian and he can control. If he wants he can control the approximations with the fundamental coupling constants and see what the errors are.

Walecka: But what happens if you take the philosophy of renormalizable nuclear matter....
Brown: We don't take that philosophy....

Lomon: I agree with your remark, except that in the face of your statement about the ultimate accuracy of the pathological high density limit, in the neutron stars one certainly must believe that the quarks play a role, but when to tackle what is another question.

Brown: We are talking about the problem of nucleon-nucleon forces and the question of working out with the consequences.

SESSION F

THE NUCLEUS CONSISTS OF MORE THAN NUCLEONS

MESONS AND ISOBARS*

Mannque Rho
Department of Physics
State University of New York, Stony Brook, NY 11794

and

Service de Physique Théorique
CEN Saclay, 91191 Gif-sur-Yvette, France

ABSTRACT

Meson and isobar degrees of freedom in nuclei are discussed in terms of the way chiral symmetry manifests itself in nuclear medium. A particular emphasis is placed on the role that pions (as the Goldstone bosons) play in enhancing and quenching the effective axial charge g_A^{eff} measured in some nuclear spin-isospin modes.

INTRODUCTION

There is no question that mesons, particularly the pion, are actually seen in nuclear processes.[1] There is also a strong indication that the isobar (Δ) has a direct influence at low-energy domain.[1] I would like to propose as I did elsewhere[2] that these meson and isobar manifestations are naturally tied in with the way chiral symmetry is operative in nuclear medium. Viewed in this way, nuclei present an intriguing facet of the complex vacuum structure of quantum chromodynamics, the candidate theory for strong interactions. I believe that this is an area of physics most relevant to the non-perturbative regime of QCD. My main theme will be that chiral symmetry is a lot more relevant aspect of QCD than quark confinement, and that pions in nuclei are the "litmus" indicator of the vacuum changes that are believed to be inherent in QCD.

Strictly speaking, it does not seem really necessary to introduce quarks and gluons for discussing mesons and isobars. Mesons and isobars treated on the same footing as the nucleon would suffice.[3] However understanding why this is so in the context of the modern picture of strong interactions will probably shed some light on just where quark-gluon degrees of freedom will come in in nuclear physics. A good example will be the quenching of the axial-vector coupling constant g_A in nuclear medium: a 30% quenching turns out to lead to a huge

*Invited talk given at IUCF Workshop on "The Interactions between Medium Energy Nucleons in Nuclei", Indiana University, Bloomington, Indiana (October 28-30, 1982). Work partially supported by U.S.D.O.E. contract DE-AC02-76ER-13001.

repulsion between two nucleons at short distance as discussed by Gerry Brown in his talk.[4]

Since the pion is the main ingredient in chiral symmetry aspects of nuclear physics, almost entire discussion will be made in terms of the effective axial charge g_A^{eff} in nuclear medium. In particular, it will be shown in what situations the g_A^{eff} can look bigger or smaller than the free space value, and how this observation reflects on the chiral structure of the nuclear medium. The discussion will be very heuristic but I am certain that some of the results obtained here will remain valid in more rigorous treatments.

PHASE TRANSITIONS AT HIGH DENSITY AND/OR HIGH TEMPERATURE

Indirect though the connection may be, the recent results obtained in lattice gauge calculations are teaching us something enlightening and fundamental in nuclear physics.[5,6] At high density and/or temperature, quarks get deconfined.[5] Furthermore massive quarks of up and down flavor become nearly massless and the Goldstone bosons, pions, disappear.[6] The former is the much conjectured nuclear matter-quark matter phase transition; the latter a chiral phase transition going from the Goldstone phase to the Wigner phase. At present, the energy scale associated with the chiral phase transition appears to be higher (so the size scale smaller) than that of the deconfinement phase transition. It is, however, not unlikely that the two will turn out to be about the same. It is a currently popular notion that confinement triggers a spontaneous breaking of chiral symmetry, in which case the two scales cannot be very different. This issue will surely be settled in the near future. What is important is that there are two phase transitions with different order parameters, the deconfinement concerned with the string tension, the chiral phase transition with the vacuum expectation value of the quark scalar density $<\bar{\psi}\psi>$

Most relevant to nuclear physics is the chiral aspect of QCD as I will argue. The confinement aspect seems to be relegated to a minor role when dealing with nuclear phenomena. In the low density-low T phase, there is the pion: $<\bar{\psi}\psi>$ is non-zero; the g_A is empirically determined to be 1.26. After the chiral phase transition (at high ρ, high T), $<\bar{\psi}\psi>$ goes to zero, pions disappear and a down quark will beta-decay to an up quark with $g_A \cong 1$. The change of g_A from 1.26 to 1 is undoubtedly a feeble one and nobody would argue for using it as a "litmus" indicator of the different chiral phases. Pisarski[7] discusses an example (e.g. $\rho \to 2\pi$ decay) where the effect can be much more dramatic; and it is possible that one can devise a clever experiment to enhance the effect. However I would argue that even at T=0 and at normal density, there is a strong indication that some selected nuclear processes can and do show the way chiral structure of the vacuum affects the

effective axial charge g_A^{eff} in nuclear medium. The fundamental
calculations referred to above clearly indicate that the crucial
degree of freedom is the pion. I will thus examine the role
that pions play in nuclear processes.

THE CHIRAL BAG PICTURE

In the chiral bag model[8], the way chiral symmetry is
realized delineates the bubble (bag) in which quarks are confined from the medium in which the Goldstone modes (pions) are
excited. Gerry Brown discusses how this picture leads to the
conventional meson exchange description of nuclear force and,
what is somewhat remarkable, to a novel description of repulsive core in the NN interaction. I will now discuss how the
same agency works to "quench" the axial charge when the matter
density increases. In fact, these two seem to be intimately
related.

In the chiral bag picture[8], the axial current $A_\mu^i(x)$ is
carried by the quarks inside the bag and by the pions outside
the bag. Consider the static situation, $\partial_t A_0(x) = 0$ or zero
energy transfer. I will assume that the axial current is
exactly conserved

$$\partial_\mu A_\mu^i(x) = 0 \ . \tag{1}$$

This means that for any baryonic states $|a>$ and $|b>$, we should
have

$$\hat{q} \cdot <b|A^i(\hat{q})|a> = 0 \ . \tag{2}$$

Consider in particular the baryonic states of n valence quarks
(n = 3 for the nucleon, = 6 for two-nucleon or dibaryon systems
etc.). Then

$$\underset{\sim}{A}^i(\hat{q}) \propto (\underset{\sim}{\sigma} - \underset{\sim}{\sigma} \cdot \hat{q}\hat{q}) \frac{\tau^i}{2} \tag{3}$$

where $\underset{\sim}{\sigma}$ is the (quark) spin Pauli matrix, and τ the (quark)
isospin Pauli matrix. The $g_A^{(n)}$ for n-quark system we want is
related to the constant that multiplies this operator, Eq. (3).
Since the term proportional to $\sigma \cdot \hat{q}\hat{q}$ comes uniquely from the
exterior pion part of the axial current of the chiral bag
model and not from the quarkish component, $g_A^{(n)}$ can be computed
entirely from the pionic piece alone. (The quarkish contribution proportional to σ must be in such a proportion that when
combined with appropriate pionic contribution Eq. (2) is
satisfied.) A detailed model calculation[9] shows that what
matters is the derivative of the pion field at the bag surface
and that when normalized to the value of g_A for the three-quark
system $[g_A^{(3)} \equiv g_A]$ the result is more or less independent of
such details as bag size, coupling constants, etc. The result

is (for n ≤ 12)

$$\frac{g_A^{(n)}}{g_A} = 1 - \frac{n-3}{9} \tag{4}$$

Note that this has a correct feature for n=12, i.e. the spin-isospin-color saturated system, for which the axial charge vanishes. For a 6-quark state, the axial charge is quenched by about 30%,

$$\frac{g_A^{(6)}}{g_A} \cong 2/3 . \tag{5}$$

Thus as two nucleons come together – and so the bags overlap – pions couple more weakly to the merged system than to the individual bags. This feature leads to a strong repulsion between two nucleons.[4] It also provides a novel interpretation of the celebrated Lorentz-Lorenz effect in π-nuclear inter-tion.[10]

THE CHIRAL FILTER

It seems highly plausible that the more pions an axial-vector probe "sees" the bigger the effective axial charge will be. Can one verify this by experimentally "filtering" the pion cloud? The answer seems to be yes.

Imagine having a complete control of the vector (V_μ^i) and the axial-vector (A_μ^i) currents. When can these operators "see" the maximum effect of the pion cloud? This question was answered some years ago[11] and the reasoning goes as follows: Suppose two bags of nucleons are sufficiently far apart that a soft-pion is exchanged between the two. Now send in V_μ^i or A_μ^i and see how the system responds. One can obtain

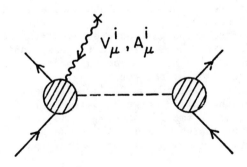

Fig. 1.   Probing soft pions

354

the necessary information by looking simply at the left-hand
vertex of Fig. 1. When the pion is soft, this vertex is
unambiguously given by a soft-pion theorem and as long as it
is not suppressed by kinematics or symmetry, the soft-pion
term will dominate. One learns from this line of reasoning
that the appropriate probe must then be the spatial component
of the vector current V^i and the time component of the axial
current (axial charge density) A_o^i.[1,11] A beautiful case for
the former is the process d(e,e')pn at small energy transfer
and large momentum transfer, much discussed recently in conferences and review articles. (See Ref. 1 for a complete list
of references). A recent treatment of this process in terms of
the chiral bag model[12] reinforces the picture that the "pion
presence" is seen up to a very large momentum transfer.

Although much less studied, the latter, the axial charge
density, has a direct bearing, perhaps more so than the M1
operator relevant for d(e,e')pn, on the working of pions in
nuclear medium. A simple calculation shows that the presence
of the exchanged soft pion of the type Fig. 1 increases the
matrix element of A_o^i or equivalently the axial charge* by a
large amount, e.g.,[11,13]

$$\frac{"g_A^{eff}"}{g_A} = 1.4,\ 1.5,\ 1.6 \quad \text{for } \rho/\rho_o = 0.5, 0.7, 1 \quad (6)$$

where ρ/ρ_o is the density of nuclear medium relative to that of
normal nuclear matter, $\rho_o = 0.17$ fm^{-3}. The effect of this
increased "charge" on the particularly faborable transition
$^{16}N(0^-,T=1) \rightarrow {}^{16}O(0^+,T=0) + e^- + \bar{\nu}_e$ has been studied in detail
by Towner and Khanna [14]. The prediction that the enhanced
charge would increase the β-decay rate by a factor of ~ 4 has
recently been confirmed by a precise measurement of the rate
by Garvey and coworkers[15]. Although this is not yet a definitive
confirmation because of various nuclear structure complications
that make the analysis model dependent, the notion of increased
charge due to soft pions seems inescapable and should be given
a closer attention than done so far.

Up to this point, I repeated what has been known since
some time. In order to make a connection to the next topic,
namely, quenching of the axial charge in nuclear medium, I now
turn to a novel interpretation of the result (6).

As is well known, the axial charge Q_5^i formally defined as
the spatial integral of the axial charge density $A_o^i(x)$ is not
Lorentz invariant.[16] It is frame dependent. (This is quite
different from the vector charge Q^i which is a bona-fide
Lorentz invariant quantity.) The $A_o^i(x)$ for a single particle

---

*As I will discuss later, the axial charge is not Lorentz in-
variant quantity in the sense used here.

transition is of the form (in momentum space)

$$A_o^i = g_A \frac{\tau^i}{2} \left[\frac{\underset{\sim}{\sigma} \cdot \underset{\sim}{p}}{E_p + M} + \frac{\underset{\sim}{\sigma} \cdot \underset{\sim}{p}'}{E_{p'} + M} \right]$$

where p and p' are, respectively, the initial and final momenta of the nucleon and for zero momentum transfer

$$A_o^i = g_A \frac{\tau^i}{2} \; \underset{\sim}{\sigma} \cdot \underset{\sim}{p} / \tfrac{1}{2}(E_p + M) \; . \tag{7}$$

Clearly for a nucleon at rest, A_o^i vanishes. But in the infinite momentum frame ($|p| \to \infty$), it is non-vanishing and finite

$$A_o^i \to g_A \tau^i \; \underset{\sim}{\sigma} \cdot \hat{p} \tag{8}$$

which is just the nucleon helicity. The coefficient multiplying $\underset{\sim}{\sigma} \cdot p$, g_A, can be taken to be the <u>invariant charge</u>. Recall that the charge commutation rule $[Q_A^i(o), Q_A^j(o)] = i\varepsilon_{ijk} Q_A^k(o)$ leads to an infinite set of sum rules for each momentum p and only in the $|p| \to \infty$ limit does one get a sum rule (Adler-Weisberger sum rule)[16] for the invariant charge g_A. Now look at the Gamow-Teller matrix element for which the operator is

$g_A \frac{\tau^i}{2} \underset{\sim}{\sigma}$, in non-relativistic limit. This is the same as (8) when dotted into with $\hat{p}$. The coefficient g_A is clearly the invariant charge and hence the <u>Gamow-Teller matrix element</u> (for zero momentum transfer) measures the <u>invariant axial charge</u>.

Let us now see what one finds when this argument is applied to a two-nucleon system. Look particularly at what happens to the pion cloud at ∞ momentum frame: Let each of the two nucleons in Fig. 1 carry half of the total momentum p. As $|p| \to \infty$, the exchanged soft pion is shaken off and its effect vanishes as

$$\underset{\sim}{\sigma} \cdot \underset{\sim}{q}/E_p \xrightarrow[]{|p| \to \infty} 0 \; . \tag{9}$$

This argument applies to all soft mesons that are exchanged. (This point is extensively discussed in Ref. 16.) Therefore the soft-pion effect which was seen to increase the effective axial charge in the $0^+ \leftrightarrow 0^-$ transition (i.e. the laboratory frame) will no longer be operative <u>when an invariant charge is measured</u>. Furthermore, the only terms that will survive in the ∞ momentum frame are expected to be short-ranged ones, namely

356

a δ-function type interaction in configuration space. I will describe this pictorially as Fig. 2,

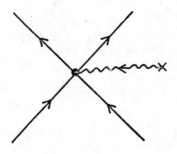

Fig. 2   Contact interaction surviving in
          infinite momentum frame

where the probe is A_o^i at $|p| \to \infty$ or equivalently A^i at zero momentum transfer. Thus ~the effective (invariant) axial charge may be obtained from two nucleons at very short distance (i.e. short-range correlation) or from a bag configuration with 6 quarks inside the bag. From Eqs. (4) and (5), we conclude immediately that g_A^{eff} can be quenched <u>as much as</u> $\sim 30\%$.

Although this line of argument can perhaps be literally taken over to many-nucleon systems, details remain still to be worked out. Nevertheless it is clear that when pion effects are <u>suppressed</u>, the axial charge gets <u>reduced</u>. My assertion is that this is just the feature that <u>reflects</u> on the modification of the chiral structure of the vacuum when matter (i.e. quarks) is introduced. As Gerry Brown discusses in his talk, this weakening of the invariant axial charge can lead to an enormous reduction in net attraction (attributed to the pion cloud), resulting in a several GeV repulsion between two nucleons at short distance.[4] It is then not surprising[2] that there can be an intimate connection (never mind the detailed mechanism or the magnitude) between this and the observed quenching[17] in giant Gamow-Teller resonances.

AXIAL WARD IDENTITY

There is another way of looking at the quenching of the invariant axial charge in nuclear medium that appears at first sight quite different from the chiral bag picture. I tend to believe that it is the same physics and hence there must surely be a simple connection, although I have neither a proof nor a convincing argument for this belief. (I have not worked out full details of the argument, and so some parts may turn out to be wrong. However I find this way of looking at the thing quite instructive.) In any event, I will argue that it is certainly intimately tied in with the chiral structure of the nuclear medium.

There has been a great deal of controversy[18] as to whether non-nucleonic degrees of freedom are responsible for an important portion of the observed quenching of Gamow-Teller strengths in β-decay[19] and in (p,n) processes.[17] This controversy is not new; it was already hotly debated in 1976 Erice school;[20] and recently given a further impetus by the beautiful experimental results obtained at the Indiana Cyclotron. A recent debate is summarized in this year's Telluride proceedings.[21]

The agencies proposed to relegate the non-nucleonic degrees of freedom [more specifically the $\Delta(1232)$] to a minor role are the core polarization due to tensor force,[18] and the exchange term in the Δ-hole interaction.[22] Individually, some of these mechanisms are quite big and can sometimes account more or less for the observed effects. The trouble with this or in that matter also with the Lorentz-Lorenz mechanism alone is that there can be large cancellations between various big terms as first pointed out by Green and Shucan[23] many years ago and subsequently by others.[24] In particular, the tensor core polarization was found to be cancelled almost completely by wave function normalization and some exchange current terms involving similar energy denominators.[24,25]

It also turns out that if the Δ-hole interaction is short-ranged, the direct Δ-hole matrix element is largely cancelled by the exchange term. The proponents for conventional nuclear physics explanation would cite the latter cancellation as an argument in their favor and the proponents for non-nucleonic degrees of freedom would cite the former cancellation in their favor. Clearly the multitude of cancellations among big terms is warning us that there is some underlying symmetry at work. I propose, as I did in Ref. 2, that the underlying invariance is chiral symmetry, and the associated constraint the axial Ward identity.

As before, I will assume that the axial current is exactly conserved. Of course everybody knows it is only partially conserved, but to make matters simple, I will forget the small breaking. Setting $\partial_\mu A_\mu^1 = 0$ is consistent with the Goldberger-Treiman relation, the Weinberg-Tomozawa relation, the KFSR sum rule etc. (see Ref. 16 on these and for original references) and is presumably no worse in the present case than there. This relation will be used at the nucleon level, not at nuclear level for which there are lots of pitfalls.[26] [I would guess that the error committed here is $O(m_\pi^2/m_N^2)$] Call this Assumption I.

I will make one more assumption and that is that the axial current will be assumed to be dominated by the A_1 meson,

358

satisfying the Weinberg sum rule (e.g. $m_{A_1}^2 = 2 m_\rho^2$).[27] The validity of this assumption is hard to assess. Although it has been around for some time, it does not have the same impressive empirical backing as does the vector dominance hypothesis for the vector current. Call this Assumption II.

Now imagine doing a relativistic field theory for nuclear matter with a Lagrangian possessing chiral invariance (including A_1 meson). From the Assumption I follows the Axial Ward identity

$$(p'-p)_\mu \Gamma_\mu^{5i}(p',p) = S_F^{-1}(p') \frac{\tau^i}{2} \gamma_5 + \frac{\tau^i}{2} \gamma_5 S_F^{-1}(p) \quad (10)$$

where p,p' are initial and final four momenta of the nucleon undergoing the transition, $S_F(p)$ the full <u>unrenormalized</u> nucleon propagator in the medium, Γ_μ^{5i} the full <u>unrenormalized</u> vertex function with external legs amputated; i is an isospin index. First remove divergences by renormalizing Γ_μ^{5i}, and S_F as in free space. The quantities $\tilde{\Gamma}_\mu^{5i}$, $\tilde{S}_F$ renormalized in such a way still satisfy the Ward identity. Next do a finite (density-dependent) medium renormalization and call the renormalization constants z_A, z_2, z_3 for the vertex, the nucleon field and the A_1 field respectively. Because of the Ward identity (10) in the quantities $\tilde{\Gamma}_\mu^{5i}$, $\tilde{S}_F$, one is led to the conclusion that at zero momentum transfer the Gamow-Teller operator must be renormalized solely through the "vacuum" (or medium) fluctuation of the A_1 field, Fig. 3. I have conjectured, to arrive

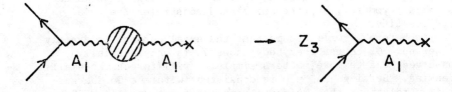

Fig. 3 A_1 field fluctuation in the medium.

at this, that the vertex renormalization is cancelled exactly by the wave function renormalization. (Note: whether the cancellation is exact or only partial when Eq. (10) holds cannot be fully verified. To do so would require a field theoretic model with chiral invariance, with N, π, ρ, ω and A_1 as minimum ingredients for a realistic treatment which I have not yet succeeded to construct to my satisfaction. It is nonetheless very plausible conjecture.)

For zero momentum transfer, the medium renormalization z_3^* is dominated by the Δ-hole bubbles, which can be summed to

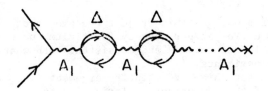

Fig. 4: Δ-hole renormalization of the A_1 field

$$z_3 = \frac{g_A^{eff}}{g_A} = (1 + g^2 \frac{m_\pi^2}{m_{A_1}^2} U)^{-1} , \qquad (11)$$

where U is the usual Δ-hole Lindhard function, g the A_1NN coupling constant and m_{A_1} the A_1 mass which is taken to be $\sqrt{2} \, m_\rho$. (There is a controversy on the A_1 mass. I take $\sqrt{2} \, m_\rho$ since it is in accordance with the Assumption II and there seems also to exist an experimental support for this mass.[28]) Equation (11) may be written in terms of the Landau-Migdal Fermi liquid parameter g_0', with

$$g_0' = g^2 \frac{m_\pi^2}{m_{A_1}^2} . \qquad (12)$$

To evaluate this, I take the value deduced by Durso , Brown and Saarela[29] from the N-N interaction

$$g^2/4\pi \cong 3\text{-}4 \qquad (13)$$

―――――――――――――――

*It is actually $(\sqrt{z_3})^2 = z_3$ that one should calculate for this, one factor of $\sqrt{z_3}$ for the A_1NN coupling constant, another factor of $\sqrt{z_3}$ relating the A_1 field to the axial current.

which implies

$$g_o' \cong 0.6 - 0.8 \ . \tag{14}$$

This is surprisingly close to the value currently accepted for g_o'. How this result can be related to the conventional model for g_o', namely the ρ exchange correlated with ω exchanges, will be commented upon below.

With the values (14), one gets*

$$g_A^{eff}/g_A \cong 0.6 - 0.7 \tag{15}$$

for nuclear matter density. This is just the amount of quenching (with due account of proper density for finite nuclei) expected in the Δ-hole mechanism[30] and consistent with what is observed in the experiments.[17]

Should one accept this Ward identity argument, what would be then the role of the tensor core polarization and the Δ-hole exchange matrix elements etc.? The answer is: they would be irrelevant. In the field theory framework, both the core polarization and the Δ-hole exchange term are vertex renormalizations. Whatever their size might be, there are other terms such as exchange currents, wave function renormalizations etc. that are just as large and at zero momentum transfer compensate those big terms. This is the constraint of chiral symmetry. In order to appreciate this point, recall that in renormalizable field theory finite vertex renormalization is completely arbitrary unless it satisfies the appropriate Ward identity. Likewise the so-called core polarization on its own would then be completely arbitrary. This is not to mean that there is no core polarization correction to the Gamow-Teller matrix element. The arbitrariness can be eliminated only when the constraint due to chiral invariance is imposed. A method is yet to be devised to incorporate, in shell-model calculations, such invariance properties. Most of the shell-model calculations performed so far on Gamow-Teller transitions have not dealt with this constraint.

(In the conventional treatment, Fig. 5a is considered as a direct term and Fig. 5b an exchange term. In the field theory language, Fig. 5a is an A_1 field renormalization and Fig. 5b a vertex renormalization. To see the analogy to QED, replace the nucleon and Δ by an electron, the A_1 by a photon and the nucleon hole by a positron. In QED, Fig. 5b along with other terms of vertex renormalization is exactly cancelled at zero momentum transfer by wave function renormalization, leaving Fig. 5a as the sole agent for charge renormalization

---

*In real nuclei, it will be less, of course.

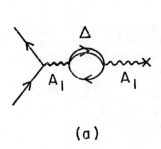
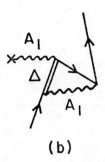

Fig. 5: Direct (a) and exchange (b) term in Δ-hole interaction.

$e = \sqrt{z_3} \, e_0$. See Ref. 31. From this point of view, it would seem utterly irrelevant whether or not Fig. 5b approximately cancels Fig. 5a as claimed to be the case for the g_0' interaction.)

(p,n) VS. NEUTRINO EXCITATION OF GAMOW-TELLER RESONANCES

What is the implication of the A_1 dominance mechanism on the Gamow-Teller resonances excited by (p,n) process?

In describing giant Gamow-Teller resonances excited by the (p,n) process, the currently successful picture[30] is the ρ exchange accompanied by an ω meson: i.e. Fig. 6a. How is

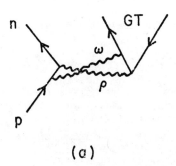
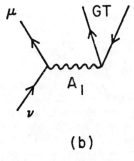

Fig. 6: Gamow-Teller states induced by (p,n) (a) and by neutrino (b).

this related to the A_1 induced excitation by neutrino (Fig. 6b), to which the Ward identity argument may apply? The answer to this is not known at the moment, but one can think of two possibilities: Either [32]

1. Although the effective operators exciting the Gamow-Teller states are of the same structure, the two processes are physically different. Thus while the process Fig. 6b is constrained by chiral invariance, the process Fig. 6a may not be. In this case, the numerical agreement between (15) [Fig. 6b] and the mechanism Fig. 6a [Ref. 30] would be purely coincidental, the old boring dispute would remain unresolved and there would be nothing fundamental about the Gamow-Teller quenching, not a very exciting possibility indeed.

2. Or they are "physically" the same, in which case Fig. 6a would be as much constrained by chiral invariance as Fig. 6b is. The question to settle would then be: are they to be added or not? (Clearly the $\omega\rho$ complex can have an A_1 quantum number, but it is not so obviously included in the extrapolating field A_1 as the $\rho\pi$ complex is.) The most appealing possibility is that they are identical, in which case the numerical agreement would have a most natural origin. If they are not the same, then again the numerical agreement would be coincidental, again somewhat unsatisfactory situation. In this case, however, there would be a clear distinction between the two in what happens to the missing GT strength. With Fig. 6a, it appears most plausible that the missing strength will be found in the Δ region[30]. However with Fig. 6b, since the axial charge is quenched by the vacuum (medium), it will be quenched in roughly the same proportion for all Gamow-Teller states (for zero momentum transfer), be that nucleon-hole or Δ-hole nature. (Of course for the Δ-hole Gamow-Teller state, momentum transfer cannot be zero, so it is not clear to what extent the concept of "charge" applies there.)

As a step towards clarifying these (exciting) issues, it would be most valuable to excite the giant Gamow-Teller resonances by neutrino. There the situation will be a lot cleaner theoretically.

CONCLUSION

When viewed in light of chiral symmetry, the "enhancement" and "quenching" of the axial charge in different axial transitions is seen to reflect on different facets of the QCD vacuum in nuclear medium. In this respect, no theoretical calculation that fails to implement this fundamental invariance property can be taken to be meaningful. The chiral structure of nucleus is, in my opinion, a lot more important issue for nuclear physics than, say, confinement, and there the pion plays a more crucial role than hitherto thought.

On the technical side, the treatment made in this talk

is very crude and rudimentary at best. Clearly a mathematical
rigor is called for. I have leaned heavily on an exact
invariance, when in nature, the symmetry is slightly broken.
This is because the argument is neater when the invariance is
exact - and also because, in the case of chiral invariance,
most of the established soft pion theorems do follow from
$\partial_\mu V^i_\mu = \partial_\mu A^i_\mu = 0$. It would be necessary to establish what the
errors are in the calculations based on it, without opening up
a Pandora's box. (See Ref. 33 for some examples on this
matter.) My bet is that I have made an error of order
$(m_\pi/m_N)^2$. Would anyone take up this bet?

ACKNOWLEDGMENTS

I am very grateful for fruitful discussion and collaboration with Gerry Brown and Vicente Vento. Some of the materials
presented here are based on the work done with them.

REFERENCES

1. See M. Rho and G.E. Brown, Comments in Particle and Nuclear
   Physics 10, 201 (1981).
2. M. Rho, in Proceedings Int. Conf. on Nucl. Structure,
   Amsterdam, Aug. 30 - Sept. 3, 1982.
3. Mesons in Nuclei, eds. M. Rho and D.H. Wilkinson (North-
   Holland, Amsterdam, 1979), Vols. I, II and III.
4. G.E. Brown, talk in this Conference.
5. L.D. McLerran and B. Svetitsky, Phys. Rev. D24, 450 (1981);
   J. Kuti , J. Polonyi and K. Szlachanyi, Phys. Lett. 98B,
   199 (1981).
6. J. Kogut et al., Phys. Rev. Lett. 48, 1140 (1982).
7. R.D. Pisarski, Phys. Lett. 110B, 155 (1982).
8. G.E. Brown and M. Rho, Phys. Lett. 82B, 177 (1979);
   G.E. Brown, M. Rho and V. Vento, Phys. Lett. 94B, 383 (1979).
9. V. Vento and M. Rho, to be published.
10. W. Weise, Phys. Lett., in press.
11. K. Kubodera, J. Delorme and M. Rho, Phys. Rev. Lett. 40,
    755 (1978).
12. J.F. Mathiot and G.E. Brown, private communication and to
    be published.
13. J. Delorme, unpublished; M. Rho, 1981 Erice lectures, in
    Prog. in Part. and Nucl. Physics (ed. by D.H. Wilkinson)
    8, 103 (1982).
14. I.S. Towner and F.C. Khanna, Nucl. Phys. A372, 331 (1981).
15. C.A. Cagliardi et al., Phys. Rev. Lett. 48, 914 (1982).
16. V. DeAlfaro, S. Fubini, G. Furlan and C. Rossetti, Currents
    in Hadron Physics (North-Holland, Amsterdam, 1973) p.467.
17. J. Rapaport, talk in this Conference.
18. A. Arima, in Spin Excitations in Nuclei, eds. F. Petrovich
    et al. (Plenum Press, N.Y.) in press; A. Arima et al.,
    Univ. of Tokyo preprint (1982); I.S. Towner and

364

F.C. Khanna, Chalk River Preprint (1982); G.F. Bertsch and I. Hamamoto, Phys. Rev. C26, 1323 (1982); J. Speth, private communication.

19. D.H. Wilkinson, 1977 Les Houches Lectures, Nuclear Physics with Heavy Ions and Mesons, eds. R. Balian, M. Rho and G. Ripka (North-Holland, Amsterdam, 1981).

20. See the lectures of A. Arima, R. Blin-Stoyle, M. Ericson and M. Rho, 1976 Erice Lectures, in Prog. in Part. and Nucl. Physics (ed. by D.H. Wilkinson) Vol. 1 (1978).

21. Proceedings of Spin Excitations in Nuclei, eds. F. Petrovich et al. (Plenum Press, N.Y.) in press.

22. A. Arima, J. Speth, I.S. Towner and F.C. Khanna, Ref. 18.

23. A.M. Green and T.H. Shucan, Nucl. Phys. A188, 289 (1972).

24. G.E. Brown, private communication (1972); M. Rho, Nucl. Phys. A231, 493 (1974).

25. E. Oset and M. Rho, Phys. Rev. Lett. 42, 47 (1979).

26. J. Delorme, in Mesons in Nuclei, eds. M. Rho and D.H. Wilkinson (North-Holland, Amsterdam, 1979) Vol. I.

27. S. Weinberg, Phys. Rev. Lett. 18, 507 (1967).

28. G. Bellini et al., Phys. Rev. Lett. 48, 1697 (1982).

29. J.W. Durso, G.E. Brown and M. Saarela, Stony Brook preprint (1982).

30. A. Bohr and B.R. Mottelson, Phys. Lett. 100B, 10 (1981); G.E. Brown and M. Rho, Nucl. Phys. A372, 397 (1981).

31. J.D. Bjorken and S.D. Drell, Relativistic Quantum Mechanics (McGraw Hill, New York, 1965) p. 168ff.

32. This part is based on a discussion with Gerry Brown.

33. T. Matsui and B.D. Serot, Annals of Physics, to be published.

EVIDENCE CONCERNING Δ-HOLE STATES*

J. Rapaport
Ohio University, Athens, Ohio 45701

ABSTRACT

Recent (p,n) experiments at intermediate energy at the
Indiana University Cyclotron have been used to study the spin-
isospin transfer response function of nuclei. In particular forward angle spectra are dominated by neutron peaks characterized
with $\Delta L = 0$ transfer, which have been identified as Gamow-Teller
(GT) 1^+ states. The GT strength function in nuclei throughout the
periodic table is presented. The sum of the total observed
GT strength is compared to a model independent sum rule to obtain
an A dependence of the quenching. General features of GT strength
distributions are presented and discussed. A especial effort
is made to estimate how much of the GT strength resides at
excitation energies between 20 - 40 MeV. The fact that less GT
strength is observed than is expected (either calculations or sum
rule) may be associated with intrinsic nucleon degrees of freedom,
in particular the excitation of Δ-hole states.

Based on empirical information obtained from medium energy
(p,n) reactions, it has been established that the sum of Gamow-
Teller strength, B(GT), observed in nuclei is less than the one
predicted from a simple sum rule. This so called "quenching" of
the B(GT) has a profound significance. Let us review some of the
facts.

The success of the experimental studies is based on the character of the isovector nucleon-nucleon amplitude which at intermediate energies is dominated by the spin transfer part (1), thus
favoring the excitations of GT states. An evaluation of the <u>total</u>
B(GT) is thus possible; this total B(GT) value is not generally
obtained from β^--decay measurements because many states are
energetically forbidden.

The sum rule is established directly from commutators of the
one-body operators $t^\pm$ and the assumption that σ_μ and $t^\pm$ are one-
body nuclear operators that can only change the direction of spin
and isospin of nucleons. This assumption treats the nucleon as
an entity with only spin and isospin degrees of freedom. Using
closure one gets the rum rule (2):

* Supported in part by the National Science Foundation.

$$S_{\beta-} - S_{\beta+} = 3(N-Z)$$

in units such that the free neutron decay has B(GT) = 3. This sum rule depends only on the neutron excess of the target nucleus and is <u>independent</u> of the assumed structure of the ground state.

In the above equation $S_{\beta-}$ is the total B(GT) in β^--decay (observed in (p,n) reactions) and $S_{\beta+}$ is the total B(GT) in β^+-decay (observed in (n,p) reactions). In many cases, especially in heavy nuclei, the β^+-decay is practically blocked by the Pauli principle and it may be assumed zero. In other cases the β^+-decay strength may be estimated using shell model calculations. In general the above sum rule provides a lower limit for the β^--decay strength which when compared with the experimental (p,n) results provides an empirical model independent value for the quenching.

A different sum rule results when internal nucleon degrees of freedom are invoked, especially coupling to the Δ-isobars (M = 1232 MeV, T = T_z = 3/2). With these new assumptions (4), it has been calculated (5) that about 65% of the classical strength should be observed in the nuclear excitation region.

In a recent paper Delorme, Ericson and Guichon (6) have extended the above classical sum rule to include explicitly Δ-isobar nucleon excitations. Using a schematic model they calculate the depopulation of the B(GT) in the low energy excitation to be consistent with the empirical results.

Approximately 60% of the sum rule limit has been observed (3). This observed quenching has been interpreted as a disagreement with the assumption that the nucleon is an entity with only two degrees of freedom. In this presentation I will discuss some experimental results on the quenching of the GT strength which may be, therefore, considered as evidence concerning Δ-hole states. The measured angular distributions are used to estimate the GT strength up to 40 MeV excitation energy.

The data to be presented have been obtained with the time-of-flight facility at Indiana University. The beam swinger (7) and two neutron detector stations have been used. With time compensated large volume plastic scintillators as neutron detectors (8), typical time resolution of 700 - 900 psec have been obtained. Energy resolution of 350 - 700 keV are obtained at intermediate proton energies (120 - 200 MeV) and 100 [m] flight paths.

At these bombarding energies the Distorted Wave Impulse Approximation (DWIA) provides a good description of the reaction mechanism. We have used as an effective interaction between the incoming nucleon and each of the target nucleons, the free N-N t-matrix as parameterized by Love and Franey (1). Exchange effects are taken into account explicitly. These calculations show the zero degree cross section for $\Delta\ell = 0$, $\Delta S = 1$ transitions are proportional to the B(GT) for the assumed single particle transition (3).

In the low momentum transfer limit ($q \sim 0$), the zero degree $\Delta\ell = 0$ (p,n) cross section in the impulse approximation may be factorized (9) as:

(1) $$\frac{d\sigma}{d\Omega} (q \sim 0) = \left(\frac{\mu}{\pi\hbar^2}\right)^2 \frac{k_f}{k_i} N_\alpha^D |J_\alpha|^2 B(\alpha) = A_\alpha |J_\alpha|^2 B(\alpha)$$

where the index $\alpha = \sigma\tau(\tau)$ denotes either a GT transition or a Fermi transition (non-spin transfer); N_α^D is a distortion factor that can be calculated; it represents the reduction in the cross section due to absorption of the incoming and outgoing waves and is calculated to extrapolate the cross section to zero momentum transfer ($q = 0$). The values J_τ and $J_{\sigma\tau}$ are volume integrals of the effective interaction and the nuclear structure factor $B(\alpha)$ at $q = 0$ becomes the square of the reduced transition probability for GT or Fermi transitions, as defined in β-decay (10).

We have measured at 160 MeV the zero degree (p,n) cross section for a number of GT and Fermi β-decay transitions throughout the periodic table. This procedure allow us to normalize the measured $0°$ (p,n) cross sections directly to a strength obtained from β-decay.

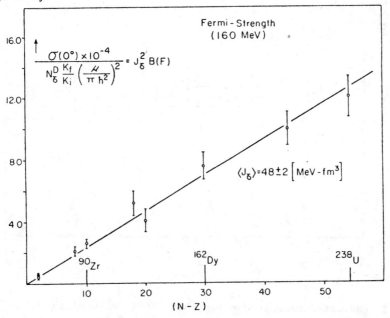

Fig. 1   The zero degree (p,n) cross sections for transitions to analog states, corrected for kinematic and distortion effects and extrapolated to $q = 0$ are plotted versus neutron excess. The data are for Ep = 160 MeV.

In fig. 1 we present the data obtained for a number of transitions to isobaric analog states (Fermi transitions). The measured 0° (p,n) cross sections at Ep = 160 MeV are divided by the quantity A (eq. 1) and plotted versus neutron excess (N-Z). The data is described by the above equation (1) with an empirical value J_{τ} = 48 ± 2 MeV-fm^3 (Ep = 160 MeV). One of the targets ^{54}Fe has a measured Fermi strength from β-decay (log ft = 3.4897; B(F) = 2T = 2).

In fig. 2 we present (p,n) data for a selected set of transitions with known β-decay log ft values. The measured zero degree (p,n) cross sections at Ep = 160 MeV, have been corrected for kinematic and distortion effects and extrapolated to momentum transfer q = 0 (eq. 1) and divided by the known B(GT) value. The constant empirical value $J_{\sigma\tau}$ = 151 ± 5 MeV-fm^2 (Ep = 160 MeV) seems to indicate that it is A independent. We note that the transition ^{42}Ca - ^{42}Sc (log ft = 3.17; B(GT) = 2.57) and the transition ^{51}V - ^{51}Cr (log ft = 5.39; B(GT) = 0.016) differ in the measured (p,n) 0° cross section in a factor larger than 150 in agreement with the β-decay results.

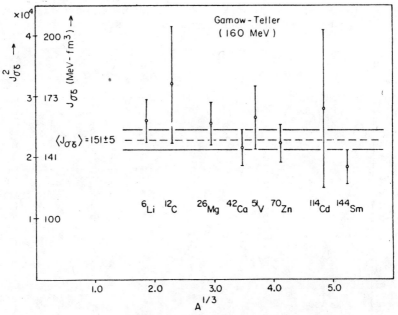

Fig. 2   Selected transitions for which B(GT) values are known
          from β-decay are used to normalize 0° (p,n) cross
          sections. The zero degree (p,n) cross sections corrected
          as in eq. (1) and divided by the known B(GT) value
          results in the empirical value $J_{\sigma\tau}$. Data are for Ep =
          160 MeV.

In fig. 3, 0° (p,n) spectra for some light nuclei are presented. The measured cross section is shown versus neutron energy. For the ^{7}Li, ^{19}Fe and ^{42}Ca nuclei the spectra are dominated by single transitions near the ground state. No evidence for other strong $\Delta\ell = 0$ transitions are found for the ^{7}Li(p,n)^{7}Be reaction. In the case for A = 19, the g.s. transition with a known B(GT) = 1.645 (obtained from β-decay log ft-value) carries a large amount of the total strength. In all other excited states with $\Delta\ell = 0$ we have determined a sum B(GT) $\sim$ 0.30. The total sum B(GT) = 1.95, is thus 65% of 3(N-Z). These low lying 1$^+$ states have predominantly particle-hole configurations with both particle and hole in the same subshell. For the case of ^{39}K we observe a splitting of the GT strength in two excitation regions. The g.s. transition corresponds to a $\nu(d_{3/2}) \rightarrow \pi(d_{3/2})$ transition while the states at Ex $\sim$ 7 MeV are $d_{3/2} - d_{5/2}$ particle-hole transitions.

In fig. 4 the 0° (p,n) spectra at Ep = 160 MeV for some medium A nuclei are presented. It is noted as the isospin of the target increases the dominance in the spectra of a giant GT resonance at about 10 MeV excitation energy. This is in sharp contrast with the spectra shown in fig. 3. The well define sharp peaks in fig. 4 correspond to the isobaric analog transition (IA).

In fig. 5 the 0° (p,n) spectra at Ep = 160 MeV for some heavy nuclei are presented. With an increase in A, we observe the disappearance of well defined 1$^+$ low-lying states. Instead the spectra has a pattern of two sets of GT resonances separated by the IA transition.

In fig. 6 we present the energy difference of these two GT resonances for the indicated nuclei. We observe a remarkable constant difference with a value of about 6 MeV. The resonance observed at lower excitation energy carries in general about 25% of the strength of the resonance at higher excitation.

In order to evaluate the total B(GT) strength we have to obtain the 0° cross section for all $\Delta\ell = 0$ transitions. This may be done without too much difficulty for the cases shown in fig. 3; however for medium to heavy nuclei the estimation of a "background" in the region of the GT states is not straightforward. An "experimentalist" background (smooth line below the resonance) may be used but as recently pointed out by Osterfeld (11), this method leads, however, to a smaller cross section. A better method is to use information from measured angular distributions. The energy spectra is divided in equally spaced energy bins and the observed angular distribution at each energy interval is then fitted with a summation of DWIA calculations with different angular momentum transfers. The cross section at $\theta = 0°$ for non $\Delta\ell = 0$ transfers provides the needed "background". This method has been used in several cases. As an example, fig. 7

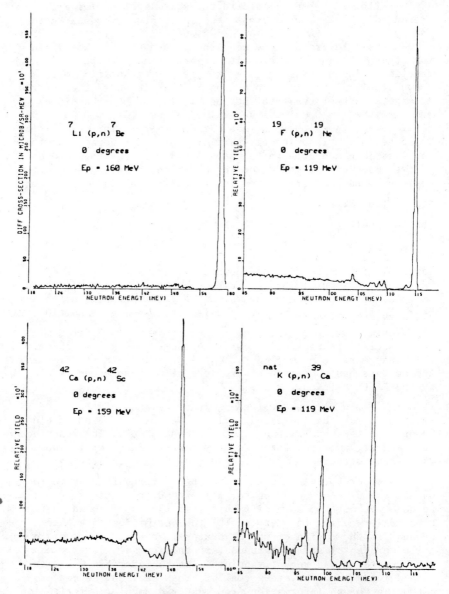

Fig. 3   Zero degree (p,n) spectra for the indicated nuclei at
Ep =´120 MeV and Ep = 160 MeV. The measured cross
section is presented versus outgoing neutron energy.

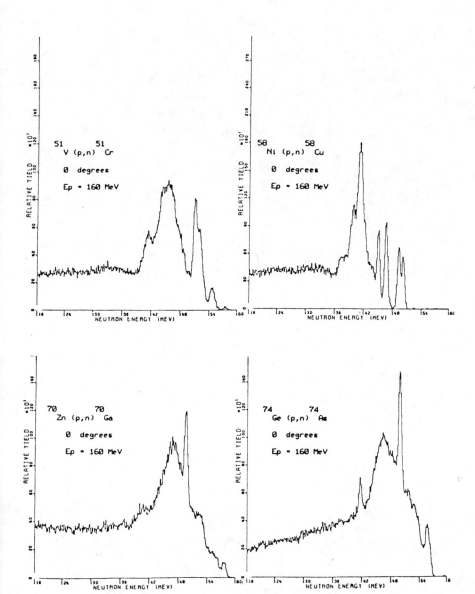

Fig. 4 Zero degree (p,n) spectra at Ep = 160 MeV for the
indicated nuclei. The measured cross section is
presented versus neutron energy.

372

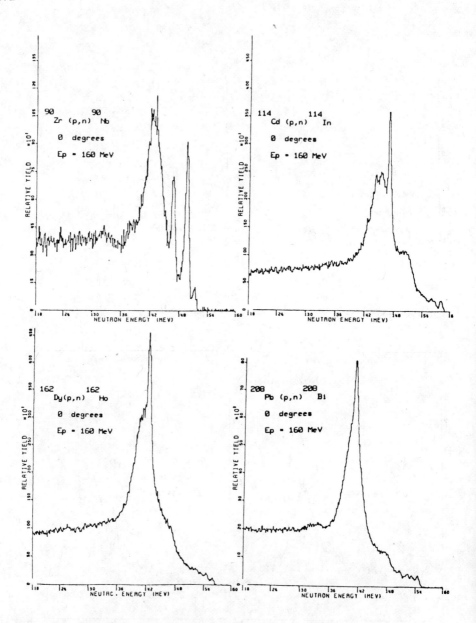

Fig. 5   Zero degree (p,n) spectra at Ep = 160 MeV for the indicated nuclei. The measured cross section is presented versus neutron energy.

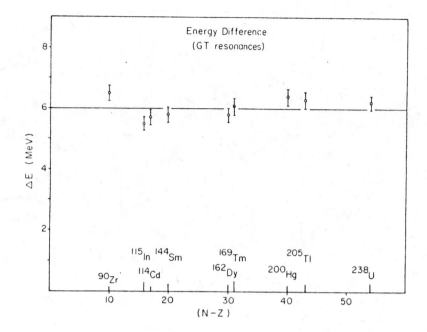

Fig. 6   Energy difference between the observed GT resonances in
          heavy weight nuclei.

shows angular distributions for the ^{58}Ni(p,n)^{58}Cu reaction at
Ep = 160 MeV. The angular distribution for 2 MeV energy bins below
10 MeV excitation may be fitted with either a pure $\Delta\ell = 0$ ($\Delta J = 1^+$)
transfer or with a combination of $\Delta\ell = 0$ ($\Delta J = 1^+$) and $\Delta\ell = 2$
($\Delta J = 3^+$) transfers DWIA calculated shapes. However, for excitation energies $10 \leq E_x \leq 20$ MeV a $\Delta\ell = 1$ ($\Delta J = 1^-$) DWIA shape
becomes predominant while the $\Delta\ell = 0$ shape characterized by a sharp
forward peak, seems to diminish. At even higher excitation energy
the $\Delta\ell = 2$ ($\Delta J = 3^+$) DWIA shape seems to fit the observed angular
distributions. The above combination of DWIA shapes used to fit
the data may not be unique. We have calculated DWIA angular distributions for $0^\pm < \Delta J^\pi < 4^\pm$ and the curves shown represent the
minimum number of $\overline{\Delta J}$ transfers that fit the observed data.
Osterfeld (11) has calculated the "background" of the
^{40,48}Ca(p,n)^{40,48}Sc spectra in a microscopic model. He was able
to reproduce the continuum part of the spectra within a factor of
1.3 with admixtures of $\Delta\ell = 1$ and $\Delta\ell = 2$ (3^+) particle-hole excitations. The reported microscopic calculations are in qualitative
agreement with the present results.

     The 0° spectrum which represents $\Delta\ell = 0$ contributions obtained
as indicated above is shown in fig. 8 for the ^{58}Ni(p,n)^{58}Cu reaction.

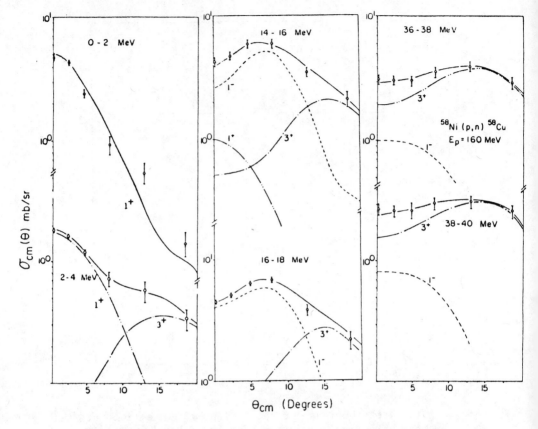

Fig. 7   Angular distributions for the ^{58}Ni(p,n)^{58}Cu reaction at
Ep = 160 MeV. 2-MeV energy bins at low, medium and high
excitation energy are shown. The curves are DWIA calculations with the indicated ΔJ transfer. The solid curve is
the sum of the calculated transfers.

This type of analysis has been completed for several cases
resulting in B(GT) values slightly larger than previously
reported (3) in which "experimentalist background" has been used.

We have been able to extract B(GT) strength in the excitation
energy region where it dominates. Uncertainties in the calculated
DWIA shapes at high excitation energy and small zero degree cross
section values for possible GT states may preclude observation of
this possible strength. We have estimated in a few cases this
possible contribution to be $\lesssim$ 20% of the observed strength. This
estimation was done in the 20 - 42 MeV excitation energy region.
Recently Bertsch and Hamamoto (12) have placed a significant part
of the "missing" B(GT) strength in this high excitation energy
region (about 50%).

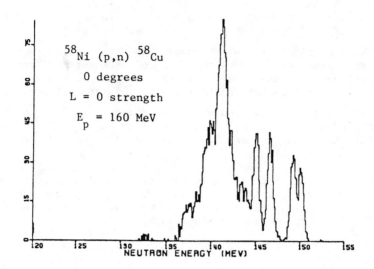

Fig. 8   Zero degree spectrum for the ^{58}Ni(p,n)^{58}Cu reaction at
          Ep = 160 MeV, representing only $\Delta\ell = 0$ contributions.

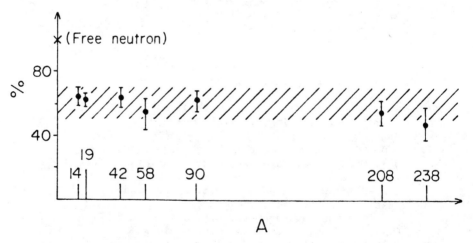

Fig. 9   Fraction of Gamow-Teller sum rule strength observed in
          (p,n) reactions. For the case of ^{58}Ni the $S_{\beta+}$ value has
          been estimated from a shell model calculation.

In fig. 9 we show a summary of the results for the GT strength
extracted from the 160 MeV data. It seems to show that independent
of A, it is observed a maximum of 65 ± 5% of the 3(N-Z) sum rule.
The data for ^{238}U should be considered preliminary. The data for
the ^{208}Pb(p,n)^{208}Bi reaction is being analyzed as indicated above.

We conclude that the observed Gamow-Teller strength is less
than the value predicted by the classical sum rule but agrees well
with the value of the modified quark spin-isospin sum rules (6) in
which explicit Δ isobars nucleon excitations are included.

The direct observation of the GT strength in the Δ region
($\sim$ 300 MeV excitation) via (p,n) reactions at 800 MeV (13) or via
(^{3}He,t) reactions at 2 GeV (14) should also provide evidence concerning Δ-hole states.

The results discussed here correspond to a collaborative
effort by a group of the following people: C.C. Foster, C. Gaarde,
C.D. Goodman, C.A. Goulding, D.J. Horen, J.S. Larsen, T.G. Masterson, E. Sugarbaker, T.N. Taddeucci and T.P. Welch.

REFERENCES

1. W.G. Love and M.A. Franey, Phys. Rev. C24 (1981) 1073 and
   references therein.
2. C. Gaarde, J.S. Larsen, M.N. Harakeh, S.Y. van der Werf,
   M. Igaraski and A. Müller-Arnke, Nucl. Phys. A334 (1980) 248.
3. C. Gaarde, J. Rapaport, T.N. Taddeucci, C.D. Goodman,
   C.C. Foster, D.E. Bainum, G.A. Goulding, M.B. Greenfield,
   D.J. Horen and E. Sugarbaker, Nucl. Phys. A369 (1981) 258.
4. M. Ericson, Ann. of Phys. 63 (1971) 562;
   M. Ericson, A. Figureau and C. Thevenet, Phys. Lett. 45B
   (1973) 19;
   E. Oset and M. Rho, Phys. Rev. Lett. 42 (1979) 47;
   H. Toki and W. Weise, Phys. Lett. 97B (1980) 12.
5. A. Bohr and B. Mottelson, Phys. Lett. 100B (1981) 10;
   G. Bertsch, Nucl. Phys. A354 (1981) 157;
   G.E. Brown and M. Rho, Nucl. Phys. A372 (1981) 397.
6. J. Delorme, M. Ericson and P. Guichon, Phys. Lett. 115B (1982)
   86.
7. C.D. Goodman, C.C. Foster, M.B. Greenfield, C.A. Goulding,
   D.A. Lind and J. Rapaport, IEEE Trans. Nucl. Sci. NS-26
   (1979) 2248.
8. C.D. Goodman, J. Rapaport, D.E. Bainum and C.E. Brient, Nucl.
   Inst. and Meth. 151 (1978) 125;
   C.D. Goodman, J. Rapaport, D.E. Bainum, M.B. Greenfield and
   C.A. Goulding, IEEE Trans. Nucl. Sci. NS-25 (1979) 577.
9. F. Petrovich, W.G. Love and R.J. McCarthy, Phys. Rev. C21
   (1980) 1718.
10. A. Bohr and B. Mottelson, Nuclear Structure (Benjamin, NY,
    1969, Vol. 1, pp. 345, 349, 411).

11. F. Osterfeld, Phys. Rev. C26 (1982) 762; also contribution to
     the 1982 Telluride Conference.
12. G. Bertsch and I. Hamamoto, 1982 reprint.
13. N.S.P. King, P.W. Lisowski, G.L. Morgan, R.E. Shamu,
     J.R. Shepard, C. Zafiratos, J.R. Ullmann and C.A. Goulding,
     Bull. Amer. Phys. Soc. 27 (1982) 720.
14. C. Gaarde, private communication (1982).

THREE-BODY FORCES

J. L. Friar

Los Alamos National Laboratory, Los Alamos, NM 87545

ABSTRACT

Three-body forces are defined and their properties discussed. Evidence for such forces in the trinucleon bound states and scattering reactions is reviewed. The binding energy defects of the trinucleon bound states, the ^{3}He charge density, the Phillips line for doublet n-d scattering lengths, and three-nucleon breakup reactions are discussed, together with the possible influence of three-body forces on these observables.

INTRODUCTION AND DEFINITIONS

Traditionally, nuclear physics has attempted to describe the nucleus as a collection of nonrelativistic nucleons interacting via two-nucleon forces.[1] These forces depend on the coordinates (and possibly momenta), as well as spins and isospins, of only two nucleons. This is a tremendous simplification which has no theoretical justification, other than a rough consistency between predictions of the theoretical (two-body) models and experiments. I use the word "rough" purposefully, because a serious impediment to the advancement of nuclear theory has been the general inability to calculate accurate nuclear wave functions for realistic nucleon-nucleon potential models. Any lack of consistency between theory and experiment is usually blamed on "poor structure calculations".

The few-nucleon systems, on the other hand, have traditionally been a testing ground for new ideas because our ability to solve the Schrödinger equation is greatest for these simple cases. Although variational techniques have long been used to obtain wave functions and energy eigenvalue bounds, the real impetus in this field was the seminal work of Faddeev.[2] Originally designed to handle the boundary conditions in scattering problems, Faddeev's approach to solving the Schrödinger equation has been very successful in treating bound states as well. Computational sophistication has improved dramatically in the past decade, to the point where we can experimentally challenge predictions of the standard two-body models.[3-8]

If this challenge proves unsuccessful, to what cause can the blame be laid? This is always a difficult question to answer, because there are many uncertainties in the conventional nuclear physics approach to calculating observables. The major possible uncertainties are threefold:

1. Three-body forces exist which depend on the simultaneous positions, momenta, spins, and isospins of three nucleons;

2. The effect of relativity is non-negligible, and relativistic corrections are important;

3.   The meson degrees of freedom and nucleon substructure make important contributions to observables.

It should be borne in mind that these 3 categories are not distinct; there is considerable overlap and in some cases the boundaries between them are completely blurred.

We will concentrate our attention on the first category. The definition of a three-body force given above seems obvious, but is incomplete. We must have some way of distinguishing 2 successive two-body forces between 3 objects and a real three-body force. Our working definition for the purpose of this talk will be: <u>Forces which depend on the simultaneous coordinates of 3 nucleons, when only nucleon degrees of freedom are taken into account</u>. The latter caveat is required, because there are few three-body force components which are in any sense "fundamental"; that is, it is possible and proper to disagree whether certain forces are three-body in nature or a complicated two-body force situation.

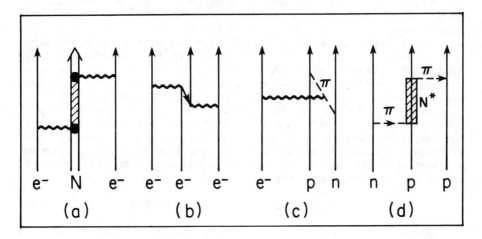

Figure 1. Various physical situations which lead to three-body forces.

The illustration above depicts 4 situations which can be understood as mediated by three-body forces. Figure (1a) illustrates 2 electrons interacting with a polarizable nucleus by means of Coulomb forces. If one chooses to treat the nucleus as an elementary particle, the force is a three-body force. If we agree to treat the nucleon degrees of freedom in the nucleus, a microscope pointed at that nucleus would reveal that the electrons were interacting with the nucleons via purely two-body forces. Thus this atomic physics conundrum is resolved by the answer to the question: Do we treat the nucleus as a "frozen" structureless object or as a collection of nucleons? Figure (1b) illustrates a classic situation, the three-electron force.[9] The middle electron is interacting via photon exchange with the outer pair of electrons. The central electron accomplishes this by emitting a photon, creating an electron-positron pair, and then annihilating the positron on the remaining electron line, which emits another photon. Everyone would agree that the lepton pair part of the diagram is special, and that without it we would be dealing with repeated two-body forces. On the other hand, a relativistic calculation which uses Dirac wave functions for the electrons and two-body forces would include this process in a natural way. In a nonrelativistic treatment, Figure (1b) depicts a three-body force; in a relativistic treatment, it is simply two two-body forces. Figure (1c) depicts a meson exchange current in a deuteron, "seen" by a photon from a passing electron.[10] This is the closest thing to a "fundamental" three-body force that I know. Figure (1d) is relevant to our real interest. A neutron emits a pion which strikes a proton, turning it into an excited state (N*), which decays by emitting a pion, later absorbed by another proton. Is this a three-body force? It is according to our definition. It is not, however, considered as such in calculations[12] which include N*'s in the nuclear wave function as identifiable components.

Thus, "freezing out" degrees of freedom always leads to three-body forces: in Figure 1(a), the nucleus; in (b), the negative-energy wave function components of the electron; in (c), the pion; in (d), the nucleon substructure (N*). I have spent too much time defining three-body forces, but I have been at meetings where the experts have argued over such matters, leaving everyone else confused. We see, however, that the litmus test is simple: degrees of freedom other than nucleons which are included in the nuclear wave function don't lead to explicit three-body forces; freezing them out produces such forces. Figure (1d) is a major (the major?) component of the nuclear three-body force.

PROPERTIES OF THREE-BODY FORCES

However we choose to define our three-body force, it will depend on the coordinates of three nucleons, including spins, isospins, etc. Thus, an obvious restriction is that one needs at least 3 nucleons for such a force to manifest itself. We will concentrate on the three-nucleon system, for reasons stated earlier. The coordinates of such a system are depicted in Figure (2); two distances, x and y, and one angle, θ, are required to completely specify the relative orientation of the three nucleons. The most important of these coordinates for our purposes is θ. It is entirely reasonable, even probable, that a three-body force would depend on this coordinate. It could, for example, be repulsive when the nucleons are in a linear configuration ($\theta=0$ or π) and attractive when the configuration is isosceles ($\theta=\pi/2$). Thus, a three-body force can select particular three-body configurations, diminishing some and enhancing others. The force will obviously depend on x and y, and the example shown in Figure (1d) depends in an essential way on spin and isospin. This means that the manifestations of such a force in the three-nucleon system and in nuclear matter could be very different, indeed, opposite!

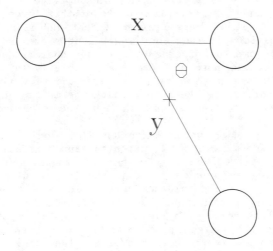

Figure 2.   Coordinates of the three-body problem.

The scales of three-body forces are best illustrated by two examples. The three-electron force (and possibly the first three-nucleon force) was developed many years ago by Primakoff and Holstein,[9] and corresponds to the diagram (1b). It is relatively easy to calculate the force if one remembers that the electromagnetic $\vec{A}^2$ term in the nonrelativistic Hamiltonian, $(\vec{p}-e\vec{A}/c)^2/2m = \vec{p}^2/2m -e\vec{A}\cdot\vec{p}/mc+e^2\vec{A}^2/2mc^2$, is due to the virtual electron-positron "pair." Thus the force we wish is obtained by connecting the $\vec{A}^2$ - vertex to two $\vec{A}\cdot\vec{p}$ vertices with photons. Schematically, this force behaves as $V_c^2 \vec{p}^2/m^3c^4$, where V_c is the ordinary Coulomb potential between two electrons. The force is momentum-dependent, and very weak in most circumstances because of the $1/c^4$ factor; it is a second-order relativistic correction and most atoms are basically nonrelativistic. To the best of my knowledge, it has never been seen experimentally.

The most popular of the three-nucleon forces is the Tucson force,[13] which continues a tradition of naming recent forces after cities (Paris, Bonn, Graz, etc.). This force is purely two-pion-exchange, the longest-ranged component of the total force. Its construction was motivated by the following argument: (1) As illustrated in Figure (1d), the primary ingredient of the two-pion-exchange-three-body-force (2PE3BF) is the off-shell pion-nucleon scattering amplitude; (2) Considerable theoretical and experimental information exists on this quantity from particle physics; (3) We need to consider only s- and p- wave π-N scattering, because the angular momentum barrier suppresses higher partial waves. With these ingredients and assumptions the Tucson force was constructed.

We wish to emphasize several aspects of this force, and similar forces. Firstly, it is very spin dependent and isospin dependent because the basic pion-nucleon ingredients are so dependent. Secondly, it is momentum dependent, although there is some evidence that the momentum-dependent terms are relatively unimportant. Thirdly, although of one-pion-range in each pair of nucleons, it has very strong short-range behavior which seems to dominate the physics. Fourthly, it does not include the exchanges of other mesons.

Schematically, the Tucson force has the form V_π^2/Mc^2, where V_π is the two-body one-pion-exchange potential and M is a mass which we can take to be the nucleon mass. The form suggests a size corresponding to a relativistic correction. Unlike the atomic case, the nuclear three-body force should be appreciable.

It should be borne in mind that the construction of such forces is a theoretical exercise with limited experimental input. How much progress on the two-nucleon problem would have been made without experimental input? A certain amount of caution should be exercised by all!

EVIDENCE FOR THREE-BODY FORCES IN BOUND STATES

One can argue that relativistic corrections in a nucleus should be a few percent <u>on the average</u>.[14] In the trinucleon system, the total potential energy is estimated to be 50-60 MeV.[3] A few percent of this would be on the order of 1 MeV. Of course, the kinetic energy largely cancels the potential energy, leaving a residue of 8.5 MeV binding for ^{3}H. Thus, a change of only a few percent in potential energy would have an appreciable effect on the binding energy. Table I shows the results of recent calculations[3,7,8,15] of the properties of ^{3}He and ^{3}H using only two-body potentials and their comparison with experiment. The binding energy refers to ^{3}H only; the binding energy difference of ^{3}He and ^{3}H is an interesting and partly unresolved problem. The Reid Soft Core, Super-Soft Core, Paris, and Argonne potentials all underbind the triton by 1.0 - 1.3 MeV and have a correspondingly too large radius. While one cannot be certain that the underbinding would not be eliminated by a judiciously rearranged two-body potential, it is <u>a priori</u> reasonable to investigate other possibilities.

Another possibility is relativistic corrections[14] to our non-relativistic two-body potential model. We have already seen that the three-body forces can be viewed in this light. Most calculations of the former effects find a small residue resulting from a cancellation of the relativistic corrections to the kinetic and potential energies. None of these calculations are sufficiently definitive to rule out the possibility that relativistic corrections to the two-body model are unimportant.

TABLE I. Comparison of calculated trinucleon properties vs. experiment. The four models correspond to the Reid Soft Core, Supersoft Core (C), Paris and Argonne (V14) potentials. The radii include the intrinsic sizes of the nucleons.

Model	$\langle r^2 \rangle^{\frac{1}{2}}_{^3\text{He}}$	$\langle r^2 \rangle^{\frac{1}{2}}_{^3\text{H}}$	E_B
RSC	2.09 fm	1.83 fm	7.2 MeV
SSC(C)	2.00 fm	1.76 fm	7.6 MeV
Paris	2.02 fm	-	7.4 MeV
Argonne (V14)	2.06 fm	1.81 fm	7.4 MeV
Expt.	1.86(3) fm	1.69(5) fm	8.5 MeV

Another possibility is that three-body forces are important and account for the discrepancies in the trinucleon bound states. At this time there exist five calculations[5,6,15,16,17] using part or all of the Tucson force and realistic two-body forces. Unfortunately, none of the calculations are identical and none get the same result. There is rough agreement that the attractive p-wave part of the Tucson force gives 1.0 - 1.5 MeV additional binding when recommended values of the coupling constants are used. The s-wave part of the interaction, however, is repulsive and estimated to decrease binding by 0.4-1.0 MeV. The net result varies from 0 to 1.3 MeV, according to the calculations. Since the two-body force models, calculational techniques, and approximations used to obtain these numbers were quite different, it is not clear what the problem is. I personally doubt that the two-body model makes that much difference, but I have no proof of this assertion. These calculations are very different and resolution of the problem will come with time.

Another interesting and controversial comparison is shown in Figure (3), which depicts the charge density of ^{3}He, with "experimental" data, and densities calculated at Los Alamos[18] using the Reid Soft Core two-body potential with and without a Coulomb interaction between the two protons. I use the word experimental advisedly, since considerable theoretical assumptions and extrapolations are required to extract this density from the experimental data. The error bars are statistical only and are considerably smaller than the sum of the theoretical uncertainties. Nevertheless, the hole reflects a longstanding problem: the lack of theoretical strength in the ^{3}He charge form factor in the region of the secondary maximum. The form factor $F(q^2)$ is simply the Fourier transform of the charge density $\rho(r)$, normalized to 1. Inverting this relationship provides a way of calculating $\rho(r)$ at any point.

384

In particular

$$\rho(o) = \frac{2}{\pi} \int_0^\infty F(q^2) \; q^2 dq \qquad\qquad (1)$$

The form factor is 1 if the momentum transfer q vanishes, decreases to 0 as q^2 increases, becomes negative, turns around and approaches 0 again. It is in the region of large negative values that theoretical calculations have always been deficient. This region of q^2 accounts at least qualitatively for the hole in ρ at the origin, which follows from the structure of the integral in (1). Various reasons have been advocated for this deficiency, including meson-exchange currents and three-body forces. The former is an exceptionally complicated business involving relativistic corrections, which we will ignore.[14] The second explanation has a particularly simple geometrical aspect, which has galvanized the theoretical community. We will assume it is correct for the purpose of exposition.

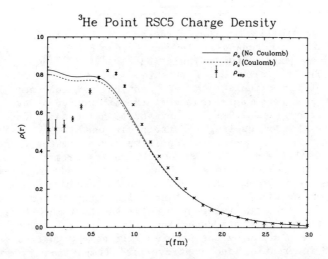

^{3}He Point RSC5 Charge Density

Figure 3. ^{3}He charge density calculations vs. experiment.

Consider Figure (2) again. The coordinate r in $\rho(r)$ is the distance from one of the protons (taken to be the bottom nucleon on the figure) and the center-of-mass of the system (the cross). Thus we have the geometrical relationship: r=2y/3. Clearly, vanishing r implies vanishing y, and this further implies a linear configuration of nucleons. Our theoretical models have too much attraction for such a configuration, implying two large a value of $\rho(o)$. We need to decrease this configuration, while increasing the binding, which depends more on the isosceles configurations. A three-nucleon force can accomplish both these goals if it behaves as follows:

We need to	Three-body force should be
A. Deplete linear configurations;	Repulsive at $\theta=0,\pi$;
B. Enhance isosceles configurations;	Attractive near $\theta=\pi/2$;
C. Increase binding.	Net attractive.

This scenario is possible, and attractive from a theoretical view-point. Time will tell if it is correct or merely a chimera.

Three further aspects of the "hole" need to be discussed before abandoning the topic. Although the hole looks very large in Figure (3), it requires less than 1% of the total charge in ^{3}He to fill it in completely. This is consistent with our earlier estimate of the size of the three-body force, and follows from the presence of r^2 in the volume element of integration. In addition, the hole has nothing to do with repulsion in the nucleon-nucleon force. The charge density is actually an integral over the wave function, which has the structure

$$\rho(r) \sim \int \Psi^2(x,3r/2,\theta)d^3x, \tag{2}$$

where we have replaced y by 3r/2 as indicated earlier. Moreover, it is easy to visualize that for y=r=0, the wave function is independent of the variable θ. Figure (4) depicts the major component (u) of the ^{3}He wave function for $\theta=0$. The deep trough I call "death valley" is due to the strong short-range repulsion when two nucleons try to overlap (y=x/2). The charge density $\rho(0)$ is given by an integral along the y=0 plane and is most strongly influenced by the large structure in the vicinity of x=1.5 fm, which is near where the nucleon-nucleon force is most attractive. Thus, "death valley" has relatively little effect on $\rho(o)$. Finally, the previously discussed inconclusive calculations using the Tucson force find a small effect on $\rho(o)$.

$$u \qquad\qquad \theta = 0°$$

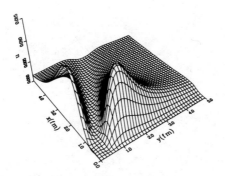

Figure 4. Principal component of the ^{3}He wave function calculated using the Reid Soft Core potential model.

EVIDENCE FOR THREE-BODY FORCES FROM SCATTERING

In principle, scattering processes should provide much more information than bound states in the search for evidence of three-body forces. In fact such evidence is lamentably scant, and is tied in a curious way to the bound state evidence.

If neutrons at vanishing energy are scattered from deuterons, the scattering lengths are measured, of which there are two. The deuteron spin of 1 couples to the neutron's spin 1/2 to form two independent total spin combinations: quartet (3/2) and doublet (1/2). The quartet scattering length is large and influenced primarily by the value of the deuteron binding energy. The doublet scattering length $a^{[2]}$ is small and strongly coupled in a somewhat mysterious way with the corresponding values of the triton binding energy.[19] This feature is known as the Phillips line[20] and is depicted in Figure (5), with 4 theoretical points[21,22] used to generate the fitted line, and the one (unfitted) experimental point. The triton binding energy defect that seems to be a universal feature of "realistic" two-body potential calculations means that $a^{[2]}$ is too large, since E_B is too small. Whatever physics accounts for the binding defect would also move $a^{[2]}$ closer to the experimental point, assuming that the Phillips line is indeed a universal feature of doublet scattering lengths. Calculations by Torre, Benayoun, and Chauvin[6] show that this general trend is followed when three-nucleon forces are included.

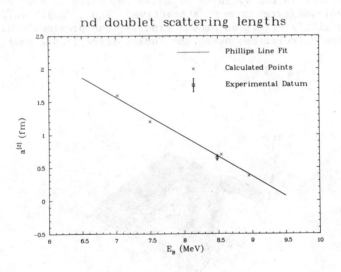

Figure 5. Calculated doublet nd scattering lengths vs. triton binding energies, and the experimental datum.

In order to exploit the additional freedom one has in scattering, one should work at slightly higher energy. Frankly, I'm not sure what types of experiments should be performed. If the geometrical feature of three-body forces we discussed earlier is correct, perhaps experiments of the the type shown in Figure (6) will show sensitivity to three-body forces. Figure (6a) depicts a bound neutron (open circle) and proton (shaded circle) colliding with a proton in the center-of-mass. Two possible final state configurations are shown, with an isosceles scheme in (b) and a collinear one in (c). These processes should show some sensitivity to three-body forces. Detailed calculations would be necessary to make this argument convincing.

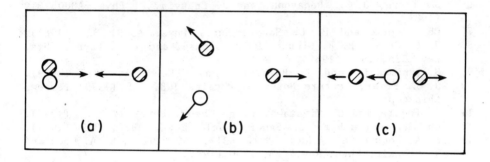

Figure 6. Various final state configurations from p+d scattering.

Finally, it has been suggested[23] that the small discrepancy between the neutron-neutron scattering lengths from the $d(\pi^-,\gamma)2n$ and $d(n,2n)p$ reactions may result from three-nucleon force effects in the final state of the latter reaction. This is an intriguing speculation which also requires more detailed calculations.

CONCLUSIONS AND SUMMARY

It is obvious that much more work needs to be performed in this rapidly developing area of investigation. Because of the inherent limitations of purely theoretical efforts to understand hadronic physics, our approaches to the three-nucleon force should be diverse; there will be many revisions of the forces. Currently there is no concrete evidence for three-nucleon forces, but several intriguing possibilities are suggestive. In recent calculations of the additional triton binding due to the Tucson force, all is chaotic. This is quite understandable in view of the complexity of the calculational problem; order will appear in time. The few-body field remains a very active area for the investigation of new ideas and fundamental nuclear mechanisms: in this case, the effect of three-nucleon forces.

388

This work was performed under the auspices of the U. S. Dept. of Energy.

REFERENCES

1. J. L. Friar, B. F. Gibson, and G. L. Payne, Comm. on Nucl. and Part. Phys. (to appear).
2. L. D. Faddeev, Zh. Eksp. Teor. Fiz. 39. 1459 (1960) [Sov. Phys. -JETP 12, 1014 (1961)].
3. A. Laverne and C. Gignoux, Nucl. Phys. A203, 597 (1973).
4. R. A. Brandenburg, Y. E. Kim, and A. Tubis, Phys. Rev. C12, 1368 (1975).
5. W. Glöckle, Nucl. Phys. A381, 343 (1982).
6. J. Torre, J. J. Benayoun, and J. Chauvin, Z. Phys. A300, 319 (1981).
7. Ch. Hajduk and P. U. Sauer, Nucl. Phys. A369, 321 (1981).
8. J. L. Friar, B. F. Gibson, D. R. Lehman, and G. L. Payne, Phys. Rev. C25, 1616 (1982).
9. H. Primakoff and T. Holstein, Phys. Rev. 55, 1218 (1939).
10. J. L. Friar, Lecture Notes in Physics 108, 445 (1979) reviews this topic.
11. L. Fujita and H. Miyazawa, Prog. Theor. Phys. 17, 360 (1957).
12. Ch. Hajduk and P. U. Sauer, Nucl. Phys. A322, 329 (1979).
13. S. A. Coon et al., Nucl. Phys. A317, 242 (1979); S. A. Coon and W. Glöckle, Phys. Rev. C23, 1790 (1981).
14. J. L. Friar, Nucl. Phys. A353, 233c (1981) reviews this topic.
15. R. B. Wiringa, submitted to Nucl Phys. A. I would like to thank Bob Wiringa for making the completely local Argonne force available before publication.
16. Muslim, Y. E. Kim, and T. Ueda, Phys. Lett. 115B, 273 (1982).
17. J. Carlson, V. R. Pandaripanda, and R. B. Wiringa, submitted to Nucl. Phys. A.
18. J.L. Friar, B. F. Gibson, E. L. Tomusiak, and G. L. Payne, Phys. Rev. C24, 625 (1981).
19. V. Efimov, Nucl. Phys. A362, 45 (1981).
20. A. C. Phillips, Rep. Prog. Phys. 40, 905 (1977).
21. G. L. Payne, J. L. Friar, and B. F. Gibson, Phys. Rev. C (Oct., 1982).
22. J. J. Benayoun, C. Gignoux, and J. Chauvin, Phys. Rev. C23, 1854 (1981).
23. I. Slaus, Y. Akaishi, and H. Tanaka, Phys. Rev. Lett. 48, 993 (1982).

QUARKS, BAGS AND THE NUCLEON-NUCLEON INTERACTION[†]

G.E. Brown

State University of New York, Stony Brook, N.Y. 11794[*]

ABSTRACT

It is argued that pionic couplings have a greater influence on the short-range nucleon-nucleon interaction than the quark degrees of freedom.

INTRODUCTION

In most calculations of the nucleon-nucleon force from quark-gluon exchange, a perturbative approach, based on asymptotic freedom is adopted.

Calculations by DeTar[1] awakened considerable interest in the above approach; we shall discuss them in some detail because they represent clearly the features of perturbative quark-gluon exchange calculations. These calculations were performed with the M.I.T. bag model[2]. Calculations in the nonrelativistic constituent quark model by Liberman[3] and many other authors following him have several characteristics in common with the bag model ones. We prefer to carry out our discussion within the framework of the bag model, because it is easy to effect chiral invariance by an extension of the bag model, whereas this isn't possible with a nonrelativistic model. As will become clear later, we view chiral invariance as the fundamental guide in constructing the nucleon-nucleon force.

DeTar's calculations gave a repulsive energy of 250-300 MeV when two nucleons came on top of each other. The mechanism for this repulsion was somewhat in the direction of the van der Waals repulsion in atomic physics. Although we now know, after the introduction of the color degree of freedom, that 12 quarks can be put into the $1s_{1/2}$ state, it is nonetheless favorable in terms of the spin-dependent interactions to put only 3 quarks in. An instructive formula derived by the M.I.T. group is the supermultiplet formula for the energies of a system of n quarks, all in the same spatial state

$$\Delta E_n = \frac{4}{3} \alpha_s \frac{\mu}{R} \left[n(n-6) + S(S+1) + 3I(I+1) \right] . \qquad (1)$$

Here S is the spin of the state, I, the isospin. The coefficient $(4/3) \alpha_s (\mu/R)$ can be chosen from the energy diference of nucleon and $\Delta(1230)$ isobar to be ~ 25 MeV.

[†]This is a summary of joint work carried out with Mannque Rho and Vicente Vento.

[*]Research supported by the U.S.D.O.E. under Contract DE-AC02-76ER13001.

390

Two separated nucleons (each with three quarks in $1s_{1/2}$
states) are lowered by an energy

$$\Delta E = 2 \times 25 \text{ MeV } [-9 + \frac{3}{4} + \frac{9}{4}] = -300 \text{ MeV.} \qquad (2)$$

The energy of a six-quark, I=0, S=1 state, with all quarks in the
$1s_{1/2}$ state, is

$$\Delta E = 25 \text{ MeV } [2] = 50 \text{ MeV.} \qquad (3)$$

Thus, this simple estimate would say that it would cost 350 MeV to
shove two nucleons on top of each other; i.e., there would be a
repulsion of ∿ 350 MeV at the origin in configuration space.

This is an overestimate; the system will adjust to lower this
energy. Chiefly, the six quarks in the overlapping nucleons will
not all remain in $1s_{1/2}$ states, but some of the quarks will go
into p-states in order to minimize the repulsion[4,5]. The nett
repulsion remaining is ∿ 100 MeV, give or take factors of 2.

Before making any estimates, one can say that the repulsion
should be of the order of the Δ-nucleon energy splitting; i.e.,
of the order of 300 MeV, because this splitting comes from the
same perturbative quark-gluon exchange mechanism that is used to
produce the repulsion in the nucleon-nucleon system. The detailed
calculations show the numerical constant multiplying this to be
rather smaller than unity.

The quark-gluon exchange calculations are difficult to carry
out in detail, because of nonlocalities, and these must be
handled properly[7]. None the less, the general magnitude of the
resulting repulsion is given by the above arguments.

We shall show that effects connected with coupling of pions
to nucleons give an order-of-magnitude greater repulsion, so that
the above perturbative quark-gluon exchange effects are only
corrections to this mechanism. In order to use the pion as we
shall do, we first discuss its nature.

THE PION

The pion has a dual nature. On the one hand, it has a
quark-antiquark substructure. On the other hand, it is the
Goldstone boson associated with chiral symmetry. The latter,
which we call its chiral nature, is the more important of the
two for nuclear physics, as we shall discuss.

The pion is a coherent superposition of quark-antiquark
states. The closest analogy in nuclear physics is a very
coherent particle-hole state; i.e., a vibration. In fact, the
spurious 1-state is, in three dimensions, the zero-energy mode
associated with the restoration of translational invariance,
just as the pion is the zero-mass mode associated with the
restoration of chiral invariance, in four dimensions. It is
well known in nuclear physics that the more collective the mode,

the closer together the particle and hole become[9]. A similar argument can be made for quark and antiquark[8], in the case of the pion.

Recently a similar argument for the smallness of the pion has been made by Brodsky[11]. In the language of ref. 10 (See, esp. §9) the physical pion has a complicated Fock space:

$$|\pi\rangle_{physical} = \sqrt{Z_2}\ |q\bar{q}\rangle + a_2|qq\bar{q}\bar{q}\rangle$$
$$+ b_2\ |q\bar{q}\ glue \rangle + \ldots \tag{4}$$

Normalization gives

$$Z_2 + |a_2|^2 + |b_2|^2 + 1 \ldots = 1. \tag{5}$$

The weak pion decay constant f_π = 93 MeV. The pion weak decay proceeds only from the first term on the right-hand side of eq. (4); using a bag-model wave function for this term

$$f_\pi \cong 0.5\ \frac{\sqrt{Z_2}}{R}\ . \tag{6}$$

Bag models often leave out the terms other than the $|q\bar{q}\rangle$ one, and this would amount to approximating $Z_2 \cong 1$, which would then give R $\sim$ 1 fm in order to fit f_π. (Note that energies must go as some constant times R^{-1} in the bag model, since R^{-1} is the only quantity one has with dimensions of energy.)

The point of Brodsky[11] and of ref. 10, is that the Fock space wave function of the pion is quite complicated, so that $Z_2 < 1$, implying that

$$R_\pi < 1\ fm. \tag{7}$$

A simple way of stating this is that if the $q\bar{q}$ pair, with quantum numbers of the pion, is moved by the interaction so far in energy – down to zero energy – then the interaction is large enough to admix a lot of more complicated configurations. In any case, it should be realized that it should be $\sqrt{Z_2}/R$, not R^{-1}, that should be equated to the measured f_π. Thus, the more complex the pion, the smaller it will be. Brodsky finds the rms pion radius to be 0.42 fm., by fitting, in addition to f_π, the process

$$\pi_o \to \gamma + \gamma\ . \tag{8}$$

The value of the radius of the $|q\bar{q}\rangle$ part of the pion is assumed to be small, in accordance with eq. (7); the precise value will be adjusted phenomenologically.

THE NUCLEON

The energy of the nucleon, as function of the adius, will be taken[13] from the chiral bag model to be[*]

$$E(R) = \frac{3\Omega_o}{R} + \frac{4\pi}{3} R^3 B - \frac{Z_o}{R} - \frac{1.71}{m_n^2} \left(\frac{g_{\pi NN}^2}{4\pi}\right) \frac{1}{R^3 + R_o^3} \qquad (9)$$

Here the first term on the right-hand side is the quark kinetic energy, with[2] $\Omega_o = 2.04$. We shall take the bag constant B to be $B^{\frac{1}{4}} = 0.137$ GeV. The term Z_o/R corrects, in a rough way, for spurious center of mass energy, with $Z_o = 0.75$. The last term comes from pion coupling. The $R_o \cong R_\pi$ gives a cut off, which we determine phenomenologically. With $g_{\pi NN}^2/4\pi = 14.6$ and $R_o = 0.33$ fm, we obtain the curve, fig. 1.

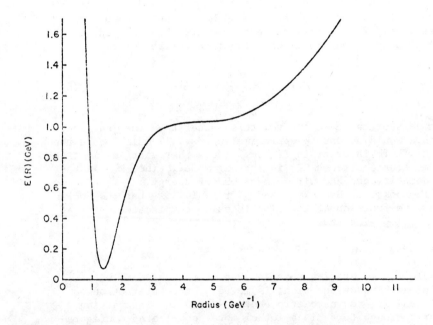

Fig. 1. Bag energy as function of bag radius R.

*The pion coupling term here includes effects from virtual intermediate Δ's as well as nucleons. In the spirit of ref. 16, only the mean-field terms, i.e., only those involving intermediate nucleons, should be used. This decreases the term by a factor of ~ 2, and means that the cutoff R_o must be decreased by a factor of $\sim 2^{1/3}$, implying a pion radius $R_\pi \cong 0.26$ fm.

Note that the minimum in the curve comes at $R \leq R_o$; the coupling to the pion provides the deep minimum.

In ref. 13 this curve is viewed as a collective potential, $U(R) = E(R)$, and the wave function $\psi(R)$, the probability amplitude for the nucleon having bag radius R, is calculated from

$$- \frac{1}{2M_{eff}} \frac{d^2\psi(R)}{dR^2} + U(R) \; \psi(R) = \varepsilon \; \psi(R). \tag{10}$$

The parameter R_o given above was set from

$$\varepsilon_o = m_n c^2 . \tag{11}$$

The M_{eff} was taken to be

$$M_{eff} = 0.48 \; m_n. \tag{12}$$

This is the mass that would follow from a collective scaling motion of breathing type.

The resulting $\psi(R)$ is shown in fig. 2. The curve tends to spread out rather far to the right; we believe that with the larger bag constant now advocated by the M.I.T. phenomenology[14], and consistent with the fits to charmonium[15] that $\psi(R)$ will drop more sharply. In the work leading to ref. 13, it was found that a relativistic treatment of the zero-point "breathing" motion also required a $B^{\frac{1}{4}}$ of $\sim$ 200 - 250 MeV, and that $\psi(R)$ was more localized about the minimum in U(R). We are presently working out the relativistic treatment.

At the minimum in U(R), fig. 1, the pionic contribution to U(R) is $- 2.18 \; m_n c^2$. Now, the point is that when two nucleons merge to form a 6-quark bag, much of the (attractive) pionic energy will be lost. The radius of the 6-quark bag will be nearly the same as that of the 3-quark ones, since the minimum in U(R) is chiefly determined by R_π. As more quarks are added to the bag, the pionic coupling must go to zero. Twelve quarks fill the $1s_{1/2}$ color and flavor states, and the pionic coupling, which goes through the operator $\sigma\tau$, is then zero. We expect the coupling to drop to zero linearly, from the three-quark bag to the 12-quark one. Thus, the six-quark bag will couple to the pion with only 2/3 the strength of the three-quark bag, for the same radius. This gives the maximum magnitude for the six-quark coupling since the 6-quark bag will have a larger R than the 3-quark one, if anything.

Whereas our above discussion was carried out within the framework of perturbative pion coupling extrapolated into the strong-coupling regime and then cut off by introducing a finite size for the pion, a nonperturbative classical solution of hedgehog type[16] exhibits the above features. We believe that

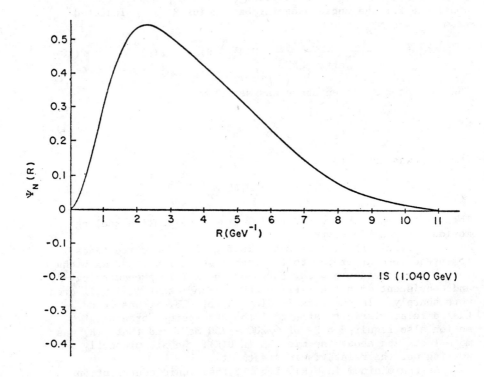

Fig. 2. Wave function for radial motion of the nucleon bag surface.

the pion cloud is nonperturbative, and the fact that our earlier discussion reproduces the results of ref. 16 is comforting.

Whereas initially each of the three-quark bags with quantum numbers of the nucleon has its energy lowered by $2.18\ m_n c^2$, according to the above estimates, by coupling to the pion cloud, the energy of the six-quark bag representing completely overlapping nucleons is lowered only $(2/3)^2$ of this, $\sim .97\ m_n c^2$. Thus, there is a nett repulsion of $3.4\ m_n c^2$ encountered in shoving the two nucleons on top of each other.

At large distances, where the two bags do not overlap, the repulsion comes from ω-meson exchange, with[17]

$$\frac{g_\omega^2}{4\pi} = 10\text{-}12 \ . \tag{13}$$

Representing the regularized repulsion in the usual way, so as to give a well-behaved potential

$$(V_\omega)_{reg} = \frac{g_\omega^2}{4\pi} \left\{ \frac{e^{-m_\omega r}}{r} - \frac{e^{-\Lambda r}}{r} \right\} \tag{14}$$

we find, taking the limit as $r \to o$,

$$\frac{g_\omega^2}{4\pi} (\Lambda - m_\omega) = 3.4 \, m_n c^2. \tag{15}$$

This would give the regularization mass Λ in the range of 1.05 - 1.1 GeV, somewhat smaller than the Λ usually used to regulate boson exchange models, but of the same general size.

As can be seen from ref. 13, considerable averaging must be done over the zero-point motion, but the above estimate should not be far wrong.

The nonperturbative hedgehog solution of ref. 16 brings the picture full circle back to the analog of the van der Waals force. In this solution, the quark orbits are "warped" by coupling to the pion field, so that only three quarks can go in the lowest (K=0) orbit. The next quarks must go into a K=1 orbit which lies substantially higher in energy. Thus, as six quarks are brought together, each threesome originally in a K=0 orbit, three of the quarks must be promoted to a K=1 orbit. This gives a mechanism for the repulsive core like that for the van der Waals interaction between atoms. Furthermore, the energies of quark excitation from K=0 to K=1 are large, so that the Born-Oppenheimer approximation implicitly assumed in our estimates should be valid. Given that some of the parameters, such as pion radius, will change as we gain more information, we none the less believe that we have a reasonable derivation of the form of the repulsion, eq. (14). This repulsion is so much greater than that found in perturbative quark-gluon exchange that it need not be handled so delicately for the purposes of low-energy nuclear physics.

BOSON EXCHANGE MODELS

We have seen in the foregoing that an understanding of the nature of the nucleon, especially of the coupling of the quarks to the pion cloud, allows us to derive the short range repulsion. At larger distances we put in the potential from ω-meson exchange. We now discuss the boson exchange potentials;

in the usual order, starting from the lightest meson, the pion.

The pion is coupled to the bag in such a way as to restore chiral invariance[12], i.e., to make the axial vector current continuous across the bag surface so that

$$\frac{\partial \vec{J}^5_\mu}{\partial x_\mu} = 0 \tag{16}$$

where $\vec{J}^5_\mu$ is the axial vector current. It was shown in ref. 12 that this prescription gives the usual Yukawa pion field asymptotically for large distances.

The σ-degrees of freedom, those corresponding to scalar, isoscalar mesons, result from two virtual pions in an S-state; the ρ-meson results from two pions in a P-state. Once the pion coupling to the bag has been effected and one has a chiral theory, then the corresponding components of the boson exchange potential can be constructed[10].

The ω degree of freedom has two components: i) the major part consists[19] of correlated ρ,π exchanges. ii) the elementary ω can be coupled as a chiral singlet[20]. The coupling of the ω-meson has not yet been worked out in detail. We used the ω-exchange potential in our earlier discussion, based largely on a strong phenomenology.

CONCLUSIONS

Perturbative quark-gluon exchange gives interactions of only modest strength at short distances; these are overwhelmed by the (nonperturbative) effects from pion clouds. The latter give a strong repulsion in the short-range nucleon-nucleon interaction. This repulsion joins on to that from ω-meson exchange at larger distances, where one has also the other components of boson exchange models.

The still somewhat crude treatment of pionic effects discussed here gives the same sort of regularization of boson exchange potentials at short distances as has usually been employed, possibly with somewhat different parameters. We believe that these effects can be included, through regularization, in the boson exchange calculations.

REFERENCES

1. C. DeTar, Phys. Rev. <u>17</u>, 302 (1978); ibid <u>17</u> 323 (1978); ibid <u>19</u>, 1451 (1979).
2. T. De Grand, R.L. Jaffe, K. Johnson and J. Kiskis, Phys. Rev. <u>D12</u>, 2060 (1975).
3. D.A. Liberman, Phys. Rev. <u>D16</u>, 1542 (1977).
4. I.T. Obukhovsky, V.G. Neudatchin, Yu. F. Smirnov and Yu. M. Tchuvil'sky, Phys. Lett. <u>88B</u>, 231 (1979).
5. M. Harvey, Nucl. Phys. <u>A352</u>, 326 (1981).

6. C.W. Wong, Prog. in Particle and Nuclear Physics $\underline{8}$, 223 (1982).

7. See, e.g., A. Faessler, F. Fernandez, G. Lübeck and K. Shimizu, Phys. Lett. $\underline{112B}$, 201 (1982).

8. G.E. Brown, Nucl. Phys. $\underline{A358}$, 39c (1981).

9. G.F. Bertsch, R.A. Broglia and C. Riedel, Nucl. Phys. $\underline{A91}$, 123 (1967).

10. G.E. Brown, Prog. in Particle and Nuclear Physics 8, 147 (1982).

11. S.J. Brodsky and G.P. Lepage, Physica Scripta $\underline{23}$, 945 (1981).

12. G.E. Brown and M. Rho, Phys. Lett. $\underline{82B}$ (1979); G.E. Brown, M. Rho and V. Vento, Phys. Lett. $\underline{94B}$, 383 (1979).

13. G.E. Brown, J.W. Durso and M.B. Johnson, Nucl. Phys. $\underline{A}$, to be published.

14. H. Hansson, K. Johnson, and C. Peterson, ITP Santa Barbara preprint, 1982; C. Carlson, H. Hansson and C. Peterson; CTP M.I.T. preprint, 1982.

15. P. Hasenfratz, R.R. Horgan, J. Kuti and J.M. Richard, Physica Scripta $\underline{23}$, 914 (1981); Phys. Letts. $\underline{94B}$, 401 (1980).

16. V. Vento, M. Rho and G.E. Brown, Phys. Lett. $\underline{103B}$, 285 (1981).

17. D.O. Riska and B. Verwest, Phys. Lett. $\underline{48B}$, 17 (1974).

18. C.G. Callan, R.F. Dashen and D.J. Gross, Phys. Rev. $\underline{D19}$, 1826 (1979).

19. J.W. Durso, M. Saarela, G.E. Brown and A.D. Jackson, Nucl. Phys. $\underline{A278}$, 445 (1977).

20. V. Vento, Phys. Lett. $\underline{107B}$, 51 (1981).

Δ(1236)-ISOBAR DEGREES OF FREEDOM
AND THE STRENGTH OF SPIN-ISOSPIN RESONANCES

J. Speth

Institut für Kernphysik, KFA Jülich, D-5170 Jülich, W. Germany, and
Lehrstuhl für Theoretische Kernphysik, Univ. Bonn,
D-5300 Bonn, W. Germany

1. INTRODUCTION

The most interesting feature of the spin-isospin modes excited
in charge-exchange reactions is the magnitude of the transition
strength. In the case of the Gamow-Teller resonances (GTR) there
exists a well established, model-independent sum rule by Ikeda et
al.) which is simply connected with the number of protons and neutrons: $S_{\beta^-} - S_{\beta^+} = 3(N-Z)$, where the l.h.s. is the difference between
the β^- and β^+ GT-strength. So far only about 60 % of this sum rule
strength has been detected experimentally. Conventional nuclear
structure effects seem to be unable to explain the "missing" GT-
strength. Several years ago already Δ-isobar degrees of freedom were
suggested to play an important role in the quenching of the axial-
vector coupling constant g_A^2). (The first study of the "quenching" of
g_A was performed by Ericson et al.[3]) without reference to the Δ-
resonance.) In this contribution I shall present a consistent calculation of different experimental quantities in the ^{48}Ca-region which
are all connected with the spin-isospin part of the particle-hole
interaction. This part of the force is of crucial importance in
connection with the "quenching".

2. SPIN-ISOSPIN DEPENDENT PARTICLE-HOLE INTERACTION

We used the generalized version of the Jülich-Stony Brook
force[4]) which includes the effects of the one-pion (V_π) and one-rho
(V_ρ) exchange potential explicitly. This force is given in the momentum space as:

$$F_{\sigma\tau}^{ph}(\vec{q}) = \int \frac{d^3k}{(2\pi)^3} \; \Omega(\vec{q}-\vec{k})\{V_\pi(\vec{k}) + V_\rho(\vec{k})\} + \delta g_0' \cdot C_0 \; \vec{\sigma}\cdot\vec{\sigma}' \; \vec{\tau}\cdot\vec{\tau}' \qquad (1)$$

Here one summarizes the effects of the other mesons in a two-body
correlation function $\Omega(\vec{q}-\vec{k})$ and a phenomenological zero range part
($\delta g_0'$) which is deduced from experiment. The quantity $C_0 = 3.02$ m/m*
[MeV fm^3] is the inverse of the density of states at the Fermi surface. It might be thought that effects from ρ-exchange should be included into $\delta g_0'$, since the ρ-mass is large and the ρ-interaction is
therefore short-ranged. However, the particle-hole interaction should
also include tensor invariants. In practice, these arise almost completely from the π- and ρ-exchange potentials. Here one has to bear
in mind that the two contributions enter with opposite sign and the
tensor force from the ρ-exchange cuts off that from the π-exchange at
short distances (large momentum transfer) as indicated in figure 1.
This is of crucial importance in connection with the precritical

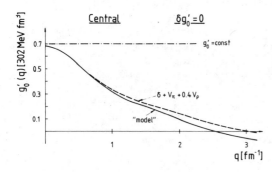

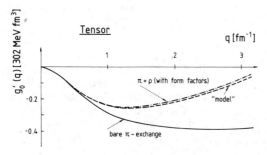

phenomena and pion condensation. The effect of the momentum dependence of the force on a given state can be discussed qualitatively by considering the corresponding form factors, since the largest (diagonal) contributions to the direct part of the ph-force are simply a double convolution of the transition density with the interaction. As we shall show in the next section, the q-dependent force gives rise to a q-dependent quenching which can be used as an experimental signature for the Δ-hole admixture.

Fig. 1. Graphical representation of the generalized spin-isospin dependent interaction of eq. (1) ("model"). It is compared with the original ansatz of ref. 4).

3. MISSING $\sigma\tau$-STRENGTH

Investigations of spin-isospin modes including the Δ-degrees of freedom have been performed by several groups. The crucial input in such an extended RPA calculation is the interaction $F_{\Delta N}^{ph}$ between a (Δ-h)-pair and a nucleon-hole (N-h)-pair, which one has to introduce in addition to the conventional ph-interaction F_{NN}^{ph}. In the case of $F_{\Delta N}^{ph}$ the exchange term due to the Pauli-principle gives rise to a large cancellation of the direct term which influences strongly the magnitude of the quenching effect. For details see the paper by Suzuki, Krewald and Speth[4]).

In the following we discuss quantities in the ^{48}Ca-region. The parameter $\delta g_0'$ in eq. (1) is fitted to the excitation energies of the first 1^+-states in ^{48}Sc. The results of ref. 5) are shown in table 1, where the results with and without the Δ-h configurations are compared. The isobar effect is found to reduce the $\sigma\tau$-strength by about 30 % in the case of the 1^+-states. It is interesting to observe that in the case of the 7^+-state practically no quenching due to Δ-h states is obtained. This effect may be explained by the q-dependence of the interaction because the form factor of the 7^+-state is peaked around $q\sim2$ $[fm^{-1}]$. Hence the coupling of a state with large multipolarity to Δ-h configurations is expected to be small. Therefore the angular momentum dependence is a characteristic experimental signature of the Δ-h model of the quenching effect.

Table 1 "Pionic" states in ^{16}O, ^{48}Ca and ^{48}Sc

Nucleus	J^π	E_{exp} (MeV)	B_{exp}	$\delta g'_0 = 0.4$		$\delta g'_0 = 0.5$		$\dfrac{B_{N+\Delta} - B_N}{B_N} \times 100$
				E_N (MeV)	B_N	$E_{N+\Delta}$ (MeV)	$B_{N+\Delta}$	(%)
^{48}Sc	1^+	3.02		3.02	2.42	2.96	1.88	22
^{48}Sc	1^+	broad structure centered around 11 MeV		11.08	21.09	10.86	14.97	29
^{48}Sc	7^+	1.6		1.41	36.62	1.65 (1.40)	33.2 (36.41)	9 0.5
^{48}Ca	1^+	10.23	4.0 ± 0.3 $\Sigma B(M1) \approx 5.2$ c)	10.21 ($g_0 = 0.3$)	8.20	10.16 ($g_0 = 0.3$)	5.29	36
^{16}O	0^-	12.80		13.17 (13.10)	0.13 (0.13)	13.35	0.12	8 (0)
	2^-	12.97	197 ± 37	13.16 (13.02)	168 (154)	13.13	126	25 (8)
		19.04 20.36	338 ± 68 467 ± 156 }	20.25 (20.14)	915 (859)	20.48	813	11 (6)
	4^-	18.98		19.41 (19.38)	5.60×10^5 (5.46)	19.75	5.14×10^5	8 (2.5)

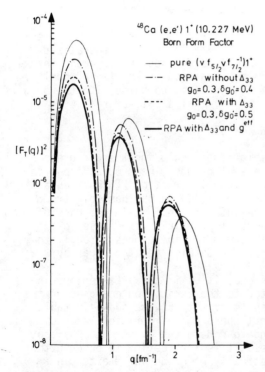

Fig. 2. Theoretical form factors of the inelastic electron scattering to the 1^+, 10.23 MeV state in ^{48}Ca, calculated in Born-approximation.

In figure 2 the form factor for inelastic electron scattering to the 1^+-state in ^{48}Ca is shown[5]). From the strongly q-dependent force one would expect naively a strongly q-dependent quenching effect. It should be largest at q=0 and disappear roughly around $q \sim 1$ [fm^{-1}]. This does not follow from our calculation. A detailed explanation is given in ref. 5). It is interesting to see that the form factor in the third maximum is no longer quenched but increased compared to the pure shell model value. This is due to the admixtures of the higher ph-configurations.

Using the RPA wave function mentioned above, Osterfeld et al.[6]) have calculated (p,n) cross sections. In figure 3 the cross sections of the three 1^+-states in ^{48}Sc are shown. Here one compares results obtained either with pure particle-hole, standard RPA or generalized RPA wave functions (including Δ-isobars) with experimental data. The in-

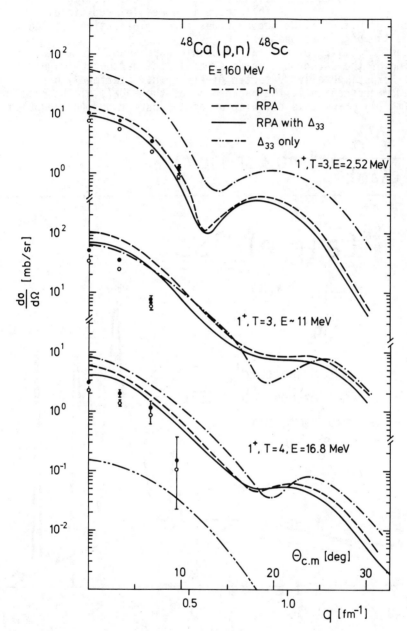

Fig. 3. Angular distributions of the ^{48}Ca(p,n) reactions to the three 1^+ GT-states in ^{48}Sc. The open circles are the data of ref. 7), the full points are the same data but with the normalization of ref. 8).

clusion of (Δ-h)-configurations reduces the cross sections at 0° by about 30 %. It is obvious that there is space for additional "quenching" effects like second order core polarization (at least for the two higher resonances).

One serious problem in all hadronic experiments is the subtraction of the background which strongly influences the magnitude of the "experimentally" detected GT-strength. Osterfeld[9]) has calculated microscopically the background of the $^{48}Ca(p,n)^{48}Sc$ reaction (see fig. 4). From this calculation of the background (dash-dotted

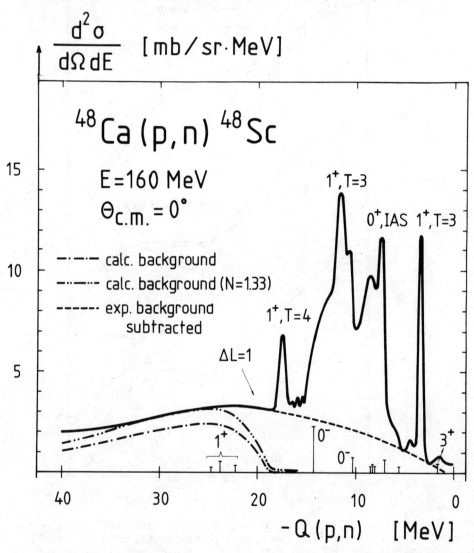

Fig. 4. Zero degree spectra for the reaction $^{48}Ca(p,n)^{48}Sc$. The data (thick full line) are taken from ref. 6). The discrete lines are calculated cross sections.

line) he concluded that most of the background subtracted in refs. 6, 7) (dashed line) is GT-strength. This would increase the GT cross section at 11 MeV by 25 % and the totally observed GT-strength from 43 % to 51 %.

4. CONCLUSION

From the discussion in the last section we conclude that there exists definitely a quenching of the GT-strength due to the coupling of the low-lying ph-states to the high-lying (Δ-h)-configurations, the magnitude of which, however, depends sensitively on the ph-interaction. The momentum dependence of the force gives rise to a momentum (angular momentum) dependent quenching which might be used as an experimental indication of the Δ-h quenching.

This work has been done in cooperation with Drs. S. Krewald, F. Osterfeld and T. Suzuki.

REFERENCES

1. K. Ikeda, S. Fujii and J.I. Fujita, Phys. Lett. 3, 271 (1961).
2. M. Rho, Nucl. Phys. A231, 493 (1974);
   K. Ohta and M. Wakamatsu, Nucl. Phys. A234, 445 (1974).
3. M. Ericson, A. Figureau and C. Thévenet, Phys. Lett. 45B, 19 (1973).
4. J. Speth, V. Klemt, J. Wambach and G.E. Brown, Nucl. Phys. A343, 382 (1980).
5. T. Suzuki, S. Krewald and J. Speth, Phys. Lett. 107B, 9 (1981).
6. F. Osterfeld, S. Krewald, J. Speth and T. Suzuki, Phys. Rev. Lett. 49, 11 (1982).
7. B.D. Anderson et al., Phys. Rev. Lett. 45, 699 (1980).
8. C. Gaarde, private communication.
9. F. Osterfeld, Phys. Rev. 26C, 762 (1982).

EFFECT OF THE Δ(1236) ISOBAR ON THE THREE-NUCLEON BOUND STATES

Ch. Hajduk, P.U. Sauer and W. Strueve

Theoretical Physics, University of Hannover, 3000 Hannover, Germany

Two-nucleon potentials, e.g., Paris[1], are unable to account simultaneously for the binding energy and the e.m. form factors of the three-nucleon bound states (dashed curves in figs.). If the force is allowed to yield explicit Δ-isobar excitation, the two-nucleon interaction gets renormalized and a three-nucleon interaction and e.m. exchange-currents (EC) arise, which are calculated (solid curves). The single-Δ wave-function components are generated nonperturbatively in a coupled-channel approach by π- and ρ-exchange transition potentials in the isospin-triplet partial waves up to angular momentum I=2, the isospin-singlet ones up to I=2 are taken from[1]. In the model, the Δ improves the ^{3}H binding energy by 0.3 MeV and simultaneously the ^{3}He charge form factor for momentum transfers Q>4 fm^{-1}. The point-proton density equivalent to the one extracted in [2] from experiment shows some structure, which potentials without isobars are unable to yield. In order to improve the magnetic form factor, isovector π- and ρ-exchange mesonic EC between nucleonic configurations are added.

1. M. Lacombe et al., Phys. Rev. C21 (1980) 861
2. I. Sick, Lecture Notes in Physics 86 (Springer 1978) p. 300

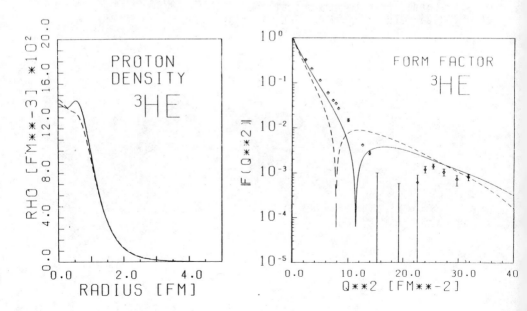

ISOVECTOR M1 TRANSITION MATRIX ELEMENTS

I.S. Towner and F.C. Khanna

Atomic Energy of Canada Limited, Chalk River Nuclear Laboratories,
Chalk River, Ontario, Canada K0J 1J0

Corrections to isovector M1 transition matrix elements are calculated in a model that incorporates the important dynamical aspects of nuclear structure (core polaristion, CP), meson exchange currents (MEC) and isobar currents (IC). The residual interaction is approximated by a one-boson-exchange potential and a short-range correlation function is introduced. The results are shown in the table. Note that: i) the contribution of MEC is significant and can be approximated quite well by that obtained in the soft-pion limit. The effect of heavier mesons is compensated by the introduction of monopole form-factors at the meson-nucleon-nucleon vertices. ii) isobar currents (RPA with direct matrix elements) provide more than half of the calculated quenching to the M1 transition matrix element. However including the exchange matrix elements reduces the quenching significantly. iii) core polarisation has a large quenching effect and a major fraction of this may be attributed to the strong tensor force. Core-excited states with energy greater than $2\hbar\omega$ contribute more than half of the total quenching from CP.

Using our results for ^{41}Ca and adding the configuration mixing calculation of McGrory and Wildenthal[1]), the B(M1) in ^{48}Ca is estimated to be ~6 μ_N^2 compared to an experimental value of ~5.2 μ_N^2. In order to avoid uncertainties of configuration mixing it will be useful to carry out measurements of B(M1) in some of the nuclei that are calculated here.

Table: Corrections to the M1 matrix element between $j=\ell+1/2$ and $j=\ell-1/2$ states of the same orbital structure expressed as a percentage of the single particle matrix element.

	A=15	A=17	A=39	A=41
CP	-12.4	-10.0	-15.4	-12.7
IC(RPA,direct)	- 8.3	- 6.6	-10.0	- 8.2
IC(RPA,Exch.)	4.6	2.0	4.7	2.6
MEC	7.5	2.4	6.9	3.4
SUM	-8.6	-12.4	-14.1	-15.3

---

1. J.B. McGrory and B.H. Wildenthal, Phys. Lett. 103B(1981)173.

GAMOW–TELLER TRANSITION MATRIX ELEMENTS

F.C. Khanna and I.S. Towner

Atomic Energy of Canada Limited, Chalk River Nuclear Laboratories,
Chalk River, Ontario, Canada K0J 1J0

Corrections to the off-diagonal Gamow-Teller (GT) transition matrix elements are calculated in a model similar to the one used[1]) to calculate the diagonal GT matrix elements. The residual interaction is approximated by a finite-range one-boson-exchange potential and a short-range correlation function is introduced. The calculation incorporates the effect of core polarisation (CP) to second order in perturbation theory, meson exchange currents (MEC) and isobar currents. Each of the MBB vertex (M is a meson and B is a nucleon or an isobar) has a monopole form factor. The results of the calculation are shown in the table. Note that: i) the contribution of MEC and the relativistic correction to the one-body operator together are quite small; ii) isobar currents (RPA with direct matrix element) gives about 40% of the calculated quenching of the matrix element. However including the effect of the exchange matrix element reduces the quenching; iii) core polarisation contributes a large part of the quenching and a major fraction of this can be attributed to the tensor force that excites particle-hole states with energy greater than $2\hbar\omega$.

Total quenching of off-diagonal GT matrix elements, for the four nuclei studied, is about 15-20%. In a GT transition like ^{48}Ca(p,n), configuration mixing can quench the matrix element still further by ~20% thereby quenching the Gamow-Teller transition by a factor of 3-4. In order to avoid the complication of configuration mixing it will be useful to carry out measurements of GT strength in the nuclei studied here.

Table: Corrections to the Gamow-Teller matrix element between $j=\ell+1/2$ and $j=\ell-1/2$ states of the same orital structure expressed as a percentage of the single-particle matrix element.

	A=15	A=17	A=39	A=41
CP	-15.5	-11.2	-18.4	-14.3
IC(RPA,direct)	- 7.4	- 5.9	- 8.9	- 7.3
IC(RPA,exch.)	4.1	1.8	4.2	2.3
MEC	1.3	0.8	2.0	1.3
Relativistic	- 0.9	- 1.2	- 1.0	- 1.3
SUM	-18.5	-15.8	-22.1	-19.4

1. I.S. Towner and F.C. Khanna, Phys. Rev. Lett. 42(1979)51.

THE KAONIC HYDROGEN ATOM IN MOMENTUM SPACE

R.H. Landau[*]
Department of Physics, University of Surrey, Guildford, Surrey
and
Department of Physics, Oregon State University, Corvallis

It is well known that the strong absorption by a heavy nucleus of a negative kaon in a bound atomic state leads to unusual and sometimes unexpected dependence on the nuclear potential. In order to obtain a theoretical understanding of this problem, and try to understand recent experimental results, we have studied Kaonic Hydrogen (K^-p), a conceptually simpler system which nonetheless displays much of the same physics. To do this, we have applied new momentum space methods to solve the combined Coulomb plus nuclear interactions problem. Our procedure is an extension of the Kwon-Tabakin [1] technique of removing the logarithmic singularity of the Coulomb potential via a subtraction procedure suggested by Landé.

We account for "absorption" by explicitly including the coupled K^-p, $\bar{K}^0n$, $\Sigma\pi$ and $\Lambda\pi$ channels with a variety of separable potentials chosen to demonstrate the model dependence of our results. In this way our procedure is similar to that of Barrett [2] - except that we treat the Coulomb force exactly and are not limited to the 1S level. We have calculated the 1S-3S level shifts and widths and some of the properties of the subthreshold Y_0^* resonance. Our results are sensitive to the channel coupling and at present appear somewhat closer to the experimental results than other calculations - yet still with a "less bound" shift for hydrogen. Further potentials and couplings are still under investigation.

A related study with R.C. Barrett [3] investigates the properties of K^--Helium in a coupled channels, p space calculation employing the computer code LPOTT.

Further studies of $\bar{p}$ and Σ atomic and nuclear bound states are anticipated.

1) Y.R. Kwon and F. Tabakin, Phys. Rev. C18 (1978) 932.
2) R.C. Barrett, J. Phys. G, 8 (1982) L39.
3) R.C. Barrett and R.H. Landau, to be published.

[*] Supported in part by the National Science Foundation (U.S.A.) and the Science and Engineering Research Council (U.K.).

408

THE NUCLEUS CONSISTS OF MORE THAN JUST NUCLEONS, DELTAS AND MESONS

Gerald A. Miller[*]
CERN, Geneva 23, Switzerland

ABSTRACT

The possibility that quark degrees of freedom have significant effects in Nuclear Physics is discussed.

INTRODUCTION

I would like to use my prerogative as one of the reviewers of this session to mention some ideas that have not been discussed by the speakers, but which are, in my opinion, extremely relevant for debates about constituents of nuclei. The main point is that it may be more accurate to treat nucleon-nucleon wavefunctions at small separation distances as collections of six quarks instead of two distinct nucleons (or e.g. two baryon resonances). This results from simple considerations of nucleon sizes[1]. If one takes the three quarks in a nucleon to be confined in a spherical region (e.g. bag[2]) of radius about w fm, two such bags overlap when their separation distance is 2 fm. This distance is roughly the same as the average spacing (1.8 fm) between nucleons in nuclei, so that the overlapping of nucleons is not so unusual. When two three-quark bags overlap one six quark bag is formed[3], and such a system is likely to behave differently than two distinct nucleons[4]. A variety of interesting examples already exists; and a partial list is given in Refs. 3,5-11. Six quark bag treatments of the short distance part of the nucleon-nucleon wave function have been made to nucleon-nucleon scattering[3,10], electron nuclear interactions[5-7], the pp → dπ[+] reaction[8], the weak proton-proton force[9], and the ^{3}H-^{3}He mass difference.[11] Non-bag-model quark treatments of nuclear physics have also been made, see e.g. Ref. 12. In many cases good agreement with experimental data is achieved and evidence for the significance of quark degrees of freedom in the nucleus is beginning to accumulate.

THE DEBATE

It is clear that a crucial element in determining the influence of quarks in nuclear physics is the size of the nucleon bag. Bags of small radius[13], R, do not overlap as much as bags of large[2], or medium values[11,14] of R. Equivalently, one can also discuss the importance of pionic effects in the structure of the nucleon. Small values of R are associated with large pionic effects and conversely[15]. This can be understood in several ways. It is well known that the electromagnetic size of a proton is characterized by a root-mean square radius of about 0.8 fm. A proton made of a bag with R much smaller than that would be

[*]Permanent Address, Physics Department, FM-15, University of Washington, Seattle, WA 98195

required to have a very significant cloud of charged mesons to fill up the volume imposed by experiment. Another way of obtaining an inverse relation between R and the importance of pionic effects is to consider, in lowest order perturbation theory, the nucleon expectation value of the pion number operator, $N\pi$. This is of second order in the pion–nucleon coupling constant. Thus it goes like $1/f_\pi^2$ (where f_π is the pion decay constant, 93 MeV) since $1/f_\pi$ is proportional, via the Goldberger–Treiman relation, to the pion–nucleon coupling constant. The expectation value of $N\pi$ is dimensionless, and the only significant dimensional parameter, besides $f\pi$, is R. Thus $\langle N_\pi \rangle$ is proportional to $1/(f_\pi^2 R^2)$ and it is clear that as R decreases, $\langle N_\pi \rangle$ grows. In a similar fashion one can show that the point-like pion's contribution to the nucleonic mass goes like $-1(f_\pi^2 R^3)$; the negative sign is universal for second order contributions to energies.

G.E. Brown has argued[16] forcefully that pionic effects in the nucleon are so significant that the modification associated with pions of the energy caused by bringing two nucleons together is sufficient to account for 3.4 GeV of repulsion. With such a large barrier to overcome, nucleons could not overlap significantly, and 6-quark-bag effects would not be relevant in nuclear physics.

The salient evidence for small bags (large pionic effects) is presented in Fig. 1 of Ref. 15. Shown therein is a very deep minimum at small R in the energy of the nucleon, E(R). This is caused by the very strong, negative contributions of the <u>pions</u> to E(R) for small values of R. Should this (curve and its interpretation as a potential energy to be used, along with a kinetic energy $-(1/2M_{eff})d/dR^2$ with M_{eff} about half a nucleon mass in a Schroedinger equation that determines the probabilty amplitude for the nucleon having a radius R) be correct, the conclusions of Brown and co-workers would result. However, there are reasons to doubt the accuracy of their Fig. 1, as well as the value of M_{eff}. First, there are some technical worries. In using that curve it is assumed that, as the bag surface vibrates, the only change in the quark wave functions is the alteration of R, and that this occurs instantaneously. Since quarks carry a fairly large constituent mass it is difficult to understand how this instantaneous rearrangement of quark orbitals can occur. Perhaps a simpler way to account for surface oscillations is to use the solition bag model.[17,18] In that model, oscillations of the confining sigma field are treated[18] in a quantum field theoretic fashion, analogous to the Bohr–Mottelson treatment of collective nuclear vibrations. The result of Goldflam and Wilets[18] is that the energy associated with the quantum fluctuation of the bag surface is quite high, about 1 GeV. Thus, one would not expect such excitations to play a dominant role in determining the properties of baryonic ground states. In the language of Ref. 16, this means that M_{eff} is much larger than half a nucleon mass.

Another worry about using Fig. 1 of Ref. 16 is possibly more fundamental. It is natural to expect that the forces that cause confinement of quarks are due to gluonic rather than pionic exchanges. QCD effects rather than old fashioned field theory cause quarks to be bound with infinite separation energy. Thus, quarks and gluons, not pions, determine the size of the confinement region. Indeed if one neglects pions, and obtains a minimum of $E(R)$ at a radius (R_0) of about 1 fm, one can show[19] that the lowest order pionic contribution to E (R) does not change the value of R_0! Furthermore, the perturbation theory treatment of pionic effects is very well justified[20] for bag radii of about 1 fm.

The result of these considerations is that I can see no reason to believe that the bag is very small and that pionic effects are dominant. Inclusion of pions in bag models is necessary in order to maintain the chiral symmetry of the Lagrangian. But cloudy-bag model calculations[14,21,22] show that pionic effects give significant, but not dominant, corrections to bag model computations of observables. (An exception is the root-mean square charge radius of the neutron, which would be essentially zero for bag models without pions.) With R in the neighborhood of 1 fm, the cloudy bag model gives very good results for pion-nucleon scattering[14], charge and magnetic moments of the nucleon[14], magnetic moments of the baryon octet[21], and the strength[14] and momentum dependence[22] of the axial vector form factor. Furthermore, $\langle N_\pi \rangle \approx 0.3$ so the schematic picture of the nucleon as three quarks in a bag of radius of about 1 fm is qualitatively valid.[22] It is also reasonable to expect that nucleons do overlap (at least sometimes) in nuclei.

HOW DOES THE NUCLEON-NUCLEON REPULSION ARISE?

There is no definitive answer to the above question at present, in my opinion. There have been a variety of interesting attempts at finding an answer that employ 6-quark wave functions and interactions between quarks. To my knowledge, the most recent is that of Ref. 10. A repulsive shift characteristic of a hard core of radius ≈ 0.3 fm is obtained from resonating group method computations. A phenomenological quark-quark force constrained by baryonic properties, rather than the single gluon exchange of earlier computations, is used. There is no large potential barrier resulting from these calculations[10]. Instead, the repulsion arises from a combination of Pauli principle and interaction effects. Of course, these calculations are not definitive. For example, long range effects of one and two pion exchanges between nucleons are absent in Ref. 10. The non-relativistic treatment of quark dynamics could also be a problem. Nevertheless, it seems that quarks could indeed give significant repulsive effects.

A PHENOMENOLOGICAL ALTERNATIVE

I would like to mention a simple procedure [8,9,24] which allows one to include 6 quark effects along with a correct description of the nucleon-nucleon scattering data. The idea is to use a spatial

separation in treating the nucleon-nucleon wave function. For
separation distances, r, larger than some fixed value r_0 ($r_0 \approx 0.9$fm)
one employs a conventional nucleon-nucleon potential, V_1 and a wave
function $\Psi(r)$. In this way, agreement with the experimental phase
shifts is guaranteed to the accuracy at which V describes the data.
On the other hand, for $r < r_0$ one resorts to a 6-quark wavefunction.
The remarkable result[24] is that $\Psi(r=r_0)$ and $\partial/\partial r \Psi(r=r_0)$ determine
the probabilty amplitudes of the 6-quark wave functions. This
conclusion is obtained by demanding conservation of the probability
current across the boundary at $r=r_0$.

CONCLUDING REMARKS

Cloudy bag model calculations, as well as simple estimates based on
the nucleon size, suggest that the bag radius is about 1 fm, and
that pions do not overwhelm quarks. With such a size for the
confinement region, it is reasonable to expect, at least
occasional, overlap of nucleons. Thus treatment of the short
distance part of the nucleon-nucleon wave function as six quarks in
a bag are reasonable. Several results of such calculations[5-13]
exist, and I believe that in the future many interesting and
significant properties of nuclei will be explained from this point
of view.

I thank E.M. Henley, L.S. Kisslinger, and A.W. Thomas for many
useful discussions of matters contained in this review.

1. R.L. Jaffe, 1982 Summer Institute on Theoretical Physics,
   Seattle, has stated that in assessing quark effects in nuclei
   one should use considerations based on energy. In this case
   the 300 MeV associated with the perturbative quark-gluon
   exchange mechanism. This is a large energy, by nuclear
   standards, but not prohibitively huge. Indeed the Δ-hole
   excitations discussed by Rapaport (Session F, this workshop)
   have this excitation energy.
2. A. Chodos et al. Phys. Rev. D9, 3471 (1974); A. Chodos et al.,
   Phys. Rev. D10, 2599 (1974); T. de Grand et al. Phys. Rev.
   D12, 2060 (1975).
3. C. Detar, Phys. Rev. 17, 302 (1978); ibid 17 323 (1978); ibid
   19, 1451 (1979).
4. Even if the MIT bag model does not apply strictly to the
   6-quark system, one would expect significant overlap effects
   when the ratio of the overlap volume to the volume of two
   nucleons is non-negligible.
5. V.A. Matveev and P. Sorba, Nuovo Cimento Lett. 20, 435 (1977).
6. H.J. Pirner and J.P. Vary, Phys. Rev. Lett. 46, 1376 (1981);
   M. Namiki, K. Okano, and N. Oshimo, Phys. Rev. C25, 2157
   (1982).

412

7. L.S. Kisslinger, Phys. Lett. 112B, 307 (1982).
8. G.A. Miller and L.S. Kisslinger, U. Washington, preprint, 1982: "Quark Contributions to the pp d $^+$ Reaction", to appear in Phys. Rev. C.
9. L.S. Kisslinger and G.A. Miller, U. Washington, preprint, 1982: "Weak Nuclear Interactions in a Hybrid Baryon-Quark Model: p-p Asymmetry, to appear in Phys. Rev. C.
10. A. Faessler, F. Fernandez, G. Lubeck and K. Shimizu, Phys. Lett. 112B, 201 (1982).
11. A.W. Thomas, TH3368-CERN, TRI-PP-82-29, to be published in Adv. in Nucl. Phys., eds. J. Negele and E. Vogt, 13 (1983).
12. D. Robson, Prog. in Particle and Nuclear Physics 8, 223 (1982).
13. G.E. Brown and M. Rho, Phys. Lett. 82B, 177 (1979); G.E. Brown, M. Rho and V. Vento, Phys. Lett 94B, 383 (1979).
13. G.E. Brown and M. Rho, Phys. Lett 82B, 177 (1979); G.E. Brown, M. Rho and V. Vento, Phys. Lett. 94B, 383 (1979).
14. S. Theberge, A.W. Thomas and G.A. Miller, Phys. Rev. D22, 2838 (1980); A.W. Thomas, S. Theberge, and G.A. Miller, Phys. Rev. D24 216 (1981).
15. R.L. Jaffe, Erice Summer School, "Etore Majorana" Lecture (1979) MIT CTP #814.
16. G.E. Brown, Session F, this workshop.
17. R. Freiedberg and T.D. Lee, Phys. Rev. D15, 1694 (1977); D16 1096 (1977); D18, 2623 (1978)
18. R. Goldflam and L. Wilets, Phys. Rev. D25, 1951 (1982).
19. C. Detar, Phys. Rev. D24, 752 (1981), ibid D24, 762 (1981).
20. L.R. Dodd, A.W. Thomas, and R.F. Alvarez-Estrada. Phys. Rev. D24, 1961 (1981).
21. S. Theberge and A.W. Thomas, CERN preprint TH-3290 (1982) to be published.
22. S.A. Chin and G.A. Miller, NSF-ITP-82-96 to appear in Phys. Lett. B; P.A.M. Guichon, G.A. Miller, and A.W. Thomas CERN preprint TH-3493 to be published (1982).
23. It is sometimes argued (e.g., G.E. Brown, Prog. in Particle and Nuclear Physics 8, 223 (1982)), that computed observables are independent of R because the reduction of quark contributions at small R is compensated by an increase in pionic contributions. However, these arguments are usually based on second-order perturbation calculations, a treatment which is not valid at small R. Sometimes exact "hedgehog" solutions are used, but these are not reliable, in my opinion since isospin conservation is strongly violated.
24. E.M. Henley, L.S. Kisslinger, and G.A. Miller, "A Theory for a Hybrid Model for the Nucleon-Nucleon System", U. Washington/Carnegie-Mellon U., preprint, 1983.

SESSION G

EXPERIMENTAL POSSIBILITIES

OF THE FUTURE

FUTURE PROSPECTS IN N-NUCLEUS INTERACTIONS

J. M. Moss

Los Alamos National Laboratory
Los Alamos, New Mexico 87545

ABSTRACT

I examine in detail two research areas, polarization observables and antiproton-nucleus reactions, which should have near-term future impact on our understanding of the interaction of medium-energy nucleons in nuclei. More speculative future experiments employing cooled beams, double spectrometer systems, and large Q-value low momentum-transfer reactions are also discussed.

INTRODUCTION

I have been asked to tell you about future possibilities, in other words, to predict what we will be or should be doing a few years from now. Webster's definition of "predict" is, "to foretell on the basis of observation, experience, or scientific reason." I can assure you that the last item has the least to do with what I will say. It is, of course, safer not to look too far into the future so I will concentrate partly on extensions of experiments that can be done now or will soon be feasible. The remainder will be more speculative and involve a heavy dose of ideas that I have gleaned from various proposals for research[1,2] to be performed or machines to be built.[3] Although I will mention possibilities for observing exotic states of nuclear matter, one subject that I will specifically omit is hypernuclei. Not that this subject is out of place here, but it has been covered in depth at other conferences. Antiproton-nucleus interactions will, however, be discussed since this subject has direct bearing on many aspects of N-nucleus physics. Interspersed through these discussions and in the final sections, I will briefly describe some more speculative areas where the machines of the future may have some impact.

POLARIZATION OBSERVABLES IN (p,p') REACTIONS

Intermediate protons offer a distinct advantage over lower energy beams in that very efficient proton polarimeters may be built. At the high resolution spectrometer (HRS) at LAMPF, the focal-plane polarimeter[4] has a scattering efficiency of about 10% and an effective analyzing power of ~ 0.4 at $E_p = 500$ MeV. In effect, this means that nearly every reaction for which a cross section is measurable can be made to yield polarization transfer (PT) observables with a bit more effort.

Spin correlation (SC) experiments are also on the near horizon when the IUCF Cooler[3] comes on-line. The high luminosity of the circulating polarized beams in the cooler will make feasible for the first time the use of jet targets of polarized nuclei. Jet targets should also allow analysis of final nuclear polarization in many cases where more conventional targetry would not. This is an exciting area for future double-spectrometer experiments.

What might a new generation of PT and SC experiments tell us about nuclei that is difficult or impossible to get by other means? To my knowledge there has not been any general analysis of what physics SC experiments would yield. Polarization transfer, on the other hand, has been examined in detail recently[5-7] and experimental evidence suggests that these new observables may be exceptionally interesting. Without delving into the theory, I will discuss two examples that should have significant future application.

In the excitation of unnatural parity states, two spin-dependent form factors enter in the inelastic scattering (or charge exchange) process, the transverse $X_T(q)$ and longitudinal $X_L(q)$. The former is similar to that measured in magnetic electron scattering. The latter is not present in (e,e'). In the eikonal single-scattering approximation, the PT observables may be used to separate these two form factors. Specifically,

$$X_L^2(q) = \sigma/4E^2\left(1 - D_{NN} + D_{SS'} - D_{LL'}\right) , \qquad (1)$$

$$X_T^2(q) = \sigma/4F^2\left(1 - D_{NN} - D_{SS'} + D_{LL'}\right) , \qquad (2)$$

where E and F are coefficients of the nucleon nucleon (N-N) scattering amplitude (as defined in Ref. 8), and σ is the differential cross section. The PT parameters are in the Ann Arbor Convention.[9] The subscripts N, S, and L denote normal, sideways, and longitudinal components of polarization referred to the incoming (unprimed) and outgoing (primed) momentum directions. Measurement of the PT observables is of special interest not only because they may provide new information, but because of the particular sensitivities of the above form factors to questions of great current interest in nuclear physics.

Specifically X_L emphasizes those aspects of the nuclear response sensitive to the pion field, whereas X_T is related to transverse fields as yielded for example by ρ-meson exchange. These sensitivities are clearly indicated in Fig. 1, which shows calculations of the longitudinal and transverse nuclear response functions by Albercio et al.[10] The precise physics input into such caluclation can have a dramatic affect on the predicted values of X_L and X_T. For example in Fig. 1, specific assumptions were made about the nature of short range N-N, Δ-N, and Δ-Δ correlations. Such issues are hardly resolved at present as the debate continues on the influence of Δ-h configurations in nuclei and the proximity of normal nuclear matter to the critical value for pion condensation.

It is interesting to note that the continuum response as calculated in Fig. 1 may actually be measured. There is no contribution from natural parity excitations (within the approximations indicated previously) to Eqs. (1) and (2). The most interesting application of Eqs. (1) and (2) might be in the (n,p) or (p,n) reactions, where isospin transfer, $\Delta T = 1$, is insured. Additionally, if

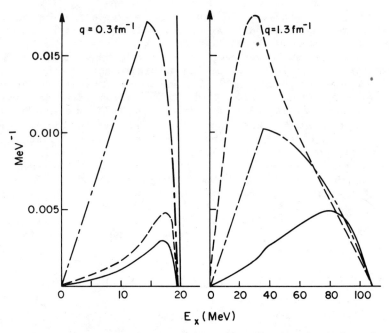

Fig. 1. Fermi gas (dot-dash), axial longitudinal (dash),
and transverse response functions from Ref. 10.

sufficient precision could be attained, such an experiment might
shed light on the questions of missing Gamow-Teller strength in the
continuum.

Another recent application of polarization analysis, which I am
confident will have significant future application, concerns the
difference between the polarization (P) and analyzing power (A) in
inelastic scattering (and charge exchange). For elastic scattering
P = A of course due to time reversal (TR) invariance. For inelastic
scattering TR does not imply P = A, however. It is easy to demonstrate that nonzero values of P-A arise from an interference
between TR odd and even terms in the N-nucleus scattering
amplitude.[11] The TR odd amplitides, which are intimately related
to the nonstatic parts of the N-N interaction, again offer the
opportunity to measure aspects of nuclear structure not easily
obtained in other experiments. At E_p = 150 MeV Carey et al.[12] have
shown that P-A for the 1^+ T = 1 state in ^{12}C can be explained only
by a transition density that includes a large term with orbital,
spin, and total angular momentum $\ell s j$ = 111 transfered to the nucleus.
The Cohen-Kurath[13] wave functions, which contain an important term
of this type, are very successful in reproducing the data (Fig. 2).
When the $\ell s j$ = 111 term is removed, (P-A)$\ast\sigma$ remains near zero at all
angles (P-A itself may still be large at diffraction minimum as is
seen in Fig. 2). The $\ell s j$ = 111 term for 0^+ to 1^+ excitation is
determined entirely by the combination of density-matrix elements
$j^{-1}_{<j>} + j^{-1}_{>j<}$ to which (e,e') is completely insensitive. It is likely

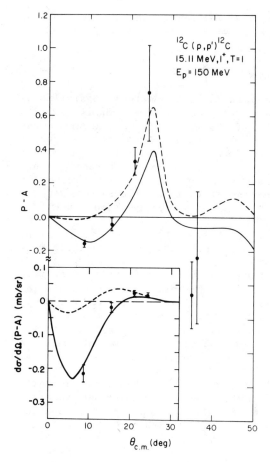

$^{12}C\,(p,p')\,^{12}C$
15.11 MeV,I^+,T=1
E_p = 150 MeV

Fig. 2. P-A and (P-A)dσ/dΩ. The solid curve is a DWIA calculation with the Cohen-Kurath wave function. The dashed curve has the $\ell\,s\,j$ = 111 term removed.

that studies of P-A for other M1 states may provide unique experimental evidence concerning the effects of one particle, one hole ground-state correlations in reducing the strength of M1 transitions.

Without pretending that all the theoretical problems associated with nucleon-nucleus scattering have been solved, it is clear that much new nuclear structure physics should result in the coming years from polarization-transfer and spin-correlation experiments. To be even more speculative, there may well be an analogous application of spin observables in the excitation of nucleons when (and if) a new generation of higher energy polarized beams becomes available. As an example consider the excitation of the Δ^{++} by the reaction

$$p + p \rightarrow \Delta^{++} + n \quad .$$

What might one learn by measuring the longitudinal and transverse form factors or P-A for such a reaction? Clearly one needs theoretical input into this problem. The nuclear structure view that this reaction is $p(p,n)\Delta^{++}$ is of course incomplete. The probe and target are not distinct entities. Nevertheless, polarization observables in the excitation of nucleons would seem like a profitable ground for extension of our present and near future experimental techniques.

As a final note on polarization observables, the most intelligent applications of PT and SC techniques will require a deep understanding of the connections between the basic operators entering in nucleon scattering and those encountered in electromagnetic and semileptonic weak processes. Considerable progress has been made in laying out these connections[14] in terms of a common formalism; however, much remains to be done. It may turn out that guidance in the interpretation of exotic processes such as neutrino inelastic scattering could be provided by nucleon polarization observables.

ANTIPROTON NUCLEUS INTERACTIONS

Antiproton nucleus interactions promise numerous possibilities for new and exciting physics in the near future when the LEAR facility at CERN comes on-line. Describing with any accuracy the nature of this new physics is another matter -- considerable uncertainity exists concerning even the basic features of the $\bar{N}$-N system. Imbedding various possible scenarios for the $\bar{N}$-N interaction into nuclear matter is also not straightforward. Nevertheless with the aid of a few ideas gleaned from the theoretical literature and two experimental proposals to LEAR, I will speculate about what the future might hold.

One feature of $\bar{N}$-N system, which has a profound impact on $\bar{N}$-nucleus interactions, is the large cross section for annihilation. The dominant annihilation channel results in the production of roughly five pions. The mean pion energy is near that appropriate for the Δ_{33} resonance. That means that the pions themselves should have a short mean free path inside a nucleus. Thus if one could implant such a catastrophic event (releasing nearly 2 GeV of rest-mass energy) inside a nucleus, conditions favorable for the production of nonnormal phases of nuclear matter might well be achieved. But how does one implant a probe that itself has an extremely high probability of interacting even in the low-density matter in the nuclear surface? A group at Los Alamos has recently suggested that the properties of the $\bar{N}$-N interaction at moderate energies conspire to give antiprotons a better chance of depositing their energy in the nuclear interior than might be expected.[2,15] Extensive calculations using the intranuclear cascade model (INC) reveal that 175-MeV antiprotons have a high probability for depositing in excess of 1 GeV in a ^{238}U nucleus (Fig 3). A key feature in these calculations is a transparency conferred on the $\bar{N}$-nucleus system by the combination of a strongly energy-dependent annihilation cross section and an attractive $\bar{N}$-nucleus potential.

The Los Alamos National Laboratory group intends to use a very large, solid-angle spectrometer at LEAR in search for possible exotic effects induced by the explosion of antiprotons inside nuclei. Pion multiplicities and correlations will be employed in the initial phase of this effort. In the examination of nuclear matter under extreme conditions, the hope is that the relatively small momentum-transfer $\bar{N}$-nucleus reactions will complement relativistic heavy-ion collisions.

From the quieter side of $\bar{N}$-nucleus interaction, what might elastic or inelastic $\bar{p}$ scattering tell us about nuclei? The Saclay-Strasbourg - Tel Aviv collaboration[1] proposes to use the SPES II spectrometer at LEAR to carry out such studies. What types of states might be strongly excited in $(\bar{p},\bar{p}')$? At present one can only speculate since few of the experiments necessary to construct the $\bar{N}$-N scattering amplitude have been performed. We are not totally in the

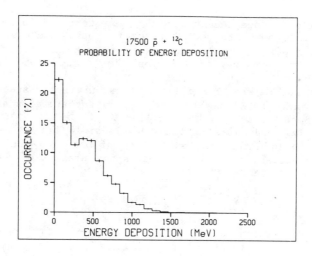

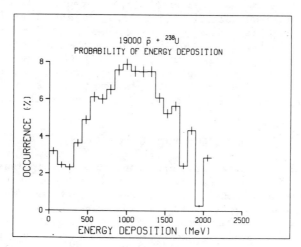

Fig. 3. Energy deposition probabilities for
antiprotons from Ref. 15.

dark, however. T-channel meson exchange models of the N-N interaction may be used in conjunction with the G-parity transformation to
construct the long-range real part of the N̄-N interaction. An
imaginary term is normally added to the real part to take annihilation into account phenomenologically. Several calculations along
these lines have recently been published.[16-18] Among the
interesting results is a strong isoscalar tensor interaction, which
profoundly influences the spin observables in the p̄p system.[17] Such
strong spin-dependences are suggestive of interesting physics for the
N̄-nucleus system.

Given a free N̄-N interaction one is still a long way from
knowing what the effective interaction will be inside a nucleus.

420

Medium effects, to use recent parlance, may turn out to be enormous. To some extent, however, what happens to the $\bar{N}N$ interaction may be irrelevant since few $\bar{N}s$ are likely to make it to the nuclear interior. In the surface region "$t\rho$" may not be a bad approximation.[19]

In a recent talk on $\bar{N}$-nucleus physics, Garreta[20] speculates that one-pion exchange (OPE) may dominate (p,p') spectra at small momentum transfer. Such states as the 15.11-MeV, 1^+,T=1 of ^{12}C would be prominent. This speculation is very plausible but it would be a mistake to conclude that the excitation of one-pion-like states is therefore uninteresting. The spin-isospin interaction for the N-N system may be written as

$$V_{\sigma\tau}(q) = V_{OPE}(q) + g_o' \quad ,$$

where g_o' is independent of q. At q = 0, $V_{OPE} \rightarrow 0$ and the cross section to pion-like states is governed by g_o'. This quantity is large as can be seen in a recent 0° (p,p') spectrum from the HRS at the Los Alamos Meson Facility (Fig. 4). The effect of short range interactions that determine g_o may be completely different in the $\bar{N}$-N system. Finally the SPES II collaboration proposes to examine antiprotonic nuclei by using the recoilless ($\bar{p}$,p) reaction. Such nuclei might exhibit relatively long-lived states if a sufficiently strong attractive potential existed[21] for a high-spin state whose wave function was largely outside the range of the annihilation potential.

NEW SPECTROMETER TECHNIQUES

As future accelerators are constructed (at least in our dreams) it is important to ask what instrumentation is required to put the beam to maximal use. Magnetic spectrometers have been vital in medium-energy charged-particle work and will undoubtedly continue to be so in the future. They are essential not only in providing very precise momentum resolution and particle

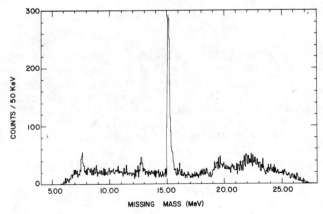

Fig. 4. Spectrum of the ^{12}C (p,p')^{12}C reaction at Ep = 497 MeV taken at zero degree.

identification, but also in the spatial separation of high-cross-section and low-cross-section reactions allowing the low background observation of the latter. The dual spectrometer system for IUCF[22] should provide much of the capability needed to use the unique properties of the electron-cooled beams. One spectrometer will have exceptional resolving power, with the promise of ~ 10-keV resolution at 200 MeV. The other spectrometer, to be used in coincidence, will detect a much broader range of momenta with a sizable solid angle. It is plausible to foresee that dual spectrometer coincidence studies will assume an increasing importance in future investigations using medium-energy nucleons.

As a very speculative example of a two-spectrometer cooled-beam coincidence experiment consider the excitation of an important but weakly excited giant resonance. It is unobservable in a single experiment because of the large continuum. Once excited, however, it remembers its strong connection with the ground state and decays by gamma emission with perhaps a 10^{-4} branching ratio. Excitation of the continuum, on the other hand does not produce the original target nucleus. In the $A(p,p')A^*$ reaction, one measures p'-A coincidences to enhance the giant resonance signal. Many other possibilities undoubtedly exist for recoil coincidence detection.

Another interesting application of spectrometers that should see even more future action is inelastic scattering at $0°$. Zero-degree scattering of medium-energy protons yields momentum transfers that are extremely small (limited only by the reaction Q- value) and are hence ideal for the study of low spin states such as 0^+ and 1^+.

Zero-degree scattering is not only promising from the experimental point of view for maximizing the signal for $\Delta \ell = 0$ transitions, it also offers relative theoretical simplicity. Distortion effects are minimal and transition densities are closest to the "photon point".

Such studies have been in progress for about a year at the HRS at LAMPF using a specially prepared beam of 500-MeV protons. Definition of the beam is accomplished by using foil strippers instead of collimators while the ions are negative. The highly tailored H⁻ beam is then stripped of its electrons and transported to the target with no further collimation. This beam then passes through the spectrometer and misses all of the focal plane detectors. Inelastic scattering is detected for energy losses as small as 3 MeV. A typical $0°$ spectrum from 500-MeV protons on ^{12}C is shown in Fig. 4. This technique is much more difficult for heavy targets due to rescattering of Coulomb-scattered particles. Active collimators show promise of alleviating this problem for heavier systems.

HIGH-Q, LOW-q REACTIONS

Zero-degree scattering even of Fermi-Lab protons will not yield zero-momentum transfer in the region of excitation of Δ-isobar-hole configurations. Recall that it is the low q aspects of the Δ-h admixtures that are of interest in the Gamow-Teller/Ml problem.[23] Thus if we are able to examine this end of the Δ-h spectrum whose influence is so strongly exerted on the central theme of this conference, we must find another way to do it.

One method is to use the (p,d) reaction at 800 MeV.[24] The q for such conditions is near zero resulting hopefully in the production of recoilless deltas in the residual nucleus. Such an experiment was recently performed at LAMPF.[24] Deuterons with energy loss corresponding to Δ production in the $^{13}C(p,d)^{12}C*$ reaction were observed in coincidence with decay protons from the excited nucleus. The constant excitation energy observed in the coincidence spectrum is very suggestive of the kinematics of a state in ^{12}C decaying rather than of a quasi-free $p+p \rightarrow d+\pi^+$ mechanism. Continued investigation is required before definitive statements can be made, however.

This reaction is an example of a general class of high Q-value, low-momentum-transfer reactions that are possible avenues for producing nuclear states ranging from slightly exotic (e.g., the previous example) to profoundly different. A. Goldhaber has recently suggested,[25] that colorless (in the quark sense) nuclear matter may exist in exotic configurations not at all resembling A-nucleons, but with excitation energies perhaps only a few hundred MeV above standard nucleonic matter. He speculates that high-Q-value, low-q reactions may be appropriate for exciting exotic configurations, or perhaps more likely, for producing identifiable precursors of such states.

What reaction mechanism could ever lead to the formation of such exotica? That is of course the proverbial "fly in the ointment;" indeed even for such "conventional" reactions as $^{13}C(p,d)^{12}C_\Delta$ the mechanism is most appropriately described by a black box. On the positive side, in hadron-hadron interactions, no proposed reaction mechanism, however absurd, can be completely eliminated from consideration. In other words if fundamental conservation laws don't forbid it, it will probably occur with a cross section that is not ridiculous by weak-interaction standards.

SUMMARY

I have emphasized polarization observables in nucleon-nucleus physics and cross sections and correlation experiments in antinuclion-nucleus physics as the most promising areas for the near future. The more distant future is largely a matter of guess work. Aside from describing what might be accomplished with new machines and devices it is difficult to be specific. If we have learned anything relevent to the future from the near past in intermediate energy physics, it is that nuclear physicists can no longer be parochial. Much more, for example, can be learned about a given state by doing (p,p'), (e,e'), and (π,π') experiments than can be gotten from one reaction in isolation. In the future this trend needs to continue with attention given to the possible complementary aspects of hadron-scattering and weak-interaction experiments. Additionally, one of the most obvious challenges of medium-energy physics, determination of the necessity or lack thereof of quark degrees of freedom, is a central theme in high-energy heavy-ion physics. We will learn more if we understand the nature of other approaches to the same subject.

REFERENCES

1. LEAR Experiment 184, Saclay, Strasbourg, Tel Aviv Collaboration.
2. LEAR Experiment 187, Los Alamos, Grenoble collaboration.
3. The IUCF Cooler-Tripler Proposal to the NSF (Dec 1980).
4. J. B. McClelland et al., Nucl. Instr. and Meth. to be published.
5. J. M. Moss, Phys. Rev. C26, 727 (1982). J. M. Moss, Proc. of the International Conference on Spin Excitations in Nuclei. Proceedings to be published.
6. E. Bleszinski et al., Phys. Rev. C. (in press).
7. W. G. Love, Ref. 5 Ibid.
8. A. K. Kerman, H. McManus, and R. M. Thaler; Ann. Phys. 8, 551 (1959).
9. R. Fernow and A. Krish; Ann. Rev. Nucl. and Part. Science 31, 107 (1981).
10. W. M. Albercio et al., Nucl. Phys. A379, 429 (1982).
11. W. G. Love and J. R. Comfort, to be published.
12. T. A. Carey et al., Phys. Rev. Lett. 49 266 (1982).
13. S. Cohen and D. Kurath, Nucl. Phys. A101, 1 (1967).
14. F. Detrovich and W. G. Love, Proceedings of the International Converence on Nuclear Physics, North Holland Publishing Co. New York. Eds. R. M. Diamond and J. O. Rasmussen, p. 499c.
15. M. R. Clover et al., Phys. Rev. C26, 2138 (1982).
16. W. W. Buck, C. B. Dover, and J. M. Richard, Ann. Phys. 121, 47 (1979).
17. C. B. Dover and J. M. Richard, Phys. Rev. C 25, 1952 (1982); Phys. Rev. D 17, 1770 (1978).
18. R. Vinh Mau, Nucl. Phys. A374, 3c (1982).
19. C. Dover, private communications.
20. D. Garreta, Invited Talk at the School on Physics of Exotic Atoms; Erice, Italy, May 1982.
21. E. H. Auerbach, C. B. Dover, and S. H. Kahana, Phys. Rev. Lett. 46, 702 (1981).
22. IUCF - Maryland proposal to the NSF (Aug. 1980).
23. W. Weise, Nucl. Phys. A374, 505c (1982).
24. C. Morris et al., to be published.
25. A. Goldhaber, Invited talk at the LAMPF II Workshop, Los Alamos, June 1982 (Proceedings to be published).

424

POLARIZATION TRANSFER IN A($\vec{\text{d}},\vec{\text{p}}$X) REACTIONS

H. Sakamoto, M. Nakamura, H. Sakaguchi, H. Ogawa and S. Kobayashi
Department of Physics, Kyoto University, Kyoto 606, Japan

and

N. Matsuoka, K. Hatanaka and T. Noro
Research Center for Nuclear Physics, Osaka University, Ibaraki 567, Japan

and

S. Kato
Laboratory of Nuclear Studies, Osaka University, Toyonaka 560, Japan

Transverse polarization transfer coefficients ($k_y^{y'}$) have been measured in the reactions ^{12}C, ^{58}Ni, ^{209}Bi($\vec{\text{d}},\vec{\text{p}}$X). The experiment was performed at RCNP AVF cyclotron laboratory using a 56 MeV vector-polarized deuteron beam. Emitted protons leaving the residuals in their continuum states were selected by Q-magnet pairs according to their momenta and were focused on the analyzer targets (Si-SSD) in the Si-Ge polarimeters[1] located at 13.5° and 20.0° laboratory angles. The energy bins of the measured protons have been chosen to be 22, 26, 31 and 36±2 MeV. The $k_y^{y'}$'s obtained are displayed in Fig. 1, where it can be seen though almost data points come near the $k_y^{y'}$ =2/3 line, there is a certain tendency of the points at lower proton energies going away from the line.

1) H. Sakamoto et al., RCNP Annual Report 1980, p.28.

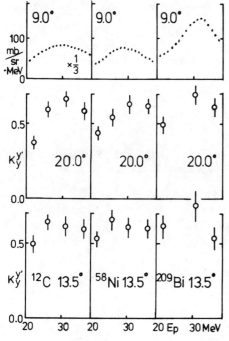

Fig. 1. Transverse polarization transfer coefficients ($K_y^{y'}$) in the reactions ^{12}C, ^{58}Ni, ^{209}Bi($\vec{\text{d}},\vec{\text{p}}$X) at Ed=56 MeV. Their corresponding energy spectra ($d^2\sigma/d\Omega dEp$) are plotted together at the top of the figure.

THE POLARIZATION SPECTROGRAPH "DUMAS"

M. Nakamura, H. Sakaguchi, H. Sakamoto, H. Ogawa, T. Ichihara, M. Yosoi
and S. Kobayashi
Department of Physics, Kyoto University, JAPAN
and
T. Noro, T. Takayama and H. Ikegami
Research Center for Nuclear Physics, Osaka University, JAPAN

Recently, it has been recognized that the polarization measurement of emitted particles in nuclear reaction may play a significant role in the exploration of nuclear reaction mechanism, where the reaction may be induced by either polarized or unpolarized beam and even heavy ion beam. The polarization measurement should cover not only the emitted particles leading to the low-lying states but also those leading to the highly excited states of residual nuclei simultaneously. The characterestics to satisfy such physical requirements, the high efficiency and high resolving power of the instrument are not always compatible with each other. However we have succeeded in design of a new separate function type polarization spectrograph[1]) with excellent performance with regard to these characteristics, by introducing combined multipole fields into the system besides the separate field elements.

The spectrograph named DUMAS (Dual Magnetic Spectrograph) is now being intalled in the experimental area for polarization measurement at RCNP and will be brought in overall test operation in the fall of 1982.

DUMAS's configuration and trajectories predicted by using the finally adopted parameters which include higher order corrections are shown in the figure. The figure also shows a tagging counter which is set at the focal plane of the first part of DUMAS to specify the momentum of emitted particles and a polarimeter is mounted at the achromatic focusing point of the second part of DUMAS, where all particles may enter the polarimeter with a small acceptance. The polarimeter consists of multi-target and multi-wire proportional counter telescope.

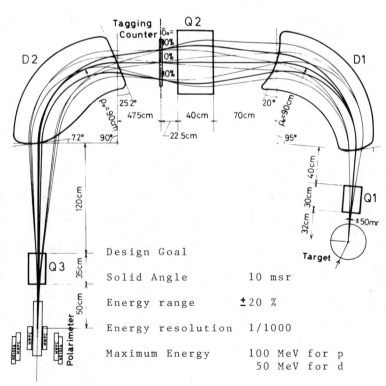

Design Goal

Solid Angle	10 msr
Energy range	± 20 %
Energy resolution	1/1000
Maximum Energy	100 MeV for p
	50 MeV for d

1) H. Ikegami, T. Noro, T. Takayama, M. Nakamura and S. Kobayashi:RCNP Annual Report 1981.

WEAK DECAY OF Λ HYPERNUCLEI

Richard Grace

Carnegie-Mellon University, Pittsburgh, PA, 15213

Abstract   Section G

A hypernucleus must ultimately decay by a weak process after the more rapid strong and electromagnetic (strangeness conserving) processes are finished. There exists four weak decay modes for a hypernucleus.

1. $\Lambda \rightarrow p + \pi^-$

2. $\Lambda \rightarrow n + \pi^o$ mesonic decay

3. $\Lambda + p \rightarrow n + p$

4. $\Lambda + n \rightarrow n + n$ nonmesonic decay

The mesonic decay modes are dominant in very light hypernuclei while the nonmesonic decay modes are dominant in heavier hypernuclei and depend on the four fermion weak interaction. This experiment is designed to test our understanding of this interaction.

We will use the Moby Dick spectrometer (at BNL) in coincidence with a range spectrometer and a TOF neutron detector to study the weak decay modes of $^{12}_{\Lambda}C$. The Moby Dick spectrometer will be used to reconstruct and tag events in which specific hypernuclear states are formed in the reaction $K^- + {}^{12}C \rightarrow \pi^- + {}^{12}_{\Lambda}C$. Subsequent emission of decay products (pions, protons and neutrons) in coincidence with the fast forward pion will be detected in a time and range spectrometer, and a neutron detector. Current tests of prototype counters have given a time resolution of $\sigma = 65$ psec. This will permit measurement of lifetimes in the range $T = 100-150$ psec. The large solid angle and low energy thresholds of the range spectrometer and neutron counters will permit branching ratio measurements between the various decay modes.

SUMMARY

SUMMARY PRESENTED AT 1982 IUCF WORKSHOP
ON
THE INTERACTION BETWEEN MEDIUM
ENERGY NUCLEONS IN NUCLEI

Herman Feshbach
Massachusetts Institute of Technology
Cambridge, Mass. 02139

Qualitatively the concept of the nucleon-nucleon interaction inside a nucleus seems clear. We expect that the nuclear environment will change the effect of one nucleon upon another from that which prevails in free space. Pauli blocking is a simple example. But in this workshop a more far reaching question is under discussion. Can the effect of the environment be given in a gross way in terms of properties like the nuclear density so that the effective nucleon-nucleon interaction will vary smoothly from nucleus to nucleus? We must bear in mind that such an interaction will be complicated, being both energy dependent and non-local, in addition of course to similar complexities which are present in the free space interaction.

As Norman Austern urged in his talk, let us make this question more precise. Formally we are concerned with the transition matrix, T_{fi} is given by

$$T_{fi} = <\Phi_f^{(-)} | \hat{T} | \Psi^{(+)}> \tag{1}$$

where

$$\hat{T} = V + V \frac{1}{E^{(+)}-H} V \tag{2}$$

and V depends upon the free space interaction between the colliding nucleons so that $\hat{T}$ is a functional of the two body interactions V_{ij}. The influence of the medium is contained in the initial and final state wave functions. The evaluation of T_{fi} generally involves the introduction of model wave functions which are related to the exact ones through a transformation

$$\Phi_f^{(-)} = U\Phi_f^{(M)(-)}$$

$$\Psi_i^{(+)} = W\Psi_i^{(M)(+)}$$

Where the superscript M has been introduced to indicate model wave functions. Introducing these into Eq. (1) T_{fi} becomes

$$T_{fi} = <\Phi_f^{(M)(-)} | U^+\hat{T}W\Psi_i^{(M)(+)}>$$

so that the effective T matrix becomes

$$T^{(M)} = U^{+}\hat{T}$$

U^{+} and W are generally complicated many body operators. Nevertheless a number of remarkable assumptions are made at this point. It is assumed that $\hat{T}^{(M)}$ is functional of two body operators $g_{ij}(\rho)$ which are functions of the nuclear density, ρ, $\hat{T}^{(M)}$ may also have an additional dependence on ρ so that

$$\hat{T}^{(M)} = \hat{T}^{(M)}\left(g_{ij}(\rho),\rho\right)$$

The effective nucleon-nucleon interaction is given by g_{ij}. But there is an additional assumption, namely that $\hat{T}^{(M)}$ and g_{ij} are independent of the initial and final states. However it is most important to note that it is not assumed, nor is it correct to do so, that $\hat{T}_M$ and g_{ij} are independent of the model used. This important point was emphasized in Norman Austern's contribution. For example, the nature of $\hat{T}^{(M)}$ and g_{ij} well depend upon whether one uses a coupled channel description or a single channel. It will depend upon the nature of the channels employed. Indeed some dependence on the states involved may enter because these may prejudice which channels are considered. For each of these cases one may very well obtain differing g_{ij}'s. It is thus most important in making statements about the effective nucleon-nucleon interaction inside nuclei that the model and the corresponding representations used in obtaining these effective interactions be specified.

In practice, the principal source of information on the effective nucleon-nucleon interaction has been obtained from inelastic scattering using the DWBA approximation. Typical data and the extraction of the interaction was presented to us by C. Olmer and J. Kelly. Olmer considered inelastic scattering to high spin states while in Kelly's experiments inelastic scattering to collective states was measured. I won't attempt to summarize their results but refer the reader to their papers.

The results obtained depend inter alia, upon the optical model wave functions employed. To obtain these, elastic scattering is fitted using an optical model potential which is then used to generate the required wave functions. It is thus no surprise that much attention was paid to the optical model during this workshop. Schwandt reviewed the proton and neutron elastic scattering data. At low energy, one uses the "standard model" potential containing a complex spin independent term and a complex spin dependent term. There are twelve parameters involved, some of which are taken to be energy dependent. The shape of the potentials is taken to be in the Woods-Saxon form. There is no a-priori justification for this choice except from very general qualitative arguments. But let me remind you of the situation in elastic electron scattering where for many years (and for reasons similar to those which prevail in nucleon scattering) the data was analyzed using a Fermi shape for the charge distribution. This is just the Woods-Saxon shape "upside down". Eventually microscopic determinations of the density were made (for example by John Negele). The shape of resultant charge distribution bore a vague resemblance to the Woods-Saxon shape and was found nevertheless to give excellent agreement with the data. I am thus not entirely surprised that the Woods-Saxon shape does not agree with the data reported to this workshop and that in fact such a shape is not predicted by the microscope models we shall discuss later in this report.

Empirically what one finds is most interesting. Focussing on the central potential, one finds, as the projectile energy increases into the several hundred MeV range, that this potential develops what is known as a wine bottle shape, repulsive at its center, attractive at longer range and of course eventually going to zero. This behavior has been incorporated into the optical model potential through the introduction of a repulsive potential whose shape is given by the square of the Woods-Saxon form:

$$-U(r) = V_{R1}f_{R1}(r) - V_{R2}\Big(f_{R2}(r)\Big)^2 + \frac{\lambda^2}{\pi}\Big(V_{SO} + W_{SO}\Big)g_{MSO}\Big(\vec{\sigma}\cdot\vec{L}\Big)$$

The square of the Woods-Saxon form drops off more rapidly than the Woods-Saxon so that the combination of the first two factors can result in an attractive region near the nuclear surface. At energies approaching 1 GeV the potential becomes purely repulsive in agreement with the multiple scattering analysis to be discussed later.

The interpretation of this important result, as result of a conversation with Fred Petrovich, is as follows. The observed energy dependence (I should and will say in what follows, momentum dependence) is in large part a representation of the non-locality of the optical model potential. That non-locality has a range, a, associated with it of the order of of the universe of the Fermi momentum, that is about 0.74 fm. As the momentum of the projectile increases and its wavelength decreases, the effect of the non-locality will begin to average out. Eventually there are enough oscillations within the range of the non-locality so that the effect of the non-local potential averages to zero. Turning to the empirical result we can interpret it as follows. A considerable fraction of the optical model potential is non-local. As the projectile energy increases it washes out leaving behind the repulsive peak which dominates the local part of the potential. This is a most interesting result which I believe is of fundamental importance.

A new initiative in the area of phenomenological analysis is the "Dirac phenomenology" presented to us at this meeting by Bunny Clark. In this analysis the Dirac equation is employed in place of the Schroedinger equation. The projectile-nucleus interaction is specified in terms of two potentials, a scalar potential, V_s, and the fourth component of the vector potential V_4. With

these assumptions a number of important consequences follow from
the structure of the Dirac equation when that equation is reduced
to an equivalent Schroedinger equation. One obtains the spin-orbit
potential, the double Woods-Saxon potential as well as some energy/
momentum dependence. One is thus able to match the experimental
data very well indeed. A particular example is the spin rotation
parameter at 497 MeV with the calculation right on the nose.

But why relativity when the projectile energy is say 100 MeV
when the nucleon mass is close to 1 GeV? The answer appears when
one looks at the empirical potentials employed. In particular, the
effect of the scalar potential is to reduce the mass of a nucleon
inside a nucleus to a fraction (~0.6) of its value in free space.
(Note this reduction in mass should not be confused with the
effective mass which is produced by the non-locality and energy
dependence of the optical potential). This change is very large and
one must look to see if there are other verifiable consequences
before we can be prepared to believe it.

There remains the question of how Dirac phenomenology enters
into consideration of reactions and transitions. This was the subject discussed by Shephard with the conclusion that this remains
an open question.

Presumably the answer can be found in the microscopic theory
which Shakin described. Shakin's model assumes the one boson model
for the nucleon-nucleon interaction and applies a relativistic self
consistent Brueckner-Hartree-Fock procedure to obtain the single
particle potential and wave-function, as well as the correlations
originating in the Bethe-Goldstone equation, taking for example
Pauli blocking into account. In my opinion, the relativistic
Bethe-Goldstone equation is not obvious. Be that as it may,
Shakin's results are in substantial agreement with the empirical
results reported by Clark. Shakin's model has the obvious merit
that its assumptions are carefully stated and the consequences
developed.

We return to the more standard procedures such as the multiple
scattering theory employed by Lannie Ray.   This theory attempts
to express the optical potential in terms of the observed nucleon-
nucleon scattering amplitude in free space. It is valid at high
energy. Ray used just the first term in the multiple scattering
expansion of the optical potential (originally used by Rayleigh!).
The second term involves correlations in the target nucleus and is
equivalent to the existence of three body potential terms. The
first order approximation does vary well at high energy. But as
the projectiles energy is reduced down to 500 MeV at begins to
fail. Ray then modifies the nucleon-nucleon amplitudes he uses in
order to obtain agreement. These changes are presumably a medium
effect which he has therefore determined empirically.

At lower energies there has been several efforts (see the
contributions of Mahaux, Pandharipande, and V. Geramb) using the
Brueckner-Hartree-Fock method, extending it from the negative
energy region with its discrete energy spectrum to the continuum.
Many approximations are involved in carrying out this difficult

432

program.

One prediction obtained is the wine bottle shape for the central potential discussed earlier. Mahaux in his paper presents a list of concerns, 19 in all, which need to be examined in order to be more confident that the results are quantitatively correct. I have picked out a few for discussion. In the first place there are higher order effects which need to be evaluated. Secondly one would like to evaluate the error associated with the local density approximation. Another emphasizes the fact that the calculations are carried out for infinite nuclear matter and do not include the effects of collectivity such as vibrational and rotational collective motions. A most important remark made by both Mahaux and Pandharipande is as follows. Let the self energy operator Σ be written $U(k,E) + iW(k,E)$ Then the optical model W is not $W(k(E),E)$ but rather

$$W_{OPT} = \frac{\tilde{m}}{m} W(k(E),E)$$

where $\tilde{m}/m$ is the "k" mass in Mahaux's nomenclature. In the nuclear interior $\tilde{m}/m$ 0.6 rising to unity for vanishing nucleon densities. This factor has not been applied in the available calculations.

This brings me naturally to the report on the nucleon-nucleon interaction presented by D. Bugg as that interaction is the primary input to the calculations just reported. He reports that the (p,p) phase shifts are well understood up to 800 MeV and that the (n,p) situation should be clarified in due course. Professor Bugg was concerned with the interpretation of the Argonne polarization experiments as a dibaryon resonance. Another explanation might be that it is an inelastic threshold for which the cross-section naturally rises as the energy increases but later is pulled down by competition. However, in my opinion, the appearance of a resonance is not in restrospect unexpected because of the existence of excited states of the nucleon such as the Δ. If the nucleon-nucleon interaction is such as to feed the incident kinetic energy into exciting the Δ so as to obtain a semi-bound state a resonance will occur. Of course there will be an associated inelastic threshold.

The existence of the Δ demonstrates that the nucleon has a structure. It is of course hardly the only supporting evidence. Regardless of the details of that structure it introduces a non-locality into the nucleon-nucleon interaction. The range of that non-locality can be estimated in various ways, e.g. $r \sim \hbar/(m_\Delta - m)c \sim 0.7$fm. There is no way of sweeping these non-local effects under the rug. They will affect not only the nucleon-nucleon force but also the nature of the optical model potential.

The particle-nucleon hole state was discussed in the report presented by J. Rapaport. In this context, considerable care must be exercized in interpreting the results. For example one can ask how much of the observed effect is a consequence of configuration mixing? But if both the Δ contribution and configuration interaction are included double counting may occur since the

Δ contributes to the residual interaction responsible for the configuration mixing. Certainly one should use the empirical information in the Δ-hole interaction available from pion-nucleus scattering. Another issue is the relation between the (p,n) process
and β decay which has been employed. That relation presumes the
validity of perturbation theory. One should ask, and indeed it
was asked by Macfarlane how good is perturbation theory? How
much of the Gamow-Teller strength is dissipated in multi-step
processes in which the Gamow-Teller operator is applied more
than once so that one obtains effectively a Fermi type transition.
The presence of a nearby isobar analog will help to confuse the
issue.

Another place where the Δ shows up is in three body forces in
which an intermediate state is formed with one nucleon replaced
in a first scattering by a Δ. Wallace used such a three body force
in considering the p-^{4}Hc elastic scattering in the 1 GeV region.

The exchange current, a phenomenon again closely related to
nuclear forces, was discussed by Friar. The important point he
made is that the exchange current operator is not positive definite.
There is therefore a substantial concellation and one must be certain that the wave functions used and the exchange current are self
consistent, i.e. stem from one theory. Friar seems to have some
concerns in this regard.

Finally Brown and Rho discussed the structure of the nucleon
in terms of the bag model. In particular, Brown discussed the
effect of the vibration of the surface of the bag. The size of
the bag has been a source of debate. However the difference
between the two camps are narrowing as the "small" bag gets bigger
and becomes surrounded by a pion cloud while the "big" bag is
found to contain not only the three quarks usually assumed but
quark-antiquark pairs as well. The latter correspond to pions of
the small bag.

Obviously a summary talk cannot be complete and refer to all
the major points raised in the course of the conference. I hope
however that it does indicate most of the issues which were
discussed and their importance for the understanding of the interaction of medium energy nucleons in nuclei.

AIP Conference Proceedings

		L.C. Number	ISBN
No.1	Feedback and Dynamic Control of Plasmas	70-141596	0-88318-100-2
No.2	Particles and Fields - 1971 (Rochester)	71-184662	0-88318-101-0
No.3	Thermal Expansion - 1971 (Corning)	72-76970	0-88318-102-9
No.4	Superconductivity in d-and f-Band Metals (Rochester, 1971)	74-18879	0-88318-103-7
No.5	Magnetism and Magnetic Materials - 1971 (2 parts) (Chicago)	59-2468	0-88318-104-5
No.6	Particle Physics (Irvine, 1971)	72-81239	0-88318-105-3
No.7	Exploring the History of Nuclear Physics	72-81883	0-88318-106-1
No.8	Experimental Meson Spectroscopy - 1972	72-88226	0-88318-107-X
No.9	Cyclotrons - 1972 (Vancouver)	72-92798	0-88318-108-8
No.10	Magnetism and Magnetic Materials - 1972	72-623469	0-88318-109-6
No.11	Transport Phenomena - 1973 (Brown University Conference)	73-80682	0-88318-110-X
No.12	Experiments on High Energy Particle Collisions - 1973 (Vanderbilt Conference)	73-81705	0-88318-111-8
No.13	π-π Scattering - 1973 (Tallahassee Conference)	73-81704	0-88318-112-6
No.14	Particles and Fields - 1973 (APS/DPF Berkeley)	73-91923	0-88318-113-4
No.15	High Energy Collisions - 1973 (Stony Brook)	73-92324	0-88318-114-2
No.16	Causality and Physical Theories (Wayne State University, 1973)	73-93420	0-88318-115-0
No.17	Thermal Expansion - 1973 (lake of the Ozarks)	73-94415	0-88318-116-9
No.18	Magnetism and Magnetic Materials - 1973 (2 parts) (Boston)	59-2468	0-88318-117-7
No.19	Physics and the Energy Problem - 1974 (APS Chicago)	73-94416	0-88318-118-5
No.20	Tetrahedrally Bonded Amorphous Semiconductors (Yorktown Heights, 1974)	74-80145	0-88318-119-3
No.21	Experimental Meson Spectroscopy - 1974 (Boston)	74-82628	0-88318-120-7
No.22	Neutrinos - 1974 (Philadelphia)	74-82413	0-88318-121-5
No.23	Particles and Fields - 1974 (APS/DPF Williamsburg)	74-27575	0-88318-122-3
No.24	Magnetism and Magnetic Materials - 1974 (20th Annual Conference, San Francisco)	75-2647	0-88318-123-1
No.25	Efficient Use of Energy (The APS Studies on the Technical Aspects of the More Efficient Use of Energy)	75-18227	0-88318-124-X